Lecture Notes in Computer Science   8901

Commenced Publication in 1973
Founding and Former Series Editors:
Gerhard Goos, Juris Hartmanis, and Jan van Leeuwen

More information about this series at http://www.springer.com/series/7407

Gopal Gupta · Ricardo Peña (Eds.)

Logic-Based Program Synthesis and Transformation

23rd International Symposium, LOPSTR 2013
Madrid, Spain, September 18–19, 2013
Revised Selected Papers

 Springer

Editors
Gopal Gupta
The University of Texas at Dallas
Richardson
Texas
USA

Ricardo Peña
Universidad Complutense de Madrid
Madrid
Spain

ISSN 0302-9743
Lecture Notes in Computer Science
ISBN 978-3-319-14124-4
DOI 10.1007/978-3-319-14125-1

ISSN 1611-3349   (electronic)

ISBN 978-3-319-14125-1   (eBook)

Library of Congress Control Number: 2014958184

LNCS Sublibrary: SL1 – Theoretical Computer Science and General Issues

Springer Cham Heidelberg New York Dordrecht London

Printed on acid-free paper

Springer International Publishing AG Switzerland is part of Springer Science+Business Media
(www.springer.com)

Preface

This volume contains papers presented at the 23rd International Symposium on Logic-based Program Synthesis and Transformation (LOPSTR) held in Madrid, Spain, on September 18 and 19, 2013. There were 21 submissions. Each submission was reviewed by at least three Program Committee members. Thirteen papers were selected for presentation at the symposium. The program also included invited talks by Peter Stuckey and Albert Rubio. All but two of the papers presented at the symposium went through another round of reviewing before being included in this volume.

We gratefully acknowledge the authors for submitting their papers, members of the Program Committee for prompt reviewing, the Easychair system in helping administer the reviewing process both for the symposium and for this volume, and the local arrangements committee for their organizational efforts. We also recognize the support of our respective institutions, the University of Texas at Dallas and Universidad Complutense de Madrid.

October 2014

Gopal Gupta
Ricardo Peña

Organization

Program Committee

Salvador Abreu	Universidade de Évora and CENTRIA, Portugal
Elvira Albert	Universidad Complutense de Madrid, Spain
Sergio Antoy	Portland State University, USA
Henning Christiansen	Roskilde University, Denmark
Hai-Feng Guo	University of Nebraska at Omaha, USA
Gopal Gupta	University of Texas at Dallas, USA
Manuel Hermenegildo	Universidad Politécnica de Madrid and IMDEA, Spain
Patricia Hill	University of Leeds, UK
Jacob Howe	City University, UK
Michael Leuschel	University of Düsseldorf, Germany
Paulo Moura	CRACS and INESC TEC, Portugal
Ricardo Peña	Universidad Complutense de Madrid, Spain
Enrico Pontelli	New Mexico State University, USA
I.V. Ramakrishnan	SUNY Stony Brook, USA
Neda Saeedloei	University of Texas at Dallas, USA
Hirohisa Seki	Nagoya Institute of Technology, Japan
Paul Tarau	University of North Texas, USA
Neng-Fa Zhou	CUNY Brooklyn College and Graduate Center, USA

Organizing Committee

Ricardo Peña	Universidad Complutense de Madrid, Spain
Clara Segura	Universidad Complutense de Madrid, Spain
Manuel Montenegro	Universidad Complutense de Madrid, Spain
Enrique Martín	Universidad Complutense de Madrid, Spain

Additional Reviewers

Ashok, Vikas	Martín-Martín, Enrique
Calejo, Miguel	Melo De Sousa, Sim ao
Chitil, Olaf	Morales, Jose F.
Garcia-Perez, Alvaro	Nampally, Arun
Gini, Maria	Ozono, Tadachika
Haemmerlé, Rémy	Proietti, Maurizio
Hunt, Seb	Son, Tran Cao
Kriener, Jael	Tarau, Paul
Lu, Lunjin	Zanardini, Damiano

Invited Talks

Search Is Dead, Long Live Proof!

Peter Stuckey

Department of Computer Science,
University of Melbourne,
Melbourne, Australia

Constraint programming is a highly successful technology for tackling complex combinatorial optimization problems. Any form of combinatorial optimization involves some form of search, and CP is very well adapted to make use of programmed search and strong inference to solve some problems that are out of reach of competing technologies. But much of the search that happens during a CP execution is effectively repeated. This arises from the combinatorial nature of the problems we are tackling. Learning about past unsuccessful searches and remembering this in the form of lemmas (or nogoods) in an effective way can exponentially reduce the size of the search space. In this sense search can be seen as a mechanism to prove lemmas, and optimization search is simply a proof that no better solution can be found, with the side effect that good solutions are found on the way. In this talk I will explain lazy clause generation, which is a hybrid constraint solving technique that steals all the best learning ideas from Boolean satisfiability solvers, but retains all the advantages of constraint programming. Lazy clause generation provides the state of the art solutions to a wide range of problems, and consistently outperforms other solving approaches in the MiniZinc challenge. Lazy clause generation allows concise lemmas to be recorded about the optimization search, and this together with methods like rapid restart mean we are no longer searching for a good solution, but instead iteratively building a proof that no better solution can be found. So search is dead, long live proof.

Program Analysis Using SMT and Max-SMT

Albert Rubio

Universitat Politécnica de Catalunya,
Barcelona, Spain

When applying the constraint-based method in program analysis the existence of good constraint solvers is key for its success. However, when the analysis requires the discovery of several properties there are two new crucial elements to be taken into account in the development of automatic tools, namely how to guide the search to and the relevant properties and how to define a good notion of progress in this process.

Our work has been focused on showing how new SMT solvers for non-linear arithmetic can improve the automatic invariant generation in imperative programs with scalar and array variables, and how to combine this invariant generation with the verification of other properties like termination. In this respect, we have observed that by considering the constraint method as a constraint optimization problem to be solved with Max-SMT, as opposed to a constraint satisfaction problem to be solved with SMT, one can devise natural notions of relevance of program properties and of progress in the analysis.

Contents

Formalization and Execution of Linear Algebra: From Theorems to Algorithms

Jesús Aransay$^{(\boxtimes)}$ and Jose Divasón

Departamento de Matemáticas y Computación, Universidad de La Rioja,
Edif. Luis Vives, c. Luis de Ulloa s/n., 26004 La Rioja, Spain
{jesus-maria.aransay,jose.divasonm}@unirioja.es

Abstract. In this work we present a formalization of the *Rank Nullity* theorem of Linear Algebra in Isabelle/HOL. The formalization is of interest because of various reasons. First, it has been carried out based on the representation of mathematical structures proposed in the HOL Multivariate Analysis library of Isabelle/HOL (which is part of the standard distribution of the proof assistant). Hence, our proof shows the adequacy of such an infrastructure for the formalization of Linear Algebra. Moreover, we enrich the proof with an additional formalization of its *computational* meaning; to this purpose, we choose to implement the Gauss-Jordan elimination algorithm for matrices over fields, prove it correct, and then apply the Isabelle code generation facility that permits to *execute* the formalized algorithm. For the algorithm to be code generated, we use again the implementation of matrices available in the HOL Multivariate Analysis library, and enrich it with some necessary features. We report on the precise modifications that we introduce to get code execution from the original representation, and on the performance of the code obtained. We present an alternative verified type refinement of vectors that outperforms the original version. This refinement performs well enough as to be applied to the computation of the rank of some biomedical digital images. Our work proves itself as a suitable basis for the formalization of numerical Linear Algebra in HOL provers that can be successfully applied for computations of real case studies.

Keywords: Linear Algebra · Verification · Code generation

Introduction

In standard mathematical practice, formalization of results and execution of algorithms are usually (and unfortunately) rather separate concerns. Computer Algebra systems (CAS) are commonly seen as *black boxes* in which one has to trust, despite some well-known major errors in their computations, and mathematical proofs are more commonly carried out by mathematicians with *pencil & paper*, and sometimes *formalized* with the help of a proving assistant. Nevertheless, some of the features of each of these tasks (formalization and computation) are considered as a burden for the other one; computation demands optimized

© Springer International Publishing Switzerland 2014
G. Gupta and R. Peña (Eds.): LOPSTR 2013, LNCS 8901, pp. 1–18, 2014.
DOI: 10.1007/978-3-319-14125-1_1

versions of algorithms, and very usually *ad hoc* representations of mathematical structures, and formalization demands more intricate concepts and definitions in which proofs have to rely on.

In this paper, we present a case study in which we aim at developing a formalization in Linear Algebra in which computations are still posible. From an existing library in the Isabelle/HOL distribution (HOL Multivariate Analysis [15], *HMA* in the sequel), which has been fruitfully applied in the formalization of major mathematical results (both in this system and also in HOL-Light, that shares a similar representation), we formalize a mathematical result, known as the "Rank Nullity theorem".

The result is of interest by itself in Linear Algebra (some textbooks name it the *Fundamental theorem of Linear Algebra*) but it is even more interesting if we consider that each linear map between *finite dimensional* vector spaces can be represented by means of a *matrix* with respect to some provided bases. Every matrix over a field can be turned into a matrix in *reduced row echelon form* (rref, from here on) by means of operations that preserve the behavior of the linear map, but change the underlying bases; the number of *non zero rows* of such a matrix is equal to the rank of the (original) linear map; the number of zero rows is the dimension of its *kernel*.

The best-known algorithm for the computation of the rref of a matrix is the Gauss-Jordan elimination method. We have implemented the algorithm over the representation of matrices in the HMA library; this representation was introduced by J. Harrison in HOL-Light and successfully applied in the formalization of Mathematics in various theorem provers, because of its succinctness and its taking advantage of the underlying type system; vectors are represented as functions over an underlying finite type; matrices as vectors of vectors. *A priori*, finite enumerable types have nice computational features, since mathematical and logical operations (traversing, epsilon operator, universal or existential quantifiers) over them can be executed. We present here some additional features, relying in previous works, that enable these possibilities in Isabelle/HOL. In this work, we link the original statement of the Rank Nullity theorem together with the Gauss-Jordan elimination algorithm, and can use both tools to produce *certified* computations of the rank and kernel of linear maps.

As we will illustrate with some examples, the performance of the algorithm is rather poor, mainly because of the data structure used to represent matrices; the executable algorithm cannot be used for real applications, but only for tests (for instance, it could be used for experimental testing or as a *reference* algorithm for more optimized versions of it). Therefore, we introduce a data type refinement that allows us to obtain a version of the algorithm performing nicely in matrices of a considerable size (but still far from specialized Computer Algebra libraries).

The paper is structured as follows; in Section 1 we describe the Isabelle features in which our development is based on. In Section 2 we present the Rank Nullity theorem, as well as its Isabelle formalization. In Section 3 we introduce the notion of rref and the formalization of the Gauss-Jordan algorithm. In Section 4 we present the choices and setup of the Isabelle code generation tool that enable to execute operations and algorithms. In Section 5 we bring together the previous ingredients and present the generated SML code from the original algorithm. Additionally, we present a refinement that enabled us to improve the performance of the certified algorithm. In Section 6 we draw some conclusions and present related works, as well as possible future research lines. The source files of the development are available from [2]; they have been developed under the Isabelle 2013 version. The previous web site also includes the SML code generated from the Isabelle specifications, and also the input matrices that have been used in the benchmarks presented in Section 5.

1   Isabelle/HOL

Isabelle [21] is a generic interactive proving assistant, on top of which different logics can be implemented; the most explored of these variety of logics is higher-order logic (or *HOL*), and it is also the one where the greatest number of tools (code generation, automatic proof procedures) are available. We do not aim to present here the fundamentals of Isabelle/HOL, just to introduce the main features that are used in our work.

The HOL type system is rather simple; it is based on non-empty types, function types ($\Rightarrow$) and type constructors κ that can be applied to already existing types (*nat, bool*) or type variables (α, β). Types can be also introduced by enumeration (*bool*) or by induction, as lists (by means of the *datatype* command). Additionally, new types can be also defined as non-empty subsets of already existing types by means of the *typedef* command; the command takes a set defined by comprehension over a given type $\{x :: \alpha . P\,x\}$, and defines a new type σ. We will refer to this new type as *abstract*, and to the underlying one as *concrete* (this terminology is particular to the context of code generation, where the abstract type cannot be directly code generated, whereas the concrete one, under precise assumptions, can be; see [8] for details).

Isabelle also introduces type classes in a similar fashion to Haskell; a type class is defined by a collection of operators (over a single type variable) and premises over them. For instance, the HMA library has a type class *field* representing the algebraic structure. Concrete types (*real, rat*) can be proved to be *instances* of a given type class (*field* in our example). Type classes can be also used to impose additional restrictions over type variables; for instance, the expression ($x :: \alpha :: field$) imposes the constraint that the type variable α possess the structure and properties stated in the *field* type class, and can be later replaced exclusively by types which are instances of that type class.

1.1   HOL Multivariate Analysis Library

The HOL Multivariate Analysis library is a set of Isabelle theories which contains a wide range of results in different mathematical fields such as Analysis, Topology or Linear Algebra. They are based on the work of J. Harrison in HOL-Light [10], which includes proofs of intricate theorems (such as the Stone-Weierstrass theorem) and has been successfully used as a basis for the Flyspeck project [11], aiming at formally verifying the proof of the Kepler conjecture by T. Hales. Among the fundamentals of the library, one of the keys is the representation of n-dimensional vectors over a given type ($\mathbb{F}^n$, where $\mathbb{F}$ stands for a generic field, or in Isabelle jargon a type variable $\alpha :: field$) taking into account that the HOL type system lacks the expressivity of dependent types. A detailed explanation can be found in [9, Section2]. The idea is to represent vectors over α by means of *functions* from a finite type variable $\beta :: finite$ to α; for proving purposes, this type definition is usually sufficient; if we need to introduce vectors of a *concrete* dimension n, β can be replaced by a (finite) type of such cardinality (we present in Section 4 a possible representation of such types).

The Isabelle type definition is as follows; the functions *vec-nth* and *vec-lambda* are the morphisms between the abstract data type *vec* and the underlying concrete data type, functions with finite domain:

```
typedef (α, β) vec = UNIV :: ((β::finite) ⇒ α) set
 morphisms vec-nth vec-lambda ..
```

The previous type also admits in Isabelle the shorter notation $\alpha \hat{\ } \beta$. The idea of using underlying finite types for vectors indices has great advantages, as already pointed out by Harrison, from the formalization point of view. For instance, the type system enforces that operations on vectors (such as addition or multiplication) are only performed over vectors of equal dimension, *i.e.*, vectors which indexing types are exactly the same (this would not be the case if we were to use, for instance, lists as vectors). Moreover, the functional flavor of operations and properties over vectors is kept (for instance, vector addition can be defined in a pointwise manner).

The representation of matrices is then derived in a natural way based on the one of vectors by iterating the previous construction (matrices over a type α will be terms of type $\alpha \hat{\ } m \hat{\ } n$, where m and n stand for finite type variables).

The HMA library already contains operations and properties of matrices defined in this way (multiplication, invertible matrices, the relationship between linear maps and matrices, determinants). Nevertheless, we missed some other standard results in Linear Algebra, that we had to introduce, such as the notion of coordinates with respect to a particular (not the canonical one) basis, the influence of changes of bases over a given matrix, or the elementary row (and column) operations over matrices (exchanging rows, multiplying a row by a constant and adding to a row another one multiplied by a constant). These elementary operations also give place to the notion of *elementary matrices*; indeed, these are the invertible matrices; each elementary matrix represents a change of bases.

Another subject that has not been explored in the Isabelle HMA library, or in HOL-Light, is the possibility to execute the previous data types and operations. As we will see in Section 4, the *finite* type class does not enable some operators over vectors and matrices to be executed, and some additional type classes have to be used.

Finally, another aspect that has not been explored in the HMA library is numerical Linear Algebra. There is no implementation of common algorithms such as Gaussian elimination or diagonalization. We aim to show that the HMA library provides a framework where algorithms over matrices can be formalized, executed and coupled with their mathematical meaning.

1.2 Code Generation

Isabelle/HOL offers a facility to generate code from specifications of data types, type classes and definitions over them, as long as these elements have an executable representation in the target languages (SML, Haskell, OCaml or Scala). The code generator is part of the trusted kernel of Isabelle [7].

As we explained before, the *vec* type is an *abstract* type, produced as a subset of the concrete type of functions from a finite type to a variable type; this type cannot be directly mapped to an SML type, since its definition, a priori, could involve HOL logical operators unavailable in SML. In the code generation process, a data type refinement from the abstract to the concrete type must be defined; the concrete type is then the one chosen to appear in the target programming language. A similar refinement is carried out over the operations of the *abstract* type; definitions over the concrete data type (functions, in our case) have to be produced, and proved equivalent (*modulo* type morphisms) to the ones over the abstract type. The general idea is that formalizations have to be carried out over the abstract representation, whereas the concrete representations are exclusively used during the code generation process. The methodology admits iterative refinements, as long as their equivalence is proved. A detailed explanation of the methodology is found in [7]; an interesting case study in [5].

In Section 5 we present two different refinements of the *vec* Isabelle type; the first one uses functions over finite domains, and is designed for simplicity. The second one uses immutable arrays (represented in the Isabelle type *iarray*) and presents a remarkable performance improvement when generated to SML.

2 The Rank Nullity Theorem of Linear Algebra

The Rank Nullity theorem is a well-known result in Linear Algebra; the following formulation has been obtained from [22, Theorem2.8].

Theorem 1 (The rank plus nullity theorem). *Let* $\tau \in L(V, W)$.

$$\dim(\ker(\tau)) + \dim(im\,(\tau)) = \dim(V)$$

or, in other notation,

$$rk\,(\tau) + null\,(\tau) = \dim(V)$$

In the previous statement, $L(V, W)$ denotes the set of linear maps between two given vector spaces V and W. It is worth noting that V must be a finite-dimensional vector space. Several textbooks impose the additional restriction of W being also finite-dimensional, but this restriction (as can be observed in the

Isabelle formalization) is only needed in the version of the theorem for matrices representing linear maps (otherwise, we would have a matrix with an infinite number of columns representing the linear map). The following formalization [1] is part of the Isabelle repository; thanks to the infrastructure in the HMA library, it comprises a total of 380 lines of Isabelle code. The Isabelle statement of the result is as follows:

```
theorem rank_nullity_theorem:
  assumes linear (f::(α::{euclidean_space}) => (β::{real_vector}))
  shows DIM (α) = dim {x. f x = 0} + dim (range f)
```

Following the ideas in the HMA library, the vector spaces are represented by means of types belonging to particular type classes; the finite-dimensional premise on the source vector space is part of the definition of the type class *euclidean-space* (in the hierarchy of algebraic structures of the HMA library [16], this is the first type class to include the requisite of being finite-dimensional). Accordingly, *real-vector* is the type class representing vector spaces over $\mathbb{R}$. The operator *dim* represents the dimension of a subset of a type, whereas *DIM* is equivalent to *dim*, but refers to the carrier set of that type.

There is one remarkable result that we did not find in textbooks, but that proved crucial in the formalization. Its Isabelle statement reads as follows:

```
lemma  inj_on_extended:
  assumes linear f and finite C and independent C and C = B ∪ W
  and B ∩ W = {} and {x. f x = 0} ⊆ span B
  shows inj_on f W
```

The result claims that any linear map f is *injective* over any collection (W) of linearly independent elements whose images are a *basis* of the *range*; this is required to prove that, given $\{e_1 \ldots e_m\}$ a basis of ker(f), when we complete this basis up to a basis $\{e_1 \ldots e_n\}$ of the vector space V, the linear map f is injective over the elements $W = \{e_{m+1} \ldots e_n\}$ and therefore its cardinality is the same than the one of $\{fe_{m+1} \ldots fe_n\}$ (and equal to the dimension of the *range* of f).[1]

The Isabelle statement of the Rank Nullity theorem over matrices turns out to be straightforward; we make use of a result in the HMA library (labeled as *matrix-works*) which states that, given any linear map f, $f(x :: real^n)$ is equal to the (matrix by vector) product of the matrix associated to f and x. The picture has slightly changed with respect to the Isabelle statement of the Rank Nullity theorem; where the source and target vector spaces were, respectively, an Euclidean space and a real vector space (of any dimension), they are now replaced by a $real^n^m$ matrix, *i.e.*, the vector spaces $real^n$ and $real^m$.

```
lemma fixes A::real^α^β
  shows DIM (real^α) = dim (null_space A) + dim (col_space A)
```

---

[1] In our opinion, this result is a typical example of a property that is unavoidable in a formalized proof, but usually skipped in paper & pencil proofs.

This statement is used to compute the dimensions of the rank and kernel of linear maps by means of their associated matrices. It exploits the fact that the *rank* of a matrix is defined to be the dimension of its *column space*, also known as column rank, which is the vector space generated by its columns; this dimension is also equal to the ones of the *row space* and the range.

3 The Gauss-Jordan Elimination Method

There are several ways of computing the dimension of the range (and consequently of the kernel) of a linear map. In our development we choose the Gauss-Jordan elimination method. The main reason is that it has several different applications. For instance, it can be used to solve systems of linear equations; Nipkow [20] has proved that the Gauss-Jordan elimination algorithm is correct in this respect; the algorithm used in that work is very succinct, but works exclusively for input square matrices with unique solution, *i.e.*, whose rank is equal to their dimension. Nipkow proves that the algorithm is *complete* (under suitable circumstances, it generates a solution) and *correct* (it generates a vector which is a solution of the linear system). The algorithm we are formalizing differs from Nipkow's since we need an algorithm capable of dealing with non-square matrices whose rank can be smaller or equal than their number of rows. We also prove a different property of the algorithm than the one he proves; namely, that the rank of the input matrix is preserved through the algorithm steps (in other words, that the rank is an *invariant* of linear maps). Another difference is implicit here; in the HMA setting, linear maps and matrices are proved to be equivalent, whereas Nipkow uses an *ad hoc* matrix data type. In this sense, our implementation of Gauss-Jordan elimination would admit further applications, such as the computation of inverse matrices (or inverses of linear maps) and the computation of determinants. The algorithm is not optimal for any of those problems, but algorithmic refinements could be used in later stages to reach better performing algorithms for each of the previous tasks, once the mathematical properties of the original algorithm are stated and proved.

The Gauss-Jordan algorithm is based on the computation of the *reduced row echelon form* of (probably non-square) matrices. The *rref* of a matrix is defined as follows (see [22]):

1. All rows consisting only of 0's appear at the bottom of the matrix.
2. In any nonzero row, the first nonzero entry is a 1. This entry is called a *leading* entry.
3. For any two consecutive rows, the leading entry of the lower row is to the right of the leading entry of the upper row.
4. Any column that contains a leading entry has 0's in all other positions.

The previous definition of rref is valid for non-square matrices. Interestingly, the rref (R) of a matrix A can be obtained by performing exclusively *row operations*, in such a way that $R = E_1 \ldots E_k A$, where E_i denote elementary matrices; since elementary operations (and elementary matrices) preserve the rank of a matrix, computing the rank of A can be reduced to computing the rank of R (its number of nonzero rows). The code in the following formalization is available from [2] in files *Elementary_Operations* and *Gauss_Jordan*.

One way to compute the successive elementary row operations that produce the rref of a matrix is through the Gauss-Jordan elimination algorithm[2]; versions of the algorithm abound in the literature; however, we preferred to introduce our own version, designed to ease the formalization. In it, the algorithm is described by means of exclusively *elementary row operations* E_i (namely *interchange_rows*, *mult_row* and *add_row*), so that the rank of a matrix A is preserved because of the previous formula $R = E_1 \ldots E_k A$. Additionally, the algorithm exploits the underlying (finite) representation of matrices, where both the indices of rows and columns are represented by *finite* types; both the types of columns and rows indices need to be *traversed*, and thus are restricted to be instances of the *enum* type class; this type class is part of the Isabelle library, and represents types for which the carrier set is *explicit*.

Algorithm 1. Gauss-Jordan elimination algorithm

Data: A is the input matrix;
$l \leftarrow 0$;   ▷ l is the index where the pivot is to be placed after each iteration;
for $k \leftarrow 0, (ncols\,A) - 1$ **do**
    ▷ Check that there is a nonzero entry over index l in column k;
    if $nonzero\,l\,(col\,k\,A)$ **then**
        $i \leftarrow index\_nonzero\,l\,(col\,k\,A)$ ▷ Let i be the index of the first nonzero entry;
        $A \leftarrow interchange\_rows\,A\,i\,l$ ▷ Rows i and l are interchanged;
        $A\,l \leftarrow mult\_row\,A\,l\,(1/A\,l\,k)$ ▷ Row l is multiplied by $(1/A\,l\,k)$;
        for $t \leftarrow 0, (nrows\,A) - 1$ **do**
            if $t \neq l$ **then**
                $A\,t \leftarrow add\_row\,A\,t\,l\,(-A\,t\,k)$ ▷ Row t is added row l times $(-A\,t\,k)$;
            end if
        end for
        $l \leftarrow l + 1$
    end if
end for

The algorithm satisfies the following properties. When applied from column 0 up to column k, the first $k+1$ columns will be in rref. Note that implicitly we are imposing additional constraints on the types indexing columns (and rows); they must be inductive, since the proofs will be performed by induction over columns' indices; we make use of an additional type class *mod-type*, which resembles the structure $\mathbb{Z}/n\mathbb{Z}$, together with some required arithmetic operations and conversion functions from it to the integers. Therefore, the underlying types used for representing the rows and columns of the input matrices must be instances of

---

[2] A somehow surprising point is that this algorithm is not even mentioned in [22], even if a detailed description of elementary operations over matrices, rref or invertible matrices is presented; this underscores our claim that algorithms and its mathematical meaning are often presented as different subjects.

the type classes *finite*, *enum* and *mod-type*, as can be noted in the Isabelle *type definition* of the algorithm:

definition *Gauss_Jordan::α::{inverse, uminus,*
    semiring_1}^columns::{mod_type}^rows::{mod_type} => α^columns^rows
where ...

In the previous algorithm definition we exclusively included the type classes required to *state* the algorithm; in the later proof of the algorithm, we have to restrict α to be an instance of the type class *field*; additionally, if we try to *execute* the algorithm (or generate code from it), the rows and columns types need to be instances of *enum*. The *finite* type class is implicit in the rows and columns types, since *mod-type* is a subclass of it.

The crucial result in the formalization of the algorithm preserving the rank of matrices is that elementary operations (*i.e.*, invertible matrices) applied to a matrix preserve its rank:

lemma fixes *A::real^'n^'m* **and** *P::real^'m^'m*
  assumes *invertible P* **and** *B = P ** A*
  shows *rank B = rank A*

As a consequence of the previous result, we also proved that linear maps are preserved by elementary operations (only the underlying bases change). Note that the previous machinery is not particular to our formalization, but could also be reused for different algorithms in numerical Linear Algebra. Additionally, we formalized a result stating that the previous algorithm produces a rref.

Moreover, the presented version of the algorithm is *executable*, as long as the rows and columns types can be generated to some type in the target languages; we present in Section 4 the details of that extraction.

4 Code Generation from Finite Types

Up to now, we have used in our development an *abstract* data type *vec* (and its iterated construction for representing matrices), for which the underlying *concrete* types are functions with an indexing type; the indexing type is instance of the *finite*, *enum* and *mod-type* type classes; these classes demand the universe of the underlying type to be finite, to have an explicit enumeration of the universe, and some arithmetical properties.

The *finite* type class is enough to generate code from some abstract data structures, such as *finite sets*, which are later mapped in the target programming language (for instance, SML) to data structures such as lists or red black trees (see [19] for details and benchmarks). Our case study is a bit more demanding, since the indexing types of vectors and matrices have to be also enumerable. The *enum* type class allows us to *execute* operations such as matrix multiplication, $A * B$ (as long as the type of columns in A is the same as the type of rows in B), algorithms traversing the universe of the rows or columns indexing types (such as operations that involve the logical operators $\forall$ or $\exists$ or the Hilbert's ϵ operator), enabling operators like "every element in a row is equal to zero" or "select the least position in a row whose element is not zero".

The standard setup of the Isabelle code generator for (finite) sets is designed to work with sets of generic types (for instance, sets of natural numbers), mapping them to *lists* on the target programming language. This poses some restrictions, since operations such as *coset* $\emptyset$ cannot be computed over arbitrary types, whereas in an *enumerable* type this is equal to a set containing every element of the enumerable type (and therefore, in the target programming language, the result of the previous operation will produce a list containing every element in the corresponding type). The particular setup enabling these kind of calculations (only for enumerable types), which are *ad-hoc* for our case study, can be found in the file *Code_Set* of our development [2].

Another different but related issue is the election of a concrete type to be used as index of vectors and matrices; we already know that the type has to be an instance of the type classes *finite*, *enum* and *mod-type*. The Isabelle library contains an implementation of *numeral types* used to represent finite types of any cardinality. It is based on the binary representation of natural numbers (by means of the two type constructors, *bit0* and *bit1*, applied to underlying finite types, and of a singleton type constructor *num1*).

```
typedef α bit0 = {0...<2 * CARD(α:: finite)}
typedef α bit1 = {0...<1 + 2 * CARD(α:: finite)}
```

From the previous constructors, an Isabelle type representing $\mathbb{Z}/5\mathbb{Z}$ (or 5 in Isabelle notation) can be used, which is internally represented as *bit1 (bit0 (num1))*. The representation of the (abstract) type 5 is the set $\{0, 1, 2, 3, 4 :: 5\}$; its concrete representation is the subset $\{0, 1, 2, 3, 4 :: int\}$. The integers as underlying type allow users to reuse (with adequate modifications) integer operations (substraction and unary minus) in the resulting finite types. As part of our development, we prove that the *num1*, *bit0* and *bit1* type constructors are instances of the *enum* type class.

```
instantiation bit0 :: enum begin
definition (enum::α bit0 list)=map (Abs_bit0'∘ int) (upt 0 (CARD α bit0))
definition enum_all P = (∀b ∈ enum. P b)
definition enum_ex P = (∃b ∈ enum. P b)
instance proof (intro_classes) ...
```

The Isabelle library already provides basic arithmetic functions for numeral types, with definitions of addition, substraction, multiplication and division. Note that, for these operations to be defined generally for arbitrary cardinalities, the cardinality of the finite type must be *computed* on demand (adding 3 and 4 in type 5 must return 2). To this aim, the Isabelle library has a type class (*card_UNIV*) for types whose cardinality is *computable*; we prove that the previous numeral types are instances of such class, enabling the computation of their cardinals (see file *Numeral_Type_Addenda* in [2] for the complete proofs).

5   Bringing It All Back Home: Formalization and Execution

In the previous section we have presented a setup that permits code generation of the vectors indexing types and their operations. Nevertheless, as we mentioned in Section 1.1, *vec* is itself an *abstract* type which also has to be *refined* to *concrete* data types that can be code generated.

We present here two such refinements. The first one consists in refining the abstract type *vec* to its underlying concrete type *functions (with finite domain)*. We expected the performance to be unimpressive, but the close gap between both types greatly simplifies the refinement; interestingly, at a low cost, an executable version of the algorithm can be achieved, capable of computing the rref of matrices of small sizes.

The second data type refinement is more informative; we refine the *vec* data type to the Isabelle type *iarray*, representing *immutable arrays* (which are generated in SML to the *Vector* structure [23]).

In order to achieve the first refinement (from abstract matrices to *functions*), the type morphisms between the type *vec* and its counterpart (functions) have to be labeled precisely in the code generator setup.

lemma [code_abstype]: vec_lambda (vec_nth v) = (v::α^β::finite)

Additionally, every operation over the abstract data type has to be *mapped* to an operation over the concrete data type (and their behavioral equivalence proved). It can be noted that because of the iterative construction of matrices (as elements of type *vec* over *vec*) each operation over matrices (as multiplication below) usually demands two lemmas to translate it to its computable version. It is also remarkable that *setsum* is *computable* as long as there is an explicit version of the *UNIV* set, and this holds since we have restricted ourselves to *enum* types.

definition
  mat_mult_row m m' f = vec_lambda(λj. setsum (λi.(mfi * m'ij)) UNIV)
lemma [code abstract]: vec_nth (mat_mult_row m m' f) =
  vec_lambda (λj. setsum (λi.(mfi * m'ij)) UNIV)
lemma [code abstract]: vec_nth (m ** m') = mat_mult_row m m'

As long as our algorithm is based on (abstract) operations which are mapped to corresponding concrete operations, the later ones will be correctly code generated. Since dealing with matrices as functions can become rather cumbersome, we also define additional functions for conversion between lists of lists and functions (so that the input and output of the algorithm are presented to the user as lists of lists).

One subtlety appears at this step; from a given list of elements, a vector of a certain dimension is to be produced; the user must add a type annotation declaring which dimension the generated vector has to be (in other words, the size of the list needs to be known in advance).

Below we present examples of the evaluation (by means of SML generated code) of the Gauss-Jordan algorithm to compute the dimension of the rank (which is also the one of the column space) and the one of the null space of given matrices of reals; the evaluation can be also performed *in* Isabelle (and therefore the code generator would not intervene):

```
value[code] rank (list_of_lists_to_matrix
    [[1,0,0,7,5],[1,0,4,8,-1],[1,0,0,9,8],[1,2,3,6,5]]::real^5^4)
value[code] dim (null_space (list_of_lists_to_matrix
    [[1,0,0,7,5],[1,0,4,8,-1],[1,0,0,9,8],[1,2,3,6,5]]::real^5^4))
```

The previous computations have been carried out with matrices represented as functions. They are almost instantaneous, but the computation of the algorithm over matrices of size 10×10 is already very slow (several minutes).

The second aforementioned refinement was designed for improving performance. The original Isabelle abstract type *vec* is mapped to the Isabelle type *iarray* (the type itself is just a wrapper of lists), which is then mapped in the code generation process to the SML *Vector* structure; the SML structure requires constant time for access operations, improving, a priori, an implementation by lists. The code equations that perform the data type and operations conversions (from type *vec* to type *iarray*) can be found in file *Matrix_To_IArray* in [2]. As in our previous example, the data type refinement demands labeling the morphisms between the abstract type (*vec*) and the concrete one (*iarray*), and introducing operations on *iarrays* that are proven equivalent to the original abstract ones. These proofs are almost straightforward, since the *iarray* and *vec* representations share a *functional* flavor (in the way of accessing elements) that can be exploited in proofs.

Our Gauss-Jordan algorithm is implemented for matrices with entries over a *field*; in our execution experiments we carry out computations over the Isabelle types *real*, *rat* (for $\mathbb{Q}$) and *bit* (an implementation of the field $\mathbb{Z}/2\mathbb{Z}$); the Isabelle type *real* admits serialisations to an SML *ad hoc* type (quotients of SML *IntInf.int* elements) and also to the SML *Real.real* type. The former offers arbitrary precision, but on a standard machine, using the optimizer compiler MLton, only (randomly generated) matrices up to 100×100 size can be computed in a reasonable time (as a matter of comparison, Gauss-Jordan algorithm in *Mathematica*® over matrices of real numbers with arbitrary precision becomes rather slow at sizes over 500×500). Table 1 shows the times used by the SML implementation Poly/ML and the optimizer compiler MLton to process and execute the Gauss-Jordan elimination algorithm generated from the Isabelle verified specification over (randomly generated) matrices whose inputs are quotients of *IntInf.int* elements. The following experiments have been carried out in a computer with an Intel Core i3-370M Processor (2 cores of 2.4 GHz) with 4GB of

Table 1. Elapsed time (in seconds) to process random $Q^{n \times n}$ matrices (with elements between -10 and 10) and computing their rrefs using the Gauss-Jordan algorithm with Poly/ML 5.5 and MLton 20100608

	Rational matrices			
	Poly/ML		MLton	
Size (n)	Processing Time (seconds)	Execution Time (seconds)	Processing Time (seconds)	Execution Time (seconds)
10	0.0	0.0	0.2	0.0
20	0.0	0.2	0.3	0.0
40	0.1	3.7	0.9	1.5
60	0.2	22.7	1.9	9.6
80	0.5	77.0	3.5	32.7
100	0.7	200.9	6.0	84.1

RAM and Ubuntu GNU/Linux 11.10. The SML code and the benchmark matrices are available from [2].

Applying profiling techniques, we detected that most of the computing time is used not in matrix operations but in the ones related to integer quotients operations (normalising quotients, computing the lcm of denominators, and the like[3]). The latter serialisation (to the SML *Real.real* type) is produced only for computing purposes, since it is inconsistent and suffers from numerical stability problems, but allows us to apply Gauss-Jordan elimination to (randomly generated) matrices up to size 700 × 700. The performance tests are presented in Table 2. The processing and execution times in Poly/ML follow a linear pattern with respect to the number of elements in the matrix (n^2).

The *rat* type is also serialised to quotients of *IntInf.int* pairs; the performance tests are therefore equal to the ones obtained for the first serialisation of type *real* and presented in Table 1.

Finally, we define our custom serialisation of type *bit* to SML; the Isabelle constants 0 :: *bit* and 1 :: *bit* are mapped in SML to 0 and 1 of type *IntInf.int*; operations over *bit* to arithmetic operations modulo 2 in *IntInf.int*. This serialisation proved empirically to perform better than other options such as the SML type *Bool*, or using *IntInf.int* with exhaustive definitions of the operations. The benchmarks of this serialisation are presented in Table 3.

With this last serialisation and Poly/ML 5.5 we get to apply Gauss-Jordan elimination, and compute the rank, of matrices of dimensions up to 2560 × 2560; computing time grows linearly on the number of matrix entries (as seen in Table 3), and thus RAM memory becomes the only practical limitation. For instance, we are able to compute the rank of the binary matrix representing the following digital image (Fig. 1), captured with a *confocal* microscope from a neuronal culture. It is worth noting that processing and computing times over matrices obtained from digital images are smaller than the ones obtained over

---

[3] Both MLton and Poly/ML make use of the GMP http://gmplib.org/ set of libraries for arithmetic.

Table 2. Elapsed time (in seconds) to process random $\mathbb{R}^{n \times n}$ matrices (with elements between -10 and 10) and computing their rrefs using the Gauss-Jordan algorithm with Poly/ML 5.2 and MLton 20100608.

	Real matrices			
	Poly/ML		MLton	
Size (n)	Processing Time (seconds)	Execution Time (seconds)	Processing Time (seconds)	Execution Time (seconds)
10	0.0	0.0	0.8	0.0
20	0.0	0.0	2.5	0.0
40	0.1	0.0	13.8	0.0
60	0.2	0.0	56.9	0.0
80	0.3	0.0	164.3	0.0
100	0.6	0.2	361.6	0.1
200	3.7	0.7	9145.4	0.5
400	20.3	5.9	-	-
600	65.8	20.5	-	-
700	98.6	44.4	-	-

Table 3. Elapsed time (in seconds) to process randomly generated $(\mathbb{Z}_2)^{n \times n}$ matrices and computing their corresponding rrefs using the Gauss-Jordan algorithm with Poly/ML 5.5 and MLton 20100608

	$\mathbb{Z}_2$ matrices			
	Poly/ML		MLton	
Size (n)	Processing Time (seconds)	Execution Time (seconds)	Processing Time (seconds)	Execution Time (seconds)
50	0.0	0.0	0.8	0.0
100	0.3	0.0	4.0	0.1
200	1.0	0.3	54.6	0.6
400	4.6	2.9	809.2	5.2
600	10.6	9.8	-	-
800	19.8	24.1	-	-
1000	31.8	45.1	-	-
1200	53.7	79.7	-	-
1400	65.6	143.0	-	-
1600	107.0	200.5	-	-

randomly generated matrices, since the first ones usually contain patterns which reduce the number of computations performed during the diagonalization.

The rank of matrices with entries in $\mathbb{Z}_2$ permits to know the number of connected components (and can be successfully applied to the computation of the number of synapses in a neuron, automating a cumbersome task previously made "by hand" by biologists) in the original image. See [13] for details about this technique. Additional benchmarks and extensive details on the previous and some other tests are presented in [3].

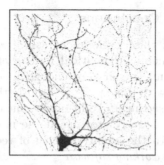

Fig. 1. Image (2048 × 2048 px.) of a neuron captured with a confocal microscope

6 Related Work and Further Work

6.1 Related Work

From the different theorem provers available in the HOL family, the ones with a better mathematical library are HOL-Light and Isabelle; this can be checked by reading through their libraries, and corroborated by informal but informative rankings such as [24]; our work here relies on the foundations that both systems share and has reused successfully the mathematical machinery that has been developed there; nevertheless, and to the best of our knowledge, both of them lack of implementations of numerical Linear Algebra; moreover, we do not know of any attempt of execution of the definitions available in that libraries. From our point of view, our work is a starting point to fill a gap between formalization and execution that aims to a greater use of these already powerful libraries.

Some other theorem provers have also formalized the computation of the rank of linear maps; for instance, the SSReflect library of Coq contains the most extensive effort to formalize finite-dimensional Linear Algebra concepts, aiming at providing a suitable library for the implementation of the classification of finite simple groups. The whole library is based upon finite-dimensional structures, and Coq itself is a constructive setting in which proofs and algorithms are intertwined, so that one would (erroneously) expect that an implementation of Gauss-Jordan elimination over matrices should be executable; as is well known [12, Sect.4], the extensive use of dependent types features in the representation of algebraic structures and matrices, which allows for relatively simple proofs, comes at a cost: these definitions have been locked to avoid the heavy computations that they would demand, since they may not finish in a reasonable amount of time. In an effort to offer executability of some of the concepts in the SSReflect library, a new library CoqEAL [4] has been carried out in which, by means of types and algorithms refinements, computable versions of, for instance, the rank of a matrix, are provided.

6.2   Further Work and Conclusions

We do not aim to present this development as a *canonical* approach to the the task of bringing together mathematical formalization and execution, but to show that proof assistants are mature enough to enable the simultaneous development of both fields with some technical effort (that once carried out, can be later reused in different settings). Additionally, one of the fields in which the Isabelle/HOL tool is more actively growing at the moment is data types and algorithms refinements, with the ambitious goal of reducing the gap between *software formalization* and *working software*.

The case study we have presented in this paper can be considered from at least two different points of view. First, as an experiment in Linear Algebra formalization, for which the HMA library has shown to be an adequate framework. With some technical effort in the code generation process, we have been capable of formalizing and executing the same "abstract" algorithm; in addition to this, we have developed tools (definitions and proofs over row and column elementary operations) that are applicable in the formalization of numerical Linear Algebra. Second, as an effort to get competitive results from a computational point of view; we have successfully applied some refinement techniques already available in Isabelle, obtaining formalized programs that can be executed over matrices of a remarkable size.

There are several directions we plan to take this work. Even if the performance of the Gauss-Jordan formalized algorithm is quite satisfying, some refinements could be thought of to reduce the number of operations that it performs; the algorithm could be implemented using *block matrices* that recursively decrease their size after each iteration of the algorithm. This would reduce the number of operations performed; on the other hand, it could demand the use of *dependent types* or *subtypes* to define submatrices (or some similar construct), falling short of the HOL type system.

Some other improvements of the algorithm are presented in the literature; for instance, instead of pivoting the first nonzero element over a given index of a column, the maximum element of the same column can be pivoted ("partial pivoting"), or even the maximum element in the whole *submatrix* ("total pivoting"); these strategies are experimentally known to improve the performance of the algorithm and specially its numerical stability. Instead of improving the performance of the Gauss-Jordan elimination algorithm, an *ad hoc* algorithm computing the rank of matrices could be implemented, and *linked* by a standard refinement technique with rank computation by Gauss-Jordan elimination.

There are further refinement techniques in Isabelle that we would like to explore as a natural continuation to our work. The work in [8] presents an infrastructure for *lifting* definitions from a *concrete* data type to an abstract one, and for *transferring* proofs from the abstract setting to the concrete one. The concept is really close to the one we have proposed in this paper, but at the moment the technology can be applied to Isabelle user defined types (as abstract type) and its underlying concrete types or quotient types. In our setting, it could have been used to lift definitions from *functions* to the type *vec*; it is also used in the code generation process of some of the fields that we used as examples. Another interesting Isabelle tool that we would like to explore is *Autoref* [18]; according to the authors, the tool automatically refines algorithms over abstract concepts to algorithms over concrete implementations; even if our underlying algebraic structures (vectors or matrices) are not completely "abstract", it could be interesting to explore the feasibility of writing down Linear Algebra algorithms in Isabelle in an imperative way (as they are usually presented in textbooks) and rely on the automatic refinement to translate these algorithms to executable ones in a functional programming setting, very much in the spirit of [17]. The previous tools and techniques could be applied to a wide range of Linear Algebra algorithms, some of them rooted in variants of Gauss-Jordan elimination.

Acknowledgments. This work has been supported by projects MTM2009-13842-C02-01 (Ministerio de Educación y Ciencia), FORMATH, nr. 243847, of the FET program within the FP7 of the European Commission, and Universidad de La Rioja, research grant FPI-UR-12.

Andreas Lochbihler provided us with great insight and invaluable ideas in how to get the right setup for code generation of sets, and also in understanding the type classes computing cardinality of types. Florian Haftmann helped us with the serialisation of the *real* Isabelle type to the *Real* SML structure. Johannes Hölzl assisted us in polishing our formalization of the Rank Nullity theorem. Julio Rubio suggested the use of *profiling* techniques to detect weaknesses in the execution experiments and commented on earlier versions of the paper. The authors also wish to thank the referees because of their valuable comments on the first version of this paper.

References

1. Aransay, J. Divasón, J.: Rank Nullity Theorem in Linear Algebra, Archive of Formal Proofs (2013). http://afp.sourceforge.net/entries/Rank_Nullity_Theorem. shtml
2. Aransay, J., Divasón, J.: Gauss-Jordan elimination in Isabelle/HOL (2013). http://www.unirioja.es/cu/jodivaso/Isabelle/Gauss-Jordan/
3. Aransay, J., Divasón, J.: Performance Analysis of a Verified Linear Algebra program in SML, Taller de Programación Funcional (TPF 2013), pp. 28–35. Lars-Åke Fredlund and Laura M. Castro (eds.) (2013)
4. Dénès, M., Mörtberg, A., Siles, V.: A Refinement-Based Approach to Computational Algebra in Coq. In: Beringer, L., Felty, A. (eds.) ITP 2012. LNCS, vol. 7406, pp. 83–98. Springer, Heidelberg (2012)
5. Esparza, J., Lammich, P., Neumann, R., Nipkow, T., Schimpf, A., Smaus, J.-G.: A Fully Verified Executable LTL Model Checker. In: Sharygina, N., Veith, H. (eds.) CAV 2013. LNCS, vol. 8044, pp. 463–478. Springer, Heidelberg (2013)
6. Formath Project: Formalisation of Mathematics. http://wiki.portal.chalmers.se/cse/pmwiki.php/ForMath
7. Haftmann, F., Nipkow, T.: Code Generation via Higher-Order Rewrite Systems. In: Blume, M., Kobayashi, N., Vidal, G. (eds.) FLOPS 2010. LNCS, vol. 6009, pp. 103–117. Springer, Heidelberg (2010)
8. Haftmann, F., Krauss, A., Kunčar, O., Nipkow, T.: Data Refinement in Isabelle/HOL. In: Blazy, S., Paulin-Mohring, C., Pichardie, D. (eds.) ITP 2013. LNCS, vol. 7998, pp. 100–115. Springer, Heidelberg (2013)

9. Harrison, J.: A HOL Theory of Euclidean Space. In: Hurd, J., Melham, T. (eds.) TPHOLs 2005. LNCS, vol. 3603, pp. 114–129. Springer, Heidelberg (2005)
10. Harrison, J.: The HOL Light Theory of Euclidean Space. J. Autom. Reasoning 50(2), 173–190 (2013)
11. Hales, T.C., Harrison, J., McLaughlin, S., Nipkow, T., Obua, S., Zumkeller, R.: A revision of the Proof of the Kepler Conjecture. Discrete & Computational Geometry 44(1), 1–34 (2010)
12. Heras, J., Coquand, T., Mörtberg, A., Siles, V.: Computing Persistent Homology within Coq/SSReflect. ACM Transactions on Computational Logic, 14(4). Article n. 26
13. Heras, J., Dénès, M., Mata, G., Mörtberg, A., Poza, M., Siles, V.: Towards a Certified Computation of Homology Groups for Digital Images. In: Ferri, M., Frosini, P., Landi, C., Cerri, A., Di Fabio, B. (eds.) CTIC 2012. LNCS, vol. 7309, pp. 49–57. Springer, Heidelberg (2012)
14. Heras, J., Poza, M., Dénès, M., Rideau, L.: Incidence Simplicial Matrices Formalized in Coq/SSReflect. In: Davenport, J.H., Farmer, W.M., Urban, J., Rabe, F. (eds.) MKM 2011 and Calculemus 2011. LNCS, vol. 6824, pp. 30–44. Springer, Heidelberg (2011)
15. Hölzl, J., et al.: HOL Multivariate Analysis (2013). http://isabelle.in.tum.de/dist/library/HOL/HOL-Multivariate_Analysis/index.html
16. Hölzl, J., Immler, F., Huffman, B.: Type Classes and Filters for Mathematical Analysis in Isabelle/HOL. In: Blazy, S., Paulin-Mohring, C., Pichardie, D. (eds.) ITP 2013. LNCS, vol. 7998, pp. 279–294. Springer, Heidelberg (2013)
17. Lammich, P., Tuerk, T.: Applying Data Refinement for Monadic Programs to Hopcroft's Algorithm. In: Beringer, L., Felty, A. (eds.) ITP 2012. LNCS, vol. 7406, pp. 166–182. Springer, Heidelberg (2012)
18. Lammich, P.: Automatic Data Refinement. In: Blazy, S., Paulin-Mohring, C., Pichardie, D. (eds.) ITP 2013. LNCS, vol. 7998, pp. 84–99. Springer, Heidelberg (2013)
19. Lochbihler, A.: Light-Weight Containers for Isabelle: Efficient, Extensible, Nestable. In: Blazy, S., Paulin-Mohring, C., Pichardie, D. (eds.) ITP 2013. LNCS, vol. 7998, pp. 116–132. Springer, Heidelberg (2013)
20. Nipkow, T.: Gauss-Jordan Elimination for Matrices Represented as Functions. Archive of Formal Proofs (2011). http://afp.sourceforge.net/entries/Gauss-Jordan-Elim-Fun.shtml
21. Nipkow, T., Paulson, L., Wenzel, M.: Isabelle/HOL: A proof assistant for Higher-Order Logic. Springer (2002)
22. Roman, S.: Advanced Linear Algebra (Third Edition). Springer (2008)
23. Gasner, E., Reppy, J.H. (eds.): The Standard ML Basis Library. http://www.standardml.org/Basis/
24. Wiedijk, F.: Formalizing 100 Theorems. http://www.cs.ru.nl/~freek/100/

Information Flow in Object-Oriented Software

Bernhard Beckert, Daniel Bruns$^{(\boxtimes)}$, Vladimir Klebanov, Christoph Scheben,
Peter H. Schmitt, and Mattias Ulbrich

Department of Informatics, Karlsruhe Institute of Technology (KIT),
Am Fasanengarten 5, 76131 Karlsruhe, Germany
bruns@kit.edu
http://www.key-project.org/DeduSec/

Abstract. This paper contributes to the investigation of object-sensitive information flow properties for sequential Java, i.e., properties that take into account information leakage through objects, as opposed to primitive values. We present two improvements to a popular object-sensitive non-interference property. Both reduce the burden on analysis and monitoring tools. We present a formalization of this property in a program logic – JAVADL in our case – which allows using an existing tool without requiring program modification. The third contribution is a novel fine-grained specification methodology. In our approach, arbitrary JAVADL terms (read 'side-effect-free Java expressions') may be assigned a security level – in contrast to security labels being attached to fields and variables only.

1 Introduction

The growing reliance of our daily lives on software systems of all kinds has increased the demand for software quality assurance. A particular concern is confidentiality of sensitive data: preventing information flow from secret (also called *high*) sources to publicly observable (also called *low*) sinks. Methods for specification and analysis of information flow play an important role in answering these concerns. Since the pioneering papers [11,13,14,21], research in information flow has grown considerably and diversified in numerous branches. This paper follows a language-based approach dealing with programs at the code level (instead of analyzing abstractions such as automata or process algebras). We will use a semantic definition of information flow, as e.g., introduced in [20]. For the analysis of information flow properties, we use a program logic, along the lines of [2,12], as opposed to the use of security type systems or dedicated analysis algorithms. Logical information flow analysis started out by investigating simple imperative programming languages and later also targeted object-oriented languages [1,9,28,32].

In imperative languages, sources and sinks are of primitive type. In an object oriented context, it is natural to consider sources and sinks of object type, too.

This work was supported by the German National Science Foundation (DFG) under project "Program-level Specification and Deductive Verification of Security Properties" within priority programme 1496 "Reliably Secure Software Systems – RS^3".

© Springer International Publishing Switzerland 2014
G. Gupta and R. Peña (Eds.): LOPSTR 2013, LNCS 8901, pp. 19–37, 2014.
DOI: 10.1007/978-3-319-14125-1_2

In this case, the usual definition of secure information flow – if a system is started in two low-equivalent states s_1, s_2 with all publicly observable values equal, then it terminates in states s_1', s_2' where all observable values are equal – is too strong. It has been replaced by object-sensitive secure information flow [3,6,9,17,18,28] that has a modified notion of low-equivalence of states: if a system is started in two states s_1, s_2 such that the observable values are related by a partial isomorphism π, then it terminates in states s_1', s_2' where all observable values are related by a partial isomorphism extending π.

A desirable outcome in developing logic-based information flow analyses is to find formalizations that allow application of existing verification tools. An instance of such reuse is [28], where the two runs of a program are encoded into a single program, allowing specifying secure information flow with JML and verifying it with the ESC/Java2 tool. The particular encoding, though, relies on ghost fields and requires instrumenting the program under investigation with ghost code. The first contribution of our paper is a formalization of object-sensitive secure information flow in Dynamic Logic that does not require any changes or additions to the investigated program. The KeY system [8] can be used to discharge the ensuing proof obligations.

To avoid loss of precision, it is reasonable to encode the partial isomorphisms of object-sensitive secure information flow explicitly in the logical formalization. This, on the other hand, holds the disadvantage that a naive encoding either increases the burden on the analysis or the burden on the user, the latter by requiring additional annotations [28]. The second contribution of this paper is an investigation into the concept of object-sensitive secure information flow itself with the aim to find alternative but equivalent formulations such that the partial isomorphisms can be restricted as much as possible. We prove (Lemma 4) that restricting the partial isomorphism π in the pre-state to be the identity still leads to an equivalent concept. We also show that additionally restricting the partial isomorphism in the post-state to newly created objects leads to a sufficient criterion for object-sensitive secure information flow (Thm. 2). Further we show that compositionality, which is considered an indispensable prerequisite for modular verification of information-flow properties and which holds for object-sensitive secure information flow only under certain conditions, holds for the sufficient criterion in general (Thm. 3). The main difference between the original property and the sufficient criterion is that the criterion admits the attacker the ability to distinguish between newly created objects and objects which already existed in the pre-state. This leads to a slightly stronger property. All three results hold the potential of significantly reducing the burden on analysis and monitoring tools. They apply under the assumption that references in Java are treated as opaque, as formalized by Postulate 1.

As a third contribution, we introduce a specification methodology where security levels (high or low) are assigned to arbitrary JAVADL terms. The set of publicly observable memory locations is thus state-dependent. This is in contrast to the typical static labeling of fields or variables only, and permits fine-grained specifications, which are especially useful for declassification.

2   Dynamic Logic for Java

In this section, we briefly review syntax and semantics of JAVADL, a Dynamic Logic for Java, as far as needed in this paper. An in-depth account can be found in [8,33]. JAVADL is an extension of classical typed first-order logic with equality (the equality symbol is denoted by $\doteq$), with which we assume the reader is familiar. The following explanations only address particularities and the modal extension.

The notion of a term in JAVADL is the same as in typed first-order logic. We assume that, among others, constant and function symbols are available for all local program variables, instance and static fields, this, result and method parameters, and operations of Java primitive data types. In addition, we make use of a special implicit program variable heap which stands for the current heap.

JAVADL formulas are inductively built up from atomic formulas using propositional operators and quantifiers, as usual. In addition

1. $\{a := t\}\phi$ is a JAVADL formula, where a is a term which refers to a location (a program variable, a static or dynamic field, or an array entry), t is a JAVADL term, and ϕ is a formula. $\{a := t\}$ is called an *update*.
2. For a JAVADL formula ϕ and any sequential Java program α, both $\langle\alpha\rangle\,\phi$ and $[\alpha]\,\phi$ are again JAVADL formulas.[1]

The basis JAVADL semantics is a structure $\mathcal{D}$ for typed first-order logic, called the *computation domain*. $\mathcal{D}$ provides the interpretation of all state-independent (sometimes also called *rigid*) function and predicate symbols. In our setup, program variables are the only non-rigid symbols. The universe D of $\mathcal{D}$ is divided into the interpretations $T^{\mathcal{D}}$ for the types T occurring in the language. In particular, we assume the existence of types Any, Obj, $Heap$, $Field$, Int with $Any^{\mathcal{D}} = D$, $Obj^{\mathcal{D}} = $ the set of all objects, $Heap^{\mathcal{D}} = $ the set of all heaps, $Field^{\mathcal{D}} = $ the set of all fields, $Int^{\mathcal{D}} = \mathbb{Z}$, $Seq^{\mathcal{D}} = $ the set of all finite nested sequences of values from D, and a subtype relation $\sqsubseteq$ such that the Java reference type hierarchy lies under $Obj \sqsubseteq Any$. Moreover, $Heap$, $Field$, and Obj are pairwise disjoint.

A *state* s is a function mapping all program variables to properly typed values in $\mathcal{D}$. By $\mathcal{D} + s$ we denote the first-order structure that interprets all, rigid and non-rigid, symbols. In most cases $\mathcal{D}$ will be implicitly understood and we write s instead of $\mathcal{D} + s$. For any state s and term t without logical variables, the evaluation t^s is as usual. If t contains logical variables, a variable assignment β is needed to evaluate the term to $t^{s,\beta}$. In the following, we will omit β whenever it is not essential. The (current) heap in a state s is completely determined by $\mathbf{heap}^s$: the value $(t.f)^s$ of a field access expression $t.f$ is obtained by $select(\mathbf{heap}^s, t^s, c_f)$. Here c_f is a constant symbol representing the Java field f,

---

[1] The definition is in fact more liberal in that α need not be a compilable program. Precisely which program sequences are allowed is explained in [8, Sect. 3.2.4]. We will nevertheless use the term 'program' synonymously.

$$\forall Int\; i((0 \le i \wedge i < maxvalue) \rightarrow \\ \{a := i\}\, \langle \alpha \rangle\, (0 \le \mathbf{r} \wedge \mathbf{r} * \mathbf{r} \le i \wedge (\mathbf{r}+1)*(\mathbf{r}+1) > i)) \tag{1}$$

$$\forall Heap\; h, h'\; \forall Int\; i, i'(\quad (select(h, \mathbf{this}, f) \doteq select(h', \mathbf{this}, f)\, \wedge \\ \{\mathbf{heap} := h\}\, \langle \mathbf{m}();\rangle\, i \doteq \mathbf{r}\; \wedge\; \{\mathbf{heap} := h'\}\, \langle \mathbf{m}();\rangle\, i' \doteq \mathbf{r}) \rightarrow i \doteq i') \tag{2}$$

Fig. 1. Two examples of JAVADL formulas

select and its counterpart *store* are state-independent functions from the theory of arrays, see [24, 29].

The recursive definition of the relation $s \models \phi$ (formula ϕ is true in state s) follows the usual pattern. Only the three modal operators need explanation. For a JAVADL formula ϕ and state s, we define:

1. $s \models \{a := t\}\phi$ iff $s' \models \phi$, where s' coincides with s except for $s'(a) = t^s$.
2. $s \models \langle \alpha \rangle\, \phi$ iff $s' \models \phi$ for some s' such that α started in s terminates in s'.
3. $s \models [\alpha]\, \phi$ iff $s' \models \phi$ for all s' such that α started in s terminates in s'.

If program α does not terminate when started in state s, then $s \models [\alpha]\, \phi$ is trivially true for all formulas ϕ, including $\phi \equiv \textit{false}$.

Let us look at the examples of Fig. 1. Formula (1) expresses that program α computes the positive integer square root for any positive input a (**result** is abbreviated by $\mathbf{r}$). Formula (2) states that the return value of method $\mathbf{m}()$ only depends on the field $\mathbf{this}.f$. Logical variables cannot occur in programs and program variables may not be quantified over. As these examples demonstrate, updates can be used as an interface between both types of variables.

We adopt the *constant domain* approach (see for instance [8, 33]), i.e., all potential objects are contained in D from the start. The generation of a new object of type T in state s is effected by changing the value of $o.\textit{created}$ from $\textit{ff}$ to $\textit{tt}$, where $o = \textit{nextToCreate}_T{}^{\mathcal{D}}(s)$, with $\textit{nextToCreate}_T{}^{\mathcal{D}}$ being the function which selects the next new object of type T to create depending on the state. In this paper computation domains $\mathcal{D}_1, \mathcal{D}_2$ will at most differ in the interpretation $\textit{nextToCreate}_T{}^{\mathcal{D}_i}$. Only functions $\textit{nextToCreate}_T{}^{\mathcal{D}}$ with $\textit{created}^{\mathcal{D}+s}$ $(\textit{nextToCreate}_T{}^{\mathcal{D}}(s)) = \textit{ff}$ and $\textit{exactInstance}_T{}^{\mathcal{D}}(\textit{nextToCreate}_T{}^{\mathcal{D}}(s)) = \textit{tt}$ are considered.

Let α be a program, $\mathcal{D}$ a computation domain and let s_1, s_2 be states. We denote "α started in $\mathcal{D} + s_1$ terminates in $\mathcal{D} + s_2$" by $\mathcal{D} + s_1 \overset{\alpha}{\leadsto} \mathcal{D} + s_2$.

3   Information Flow in Java

In Fig. 2 we reproduce a typical example of object-sensitive information flow. If low-equivalence of states required the values of x and y to be equal, method $\mathbf{m1}()$ would be rated as insecure. However, we treat object references in Java as

```
final class C {
    static C x, y;        // low variables
    static boolean h;     // high variable
    static void m1() { if (h) {x = new C(); y = new C();}
                        else   {y = new C(); x = new C();} } }
```

Fig. 2. Secure object creation

opaque, i.e., references can only be compared by the == function, cf. [23]. Thus
m1() obviously does not leak information.[2]

We describe publicly (and thus attacker-) visible parts of the program state
as sets of JAVADL terms. The attacker sees the term and the corresponding
evaluation in the pre- and post-state of a method as if they were printed on
a screen. Further, we assume that the attacker knows the program code. This
allows them to trace back the observed differences in low values in the post-state
to high values in the pre-state. In summary, an attacker *can* compare observed
values that are of a primitive type to each other and to literals (of that type) as
by using ==; *can* compare observed values of object reference type to each other
and to null as by using the == predicate and observe their (runtime) type and
the length attribute for array references; *cannot* learn more than object identity
from object references (e.g., the order in which objects have been generated
cannot be learned).

Formally, we call a sequence of JAVADL terms (which itself is a JAVADL
term), an *observation expression*. The low locations of Fig. 2, for instance, give
rise to the observation expression $\langle \text{C.x}, \text{C.y} \rangle$. Let R be an observation expression
and s a state. An attacker is able to observe the tuple (R, R^s), where $R^s =
\langle e_1^s, \ldots, e_k^s \rangle$ if $R = \langle e_1, \ldots, e_k \rangle$. Hence, they are able to deduce for any $1 \leq i \leq k$
that e_i^s is the value of the term e_i. Additionally, an attacker can learn the result of
the comparison of any two values $e_i^s = e_j^s$ and, in case of reference values, retrieve
their runtime type $type(e_i^s)$ and, for array references, their length $len(e_i^s)$.

Definition 1. *By $Obj(R^s)$ we denote the set of objects observable by R in state s,
that is, $Obj(R^s) = \{o \in Obj^{\mathcal{D}} \mid \exists i(o = R^s[i])\} \cup \bigcup_{i \in \{j \mid R^s[j] \in Seq^{\mathcal{D}}\}} Obj(R^s[i])$.*

In an object-oriented setting, what is observable may depend on the state.
For example, if *o.next.val* is observable, then it depends on the state what
object *o.next* evaluates to. Moreover, if all locations in a linked list are observed,
then the *number* of observable locations may depend on the state, since the list
length does.

Observation expressions cover such cases: JAVADL includes a sequence definition operator $seq\{i\}(from, to, e)$ with the semantics $[seq\{i\}(from, to, e)]^s =$

---

[2] In [19] it has been demonstrated that this abstraction might be broken, e.g., by
the implementation of native methods such as Object::hashCode().This potential
leakage can be dealt with by assigning a high security level to the output of native
methods or by using the security type analysis proposed in the quoted paper.

$\langle [e^{[i \to n]}]^s, \ [e^{[i \to n+1]}]^s, \ \ldots, \ [e^{[i \to m-1]}]^s \rangle$, if $from^s = n < m = to^s$ are integers. Here $e^{[i \to n]}$ is the term obtained from e by replacing all occurrences of the variable i by the literal n. Further JAVADL contains a reachability operator $e.it(f, i)$, where e is a JAVADL term of type T, f is an attribute defined in class T and also of type T and i is an integer term. The semantics of $e.it(f, i)$ is defined by $[e.it(f, i)]^s = f. \ldots .f([e]^s)$ (k times) with $k = i^s$. The observation of all elements of a linked list can be modeled by the observation expression $seq\{i\}(0, \texttt{list.len}, \texttt{list}.it(\texttt{next}, i).\texttt{val})$.

Here and in the following we abbreviate length by len. Further, we write sequences of fixed length as $\langle \ldots \rangle$ and denote the concatenation of two sequences R_1 and R_2 by $R_1; R_2$. For uniformity of notation we will frequently write $f(e_0)$ instead of $e_0.f$.

Our approach generalizes and unifies declassification of terms [4,31]. It already proved to be useful in a recent case-study [15] which uses our approach and implementation. Here, whether information is considered secret or public depends on the internal state of the system. Therefore, the information flow specifications of [15] make use of conditional terms. Another application is information flow class invariants: if a program or library has a public interface with several methods, then often it has to be ensured that any sequence of calls to those methods is secure. For this purpose it is useful to define the knowledge of the caller by a list of terms. The program is secure, if for any method of the interface the final values of those terms depend at most on their initial values. An illustrating example can be found in the companion technical report [7].

4   Isomorphisms

We assume that the reader is familiar with the concept of isomorphism for typed structures [25]. In this section we collect the results needed later on for easy reference.

We will consider isomorphisms only on the computation domain $\mathcal{D}$, and the structures $\mathcal{D} + s$ (see Section 2) for different states s. If π is an isomorphism from $\mathcal{D} + s_1$ onto $\mathcal{D} + s_2$, we will say that s_2 is isomorphic to s_1 and write $s_2 = \pi(s_1)$. We will need the following (folklore) results:

Lemma 1. *Let ρ be an automorphism of $\mathcal{D}$, s a state, ϕ a formula, e an expression. Then* $\quad s \models \phi \Leftrightarrow \rho(s) \models \phi \quad$ *and* $\quad e^{\rho(s)} = \rho(e^s)$.

Lemma 2. *Let $\mathcal{D}$ be a computation domain and π' be a bijection from X onto Y for finite subsets $X, Y \subseteq Obj^{\mathcal{D}}$ with*

1. *If null $\in X$ then $\pi'(\text{null}) = \text{null}$ and null $\in Y$ implies null $\in X$.*
2. *π' preserves the exact types of its arguments.*
3. *π' preserves the length of array objects.*

Then there is a computation domain $\mathcal{D}'$ and an isomorphism $\pi : \mathcal{D} \to \mathcal{D}'$ extending π'.

Definition 2 (Partial isomorphism w.r.t. R). *Let R be an observation expression and s_1, s_2 be two states.*

A partial isomorphism *with respect to R from s_1 to s_2 is a bijection π : $Obj(R^{s_1}) \rightarrow Obj(R^{s_2})$ such that (a) the requirements of Lemma 2 hold and (b) $\pi_{Seq}(R^{s_1}) = R^{s_2}$ where π_{Seq} is defined on sequences as $\pi_{Seq}(\langle e_1, \ldots, e_k \rangle) = \langle e_1', \ldots, e_k' \rangle$ with $e_i' = \pi(e_i)$ if $e_i \in Obj^{\mathcal{D}}$, $e_i' = \pi_{Seq}(e_i)$ if $e_i \in Seq^{\mathcal{D}}$ and $e_i' = e_i$ else.*

It will greatly simplify notation to stipulate that every partial isomorphism π is also defined on all primitive values w with $\pi(w) = w$.

If $p \in R$ for all program variables p, every automorphism extending a partial isomorphism π with respect to R according to Lemma 2 is a total isomorphism from $\mathcal{D} + s_1$ onto $\mathcal{D} + s_2$ since $\pi(p^{s_1}) = p^{s_2}$ by requirement (b).

Not every partial isomorphism can be extended to a total isomorphism, on the other hand. If q is a program variable such that q does not appear as a subterm in R, then $\pi(q^{s_1}) = q^{s_2}$ is not required.

To clarify the role of the additional condition (b) in Def. 2 let x be a program variable of type C and f a field in C, say of type integer such that $R = \langle x, f(x) \rangle$ and let s_1, s_2 be states. In this case the condition implies $\pi((f(x))^{s_1}) = (f(x))^{s_2} = f^{s_2}(x^{s_2}) = f^{s_2}(\pi(x^{s_1}))$. This amounts to the usual requirements on isomorphisms on mathematical structures.

5   Formalizing Information Flow

As mentioned before, we treat object references as opaque. This means in particular that the behavior of a Java program cannot depend on the values of references up to comparison by ==. Hence, if a program α is started in two isomorphic states, then α also terminates in isomorphic states (if α terminates.) Though this assumption is not always made explicit, it is widely used in the literature [1,3,26,28]. Opaqueness of references can be formalized in our setting as follows:

Postulate 1. *Let s_1, s_2 be states. Let α be a program which started in s_1 terminates in s_2, and let $\rho : \mathcal{D} \rightarrow \mathcal{D}'$ be an isomorphism from computation domain $\mathcal{D}$ onto computation domain $\mathcal{D}'$.*

Then α started in $\mathcal{D}' + \rho(s_1)$ terminates in $\rho'(s_2)$, where $\rho' : \mathcal{D} \rightarrow \mathcal{D}'$ is an isomorphism that coincides with ρ on all objects existing in state s_1, i.e. for all $o \in Obj^{\mathcal{D}}$ with $created^{s_1}(o) = tt$ we know $\rho(o) = \rho'(o)$. (See beginning of Sect. 4 for the definition of $\rho(s_i)$.)

The reason why we cannot assume $\rho = \rho'$, is that α may generate new objects and there is no reason why a new element o' generated in the run starting in state $\mathcal{D}' + \rho(s_1)$ should be the ρ-image of the new element o generated in the run of α starting in state $\mathcal{D} + s_1$.

5.1   Basic Information Flow Definition and Its Properties

We start with formalizing the basic object-sensitive non-interference property for Java. Apart from the more flexible assignment of security levels, this property does not yet exceed the state of the art in object-sensitive non-interference (cf. Sect. 1). We consider here the termination-insensitive case. Extensions taking termination into account, as well as differentiating between normal and abnormal termination, are straightforward.

Definition 3 (Agreement of states). *Let R be an observation expression. We say that two states s, s' agree on R, abbreviated by* agree(R, s, s'), *iff there exists a partial isomorphism $\pi : Obj(R^{s_1}) \rightarrow Obj(R^{s_2})$ with respect to R. The partial isomorphism π is uniquely determined by R, s and s'. We use the notation* agree(R, s, s', π) *to indicate that* agree(R, s, s') *is true and π is the mapping thus defined.*

Notice that because of our tacit agreement on the values of partial isomorphisms on primitive values, agree(R, s, s') entails $(e_i)^s = (e_i)^{s'}$, if e_i is a term of primitive type.

We now define what it means for a program α (when started in a state s) to allow information flow only from R_1 to R_2, a fact which we denote by flow(s, α, R_1, R_2). The intuition is that R_1 describes the low locations in the pre-state and R_2 describes the low locations in the post-state. Thus, the values of the variables and locations in R_2 in the post-state must at most depend – up to isomorphism of states – on the values of the variables and locations in R_1 in the pre-state and on nothing else.

Definition 4 (The predicate flow**).** *Let α be a program and R_1 and R_2 be two observation expressions.*

Program α allows information to flow only from R_1 to R_2 when started in s_1, denoted by flow(s_1, α, R_1, R_2), *iff, for computation domains $\mathcal{D}, \mathcal{D}'$ and all states s'_1, s_2, s'_2 such that $\mathcal{D} + s_1 \overset{\alpha}{\rightsquigarrow} \mathcal{D} + s_2$ and $\mathcal{D}' + s'_1 \overset{\alpha}{\rightsquigarrow} \mathcal{D}' + s'_2$, we have*

if agree(R_1, s_1, s'_1, π^1) *for some π^1*
then agree(R_2, s_2, s'_2, π^2) *for some π^2 that is compatible with π^1*

where π^2 is said to be compatible with π^1 if
$\pi^2(o) = \pi^1(o)$ *for all $o \in Obj(R_1^{s_1}) \cap Obj(R_2^{s_2})$ with created$^{s_1}(o) = tt$.*

In the most common case, the *low* locations before program execution will be the same as the *low* locations after program execution, i.e., $R_1 = R_2$. But, that may not be true in all cases. To *declassify* an expression e_{decl}, one would choose $R_1 = R_2; e_{decl}$.

Consider the following extension of method m1() from Fig. 2 as if defined in class C:

```
C next;
static void m2() { if(h) {x=new C(); y=new C(); x.next=y;}
                   else {y=new C(); x=new C(); x.next=x;} }
```

Whether m2() leaks information or not depends on the examined observation expression. For $R = \langle C.x, C.y \rangle$ the observation will always consist of two freshly created, distinct object references. If $\text{agree}(R, s_1, s_1', \pi^1)$, the partial isomorphism π^2 defined as an extension of π^1 by $\pi_2(x^{s_2}) = x^{s_2'}$ and $\pi_2(y^{s_2}) = y^{s_2'}$ ensures that $\text{agree}(R, s_2, s_2', \pi^2)$ and, therefore, $\text{flow}(s_1, \text{m2}(), R, R)$.

But if $R' = \langle C.x, C.y, C.x.next \rangle$ is chosen, π^2 is no longer a partial isomorphism as $\pi^2(next^{s_2}(x^{s_2})) = next^{s_2'}(x^{s_2'})$ would need to hold. But if $h^{s_1} = tt$ and $h^{s_1'} = f\!f$, the resulting heap structures are not isomorphic: $\pi^2(next^{s_2}(x^{s_2})) = \pi^2(y^{s_2})$ and $next^{s_2'}(x^{s_2'}) = x^{s_2'} = \pi^2(x^{s_2})$ which cannot be equal as π^2 is an injection. The attacker can learn the value of h by comparing x and $x.next$: $\text{flow}(s_1, \text{m2}(), R', R')$ does not hold.

For later reference we state the following lemma.

Lemma 3. *If $\text{agree}(R, s, s', \pi)$ and ρ is an automorphism on $\mathcal{D}$ then also (1) $\text{agree}(R, s, \rho(s'), \rho \circ \pi)$ and (2) $\text{agree}(R, \rho(s), s', \pi \circ \rho^{-1})$.*

Proof. Part 1: By assumption the mapping π given by $\pi_{Seq}(R^s) = R^{s'}$ is a partial isomorphism, where π_{Seq} is defined as in Def. 2. Since π is a partial isomorphism and ρ is an automorphism also $\rho \circ \pi$ is a partial isomorphism. Further, let $(\rho \circ \pi)_{Seq}$ be defined as $(\rho \circ \pi)_{Seq}(\langle e_1, \ldots, e_k \rangle) = \langle e_1', \ldots, e_k' \rangle$ with $e_i' = \rho \circ \pi(e_i)$ if $e_i \in Obj^{\mathcal{D}}$, $e_i' = (\rho \circ \pi)_{Seq}(e_i)$ if $e_i \in Seq^{\mathcal{D}}$ and $e_i' = e_i$ else. Then $(\rho \circ \pi)_{Seq}(R^s) = \rho \circ \pi_{Seq}(R^s)$, because ρ is an automorphism on $\mathcal{D}$, and $\rho \circ \pi_{Seq}(R^s) = \rho(R^{s'})$, because of $\pi_{Seq}(R^s) = R^{s'}$. Finally we derive by Lemma 1 $\rho(R^{s'}) = R^{\rho(s')}$ and thus we have $(\rho \circ \pi)_{Seq}(R^s) = R^{\rho(s')}$. Hence $\text{agree}(R, s, \rho(s'), \rho \circ \pi)$ holds.

Part 2: By symmetry from Part 1.   □

5.2   An Optimized but Equivalent Formulation

In this section, we introduce flow*, an optimized version of the flow property from Def. 4. The property flow* restricts the partial isomorphism of the pre-state to be the identity. This simplifies the formulation of verification conditions considerably (see Theorem 1 below), also making them easier to verify. Yet, it is semantically equivalent to flow.

Definition 5 (The flow* predicate). *Let α be a program and R_1 and R_2 be two observation expressions.*

We say that α allows simple information flow only from R_1 to R_2 when started in s_1, denoted by $\text{flow}^*(s_1, \alpha, R_1, R_2)$, iff, for all computation domains $\mathcal{D}$, $\mathcal{D}'$ and states s_1', s_2, s_2' such that $\mathcal{D} + s_1 \overset{\alpha}{\rightsquigarrow} \mathcal{D} + s_2$ and $\mathcal{D}' + s_1' \overset{\alpha}{\rightsquigarrow} \mathcal{D}' + s_2'$, we have

if $\text{agree}(R_1, s_1, s_1', id)$
then $\text{agree}(R_2, s_2, s_2', \pi^2)$ for some π^2 compatible with id.

Note that $\text{agree}(R_1, s_1, s_1', id)$ implies in particular $Obj(R_1^{s_1}) = Obj(R_1^{s_1'})$ since $\pi^1 = id$ is a bijection from $Obj(R_1^{s_1})$ onto $Obj(R_1^{s_1'})$.

Lemma 4. *For all programs α, any two observation expressions R_1 and R_2, and any state s_1 $\text{flow}^*(s_1, \alpha, R_1, R_2)$ ⇔ $\text{flow}(s_1, \alpha, R_1, R_2)$.*

Proof. flow(s_1, α, R_1, R_2) $\Rightarrow$ flow*(s_1, α, R_1, R_2) is obviously true. Thus it suffices to show flow*(s_1, α, R_1, R_2) $\Rightarrow$ flow(s_1, α, R_1, R_2).

To prove flow(s_1, α, R_1, R_2) we fix, in addition to s_1, states s_1', s_2, s_2' such that $s_1 \overset{\alpha}{\rightsquigarrow} s_2$ and $s_1' \overset{\alpha}{\rightsquigarrow} s_2'$, and assume agree($R_1, s_1, s_1', \pi^1$). We need to show agree(R_2, s_2, s_2', π^2) with π^2 extending π^1.

By Lemma 2, there is an automorphism ρ on $\mathcal{D}'$ extending $(\pi^1)^{-1}$. From agree(R_1, s_1, s_1', π^1) we conclude agree($R_1, s_1, \rho(s_1'), \rho \circ \pi^1$) using Lemma 3. Since ρ extends $(\pi^1)^{-1}$ we have agree($R_1, s_1, \rho(s_1'), id$). By Postulate 1. there is a state s_3' and a computation domain $\mathcal{D}''$ such that $\mathcal{D}'' + \rho(s_1') \overset{\alpha}{\rightsquigarrow} \mathcal{D}'' + s_3'$. This enables us to make use of the assumption flow*(s_1, α, R_1, R_2) and conclude agree(R_2, s_2, s_3', π^3). Furthermore, $\pi^3(o) = o$ for all $o \in Obj(R_1^{s_1}) \cap Obj(R_2^{s_2})$.

Again, appealing to Postulate 1. in the situation that $\rho(s_1') \overset{\alpha}{\rightsquigarrow} s_3'$ and considering the inverse automorphism ρ^{-1}, we obtain an automorphism ρ' such that $\rho^{-1}(\rho(s_1')) = s_1' \overset{\alpha}{\rightsquigarrow} \rho'(s_3')$ and ρ' coincides with ρ^{-1} on all objects in $\{o \in Obj^{\mathcal{D}} \mid created^{\rho(s_1')}(o) = tt\}$.

Again, using Lemma 3, this time for the isomorphism ρ', we obtain from agree(R_2, s_2, s_3', π^3) also agree($R_2, s_2, \rho'(s_3'), \rho' \circ \pi^3$). Since α is a deterministic program and we have already defined s_2' to be the final state of α when started in s_1' in the computation domain $\mathcal{D}'$ we get $s_2' = \rho'(s_3')$ and thus agree($R_2, s_2, s_2', \rho' \circ \pi^3$). Because π^2 is uniquely determined by R_2, s_2 and s_2', we have $\rho' \circ \pi^3 = \pi^2$.

Finally, we show that $\rho' \circ \pi^3$ extends π^1, i.e., for every $o \in Obj(R_1^{s_1}) \cap Obj(R_2^{s_2})$ with $created^{s_1}(o) = tt$ we need to show $\rho' \circ \pi^3(o) = \pi^1(o)$. Since $\pi^3(o) = o$ for $o \in Obj(R_1^{s_1}) \cap Obj(R_2^{s_2})$ it suffices to show $\pi^1(o) = \rho'(o)$. By the definition of isomorphic states we obtain from $created^{s_1}(o) = tt$ also $created^{\rho(s_1)}(o) = tt$. Thus $\rho'(o) = \rho^{-1}(o)$ and by choice of ρ further $\rho^{-1}(o) = \pi^1(o)$, as desired. $\square$

6   Verification Conditions

The ultimate goal is to prove information flow properties flow(s_1, α, R_1, R_2) for particular observations R_i and a program α. To this end, specialized proof rules for the flow predicate could be introduced. We pursue this approach in another paper. Here, we will show how to derive verification conditions directly from the definition. We will show how flow(s_1, α, R_1, R_2) can be expressed by a JAVADL formula – to be discharged by a standard JAVADL calculus. This exposition should also convey the idea how to obtain verification conditions with methodologies other than Dynamic Logic.

Theorem 1. *Let α be a program, and let R_1, R_2 be observation expressions. There is a JAVADL formula ϕ_{α, R_1, R_2} making use of self-composition such that*

$$\mathcal{D} + s_1 \models \phi_{\alpha, R_1, R_2} \qquad \text{iff flow}(s_1, \alpha, R_1, R_2)$$
for all computation domains $\mathcal{D}$.

We will explain here the construction of ϕ_{α, R_1, R_2} only. The complete proof of Thm. 1 can be found in the companion technical report [7].

The property to be formalized requires quantification over states. A state s is determined by the value of the heap h^s in s and the values of the (finitely many) program variables a^s in s. We can directly quantify over heaps h and refer to the value of a field f of type C for the object o referenced by the term e as $selectC(h, e, f)$. We cannot directly quantify over program variables, as opposed to quantifying over the values of program variables, which is perfectly possible. Thus we use quantifiers $\forall x$, $\exists x$ over the type domain of the variable and assign x to a via an update $a := x$. There are four states involved, the two pre-states s_1, s_1' and the post-states s_2, s_2'. Correspondingly, there will be, for every program variable v, four universally quantifier variables v, v_1', v_2, v_2' of appropriate type representing the values of v in states s_1, s_1', s_2, s_2'. There are some program variables that make only sense in pre-states, e.g., **this**, and variables that make only sense in post-state, e.g., **result**. There will be only two logical variables that supply values to them instead of four. This leads to the following schematic form of ϕ_{α, R_1, R_2}:

$$\phi_{\alpha, R_1, R_2} \equiv \forall Heap\ h_1', h_2, h_2' \forall To' \forall T_r r, r' \forall \ldots v_1', v_2, v_2' \ldots$$
$$(Agree_{pre} \wedge \langle \alpha \rangle\ \mathrm{sv}\{s_2\} \wedge \{\mathrm{in}\ s_1'\}\ \langle \alpha \rangle\ \mathrm{sv}\{s_2'\} \rightarrow \{\mathrm{in}\ s_2\}\{\mathrm{in}\ s_2'\}(Agree_{post} \wedge Ext))$$

To maintain readability we have used suggestive abbreviations: (1) $\{\mathrm{in}\ s_1'\}\ \langle \alpha \rangle$ signals that an update $\{$**heap** $:= h_1'$ $\|$ **this** $:= o'$ $\|$ $\ldots a_i := v_1' \ldots\}$ is placed before the modal operator. The a_i cover all relevant parameters and local variables. (2) The construct $\mathrm{sv}\{s_2\}$ abbreviates a conjunction of equations $h_2 \doteq$ **heap**, $r \doteq$ **result**, $\ldots$, $v_2 \doteq a_i$, $\ldots$. (3) Analogously, $\mathrm{sv}\{s_2'\}$ stands for the primed version $h_2' \doteq$ **heap**, $r' \doteq$ **result**, $\ldots$, $v_2' \doteq a_i$, $\ldots$. (4) The shorthand $\{\mathrm{in}\ s_2\}\{\mathrm{in}\ s_2'\}E$ in front of a formula is resolved by (a) prefixing every occurrence of a heap-dependent term e with the update $\{$**heap** $:= h_2\}$ and (b) every primed term e' with $\{$**heap** $:= h_2'\}$. (5) The same applies to $\{\mathrm{in}\ s_1'\}E$. Note that there is no $\{\mathrm{in}\ s_1\}$, and no quantified variables o, v since the whole formula ϕ_{α, R_1, R_2} is evaluated in state s_1.

Furthermore we use the notation $(R_i^1)'$, R_i^2, $(R_i^2)'$ for the expressions obtained from R_i by replacing each state dependent designator v by v_1', v_2, v_2' respectively. Technically, these substitutions are effected by prefixing R_i with an appropriate update. For short we use $R[i]$ instead of $seqGet_{Any}(r, i)$, $t \sqsubseteq A$ for $instance_A(t)$, and $eInst_A$ for $exactInstance_A$.

We now supply the definitions of the abbreviations used above:

$Agree_{pre} \equiv R_1 \doteq (R_1^1)'$

$Agree_{post} \equiv Agree_{type\&prim}(R_2^2, (R_2^2)') \wedge Agree_{obj}(R_2^2, R_2^2, (R_2^2)', (R_2^2)')$

$Ext \equiv Agree_{obj}(R_1, R_2^2, (R_1^1)', (R_2^2)')$

These definitions make use of the predicates $Agree_{type\&prim}$, $Agree_{obj}$ and $Agree_{obj}^2$ which are recursively defined as

$Agree_{type\&prim}(Seq\ X, Seq\ X') \equiv$
$\quad X.\mathbf{len} \doteq X'.\mathbf{len} \wedge \forall i(0 \leq i < X.\mathbf{len} \rightarrow$
$$\bigwedge_{A\ \text{in}\ \alpha}(eInst_A(X[i]) \leftrightarrow eInst_A(X'[i]))$$
$$\wedge\ (X[i] \not\sqsubseteq Obj \wedge X[i] \not\sqsubseteq Seq \rightarrow X[i] \doteq X'[i])$$
$$\wedge\ (X[i] \sqsubseteq Seq \rightarrow Agree_{type\&prim}(X[i], X'[i])))$$

$Agree_{obj}(Seq\ X, Seq\ Y, Seq\ X', Seq\ Y') \equiv$
$$\forall i (0 \le i < Y.\mathbf{len} \to (Y[i] \in Obj \to Agree^2_{obj}(X, Y[i], X', Y'[i]))$$
$$\wedge\ (Y[i] \in Seq \to Agree_{obj}(X, Y[i], X', Y'[i])))$$
$Agree^2_{obj}(Seq\ X, Obj\ y, Seq\ X', Obj\ y') \equiv$
$$\forall i (0 \le i < X.\mathbf{len} \to (X[i] \doteq Obj \to (X[i] \doteq y \leftrightarrow X'[i] \doteq y'))$$
$$\wedge\ (X[i] \in Seq \to Agree^2_{obj}(X[i], y, X'[i], y')))$$

In many cases these definitions are much simpler. Frequently it is the case that $R_i.\mathbf{length}$ is not state dependent, then quantification over index i reduces to a disjunction of fixed length. Also the exact type of an expression can often be checked syntactically and needs not be part of the formula. In other cases however, e.g., if R_i is a variable of type Seq, the full definition is necessary.

Reconsider method m1() from Fig. 2 on page 23. Let $R = \langle C.x, C.y \rangle$. Then, $(R.\mathbf{len})^s = 2$ for all states s and the exact type of both fields x, y is always C. Thus $Agree_{pre}$ equals $x \doteq x'_1 \wedge y \doteq y'_1$. $Agree_{post}$ equals $x_2 \doteq y_2 \leftrightarrow x'_2 \doteq y'_2$. The complete formula $\phi_{m3(),R,R}$ is (after some simplification)

$$\phi_{m3(),R,R} \equiv$$
$$\forall Heap\ h'_1, h_2, h'_2 \forall C\ o' \forall x'_1, x_2, x'_2, y'_1, y_2, y'_2 ((x \doteq x'_1 \wedge y \doteq y'_1\ \wedge$$
$$\langle m3() \rangle (x_2 \doteq x \wedge y_2 \doteq y)\ \wedge\ \{x := x'_1, y := y'_1\} \langle m3() \rangle (x'_2 \doteq x \wedge y'_2 \doteq y))$$
$$\to$$
$$(x_2 \doteq y_2 \leftrightarrow x'_2 \doteq y'_2\ \wedge\ x \doteq x_2 \to x'_1 \doteq x'_2\ \wedge\ y \doteq x_2 \to y'_1 \doteq x'_2\ \wedge$$
$$x \doteq y_2 \to x'_1 \doteq y'_2\ \wedge\ y \doteq y_2 \to y'_1 \doteq y'_2))$$

7 An Efficient Compositional Criterion

Though flow* from Def. 5 already simplifies the formulation of verification conditions and consequently checking for flow, we want to present another information flow property, flow**, which is still simpler to check. flow** is a criterion for flow, i.e., a sufficient but not a necessary condition. Roughly speaking, the main difference between flow and flow** is that flow** admits the attacker to distinguish between newly created objects and objects which already existed in the pre-state. This property of flow** is responsible for its compositionality (Thm. 3), which is an indispensable prerequisite for *modular* verification of information-flow properties. On the face of it, flow** takes more words to explain than the original flow property, but it is easier to prove: the partial isomorphism only differs from the identity on new objects. This reduces the effort to verify flow** considerably if only few or no new objects are created. Also, there is no obligation that one isomorphism is an extension of another.

On the other hand, an additional observation expression N_2 has to be given which exactly names the new elements of the set of objects observable in the post-state. Further it has to be proven that N_2 exactly names the new elements. However, normally it is quite an easy task to prove whether an object is newly created or not. Additionally, if a newly created object is observable in the post-state by an observation expression R_2, then there has to be a term in R_2 which evaluates to this object. Hence N_2 is normally an explicit subexpression of R_2 and can be named easily.

Definition 6 (The predicate flow****).** *Let N_2 be an observation expression such that all terms in N_2 are of object type. Let, furthermore, α be a program, R_1, R_2 observation expressions, and s_1 a state.*

The predicate flow$^{**}(s_1, \alpha, R_1, R_2, N_2)$ *is true iff, for all computation domains $\mathcal{D}$, $\mathcal{D}'$ and states s_1', s_2, s_2' such that $\mathcal{D} + s_1 \overset{\alpha}{\rightsquigarrow} \mathcal{D} + s_2$ and $\mathcal{D}' + s_1' \overset{\alpha}{\rightsquigarrow} \mathcal{D}' + s_2'$, we have*

if agree(R_1, s_1, s_1', id)

then all objects in $Obj(N_2^{s_2})$ and $Obj(N_2^{s_2'})$ are new and
   agree(N_2, s_2, s_2', π) *for a partial isomorphism π and*

   if agree(N_2, s_2, s_2', id) *then*   agree(R_2, s_2, s_2', id)

Theorem 2. *Let N_2 be an observation expression such that all expressions in N_2 are of object type. Let furthermore α be a program, R_1, R_2 observation expressions, and s_1 a state.*

1. flow$^{**}(s_1, \alpha, R_1, R_2, N_2) \Rightarrow$ flow(s_1, α, R_1, R_2).
2. *If*   *for all domains $\mathcal{D}$ such that $\mathcal{D} + s_1 \overset{\alpha}{\rightsquigarrow} \mathcal{D} + s_2$ we have $Obj(N_2^{s_2}) = \{o \in Obj(R_2^{s_2}) \mid created^{s_1}(o) = f\!f\}$ and $\{o \in Obj(R_2^{s_2}) \mid created^{s_1}(o) = t\!t\} \subseteq Obj(R_1^{s_1})$*
   then   flow$(s_1, \alpha, R_1, R_2) \Rightarrow$ flow$^{**}(s_1, \alpha, R_1, R_2, N_2)$.

For the proof of the theorem we need the following auxiliary lemma. It states that we always can find domains $\mathcal{D}_2$, $\mathcal{D}_2'$ and therefore *nextToCreate* functions such that in two runs of a program α, which are started in R equivalent states, the same new objects are chosen for those objects which are observable by R.

Lemma 5. *Let α be a program such that $\mathcal{D} + s_1 \overset{\alpha}{\rightsquigarrow} \mathcal{D} + s_2$, $\mathcal{D}' + s_1' \overset{\alpha}{\rightsquigarrow} \mathcal{D}' + s_2'$, agree$(R, s_1, s_1', id)$ and agree(N, s_2, s_2', π) hold true for observation expressions R and N. In addition we assume that all objects in $Obj(N^{s_2})$ and $Obj(N^{s_2'})$ are new.*

Then there are domains $\mathcal{D}_2$, $\mathcal{D}_2'$ and isomorphisms $\rho : \mathcal{D} \to \mathcal{D}_2$, $\rho' : \mathcal{D}' \to \mathcal{D}_2'$ such that α started in $\mathcal{D}_2 + s_1$ terminates in $\mathcal{D}_2 + \rho(s_2)$, α started in $\mathcal{D}_2' + s_1'$ terminates in $\mathcal{D}_2' + \rho'(s_2')$ and agree$(N, \rho(s_2), \rho'(s_2'), id)$ and $\rho(o) = o$, $\rho'(o') = o'$ for all o existing in state s_1 and for all o' existing in state s_1'.

We omit the proof of Lemma 5 and go for the proof of Thm. 2 instead.

Proof (Theorem 2).
   Part 1: We assume flow$^{**}(s_1, \alpha, R_1, R_2, N_2)$ and show flow$^*(s_1, \alpha, R_1, R_2)$. To this end we fix states s_1', s_2, s_2' and domains $\mathcal{D}$, $\mathcal{D}'$ such that $\mathcal{D} + s_1 \overset{\alpha}{\rightsquigarrow} \mathcal{D} + s_2$, $\mathcal{D}' + s_1' \overset{\alpha}{\rightsquigarrow} \mathcal{D}' + s_2'$ and agree(R_1, s_1, s_1', id). We need to show agree(R_2, s_2, s_2', π), where the uniquely determined partial isomorphism π is compatible with id.

   By assumption we obtain agree(N_2, s_2, s_2', σ) and we know that all objects in $Obj(N_2^{s_2})$ and $Obj(N_2^{s_2'})$ are new. By Lemma 5 there are domains $\mathcal{D}_2$, $\mathcal{D}_2'$ and isomorphisms $\rho : \mathcal{D} \to \mathcal{D}_2$, $\rho' : \mathcal{D}' \to \mathcal{D}'_2$ such that α started in $\mathcal{D}_2 + s_1$ terminates in $\mathcal{D}_2 + \rho(s_2)$, α started in $\mathcal{D}'_2 + s'_1$ terminates in $\mathcal{D}'_2 + \rho'(s'_2)$, and agree$(N_2, \rho(s_2), \rho'(s'_2), id)$. This enables us to use flow$^{**}(s_1, \alpha, R_1, R_2, N_2)$ again, now for the domains $\mathcal{D}_2$, $\mathcal{D}'_2$ in place of $\mathcal{D}$, $\mathcal{D}'$ to obtain agree$(R_2, \rho(s_2), \rho'(s'_2), id)$. Another appeal to Lemma 3 yields agree$(R_2, s_2, s'_2, \rho' \circ \rho^{-1})$. For $o \in Obj(R_1^{s_1}) \cap Obj(R_2^{s_2})$ we have $\rho' \circ \rho^{-1}(o) = o$, thus $\rho' \circ \rho^{-1}$ is compatible with id and the claim is proved.

Part 2: For the reverse implication we assume flow(s_1, α, R_1, R_2).

For the proof of flow$^{**}(s_1, \alpha, R_1, R_2, N_2)$ we consider states s'_1, s_2, s'_2 and domains $\mathcal{D}, \mathcal{D}'$ such that $\mathcal{D} + s_1 \overset{\alpha}{\leadsto} \mathcal{D} + s_2$, $\mathcal{D}' + s'_1 \overset{\alpha}{\leadsto} \mathcal{D}' + s'_2$ and agree(R_1, s_1, s'_1, id). From flow(s_1, α, R_1, R_2) we obtain agree(R_2, s_2, s'_2, π) for π compatible with id. By case assumption we know $Obj(N_2^{s_2}) = \{o \in Obj(R_2^{s_2}) \mid created^{s_1}(o) = \mathit{ff}\}$. We see that π is a partial isomorphism from $Obj(N_2^{s_2})$ onto $Obj(N_2^{s'_2})$. This already gives us agree(N_2, s_2, s'_2, π). We assume agree(N_2, s_2, s'_2, id) to verify the remaining part of flow$^{**}(s_1, \alpha, R_1, R_2, N_2)$ with the intention to show agree(R_2, s_2, s'_2, id).

By agree(R_2, s_2, s'_2, π) we already know $\pi_{Seq}(R_2^{s_2}) = R_2^{s'_2}$, where π_{Seq} is defined as $\pi_{Seq}(\langle e_1, \dots, e_k \rangle) = \langle e'_1, \dots, e'_k \rangle$ with $e'_i = \pi(e_i)$ if $e_i \in Obj^{\mathcal{D}}$, $e'_i = \pi_{Seq}(e_i)$ if $e_i \in Seq^{\mathcal{D}}$ and $e'_i = e_i$ else. It remains to be shown that $\pi(e_i) = e_i$ for $e_i \in Obj(R_2^{s_2})$. We distinguish two cases: (1) $created^{s_1}(e_i) = tt$ and (2) $created^{s_1}(e_i) = \mathit{ff}$.

In case (1) we obtain $e_i \in Obj(R_1^{s_1})$ by the assumption $\{o \in Obj(R_2^{s_2}) \mid created^{s_1}(o) = tt\} \subseteq Obj(R_1^{s_1})$. Hence $\pi(e_i) = e_i$ since π is compatible with id. In case (2) use assumptions agree(N_2, s_2, s'_2, id) and $Obj(N_2^{s_2}) = \{o \in Obj(R_2^{s_2}) \mid created^{s_1}(o) = \mathit{ff}\}$, and also arrive at $\pi(e_i) = e_i$. $\qquad\square$

The next lemma shows that the verification condition for flow** normally is much simpler than the one for flow*.

Lemma 6. *Let α be a program, let R_1, R_2, N_2 be observation expressions. Then there is a JAVADL formula $\phi_{\alpha, R_1, R_2, N_2}$ such that for all states s_1* flow$^{**}(s_1, \alpha, R_1, R_2, N_2) \quad \Leftrightarrow \quad s_1 \models \phi_{\alpha, R_1, R_2, N_2}$.

Proof. The desired formula follows a pattern similar to the one in Thm. 1.

$$\phi_{\alpha, R_1, R_2, N_2} \equiv \forall Heap\ h'_1, h_2, h'_2 \forall To' \forall T_r r, r' \forall \dots v'_1, v_2, v'_2 \dots$$
$$(R_1 \doteq (R_1^1)' \wedge \langle \alpha \rangle \text{ save}\{s_2\} \wedge \text{in}\{s'_1\} \langle \alpha \rangle \text{ save}\{s'_2\}$$
$$\to \{\text{in } s_2\}\{\text{in } s'_2\}(newIso \wedge (N_2^2 \doteq (N_2^2)' \to R_2^2 \doteq (R_2^2)')))$$

The abbreviations used above are defined as follows:

$newIso \equiv newOn(\text{heap}, N_2^2) \wedge newOn(h'_1, (N_2^2)') \wedge Agree_{type}(N_2^2, (N_2^2)') \wedge$
$\qquad Agree_{obj}(N_2^2, N_2^2, (N_2^2)', (N_2^2)')$
$newOn(Heap\ h, Seq\ X) \equiv$
$\qquad \forall i (0 \le i < X.\textbf{len} \to \ (X[i] \sqsubseteq Obj \to select(h, X[i], created) \doteq FALSE)$
$\qquad\qquad \wedge (X[i] \sqsubseteq Seq \to newOn(h, X[i]))))$

$Agree_{type}(Seq\ X, Seq\ X') \equiv$

$X.\textbf{len} \doteq X'.\textbf{len} \wedge \forall i (0 \leq i < X.\textbf{len} \rightarrow$
$$\bigwedge_{A\ in\ \alpha}(eInst_A(X[i]) \leftrightarrow eInst_A(X'[i]))$$
$$\wedge\ (X[i] \sqsubseteq Seq \rightarrow Agree_{type}(X[i], X'[i])))$$

$Agree_{obj}$ as in Thm. 1. We skip the rest of the proof, since it greatly parallels the one given for Thm. 1.   □

We now show the compositionality of flow**. To this end we need to prove that flow** implies that the set of objects, which can be observed by an attacker in the post-state, contains only objects which are newly created or which already have been observed in the pre-state.

Lemma 7. *Let s_1, s_1', s_2, s_2' be states such that $s_1 \overset{\alpha}{\leadsto} s_2$ and $s_1' \overset{\alpha}{\leadsto} s_2'$.*
flow$^{**}(s_1, \alpha, R_1, R_2, N_2)$ implies flow(s_1, α, R_1, R_2) and agree$(R_1, s_1, s_1') \Rightarrow \{o \in Obj(R_2^{s_2}) \mid created^{s_1}(o) = tt\} \subseteq Obj(R_1^{s_1})$.

Lemma 7 in combination with Thm. 2 gives an almost complete characterization of flow**. Indeed we can show that flow$^{**}(s_1, \alpha, R_1, R_2, N_2)$ also implies agree$(R_1, s_1, s_1') \Rightarrow \{o \in Obj(R_2^{s_2}) \mid created^{s_1}(o) = f\!f\} \subseteq Obj(N_2^{s_2})$ which makes this characterization tight. This characterization shows in particular that the main difference between flow and flow** is that, roughly speaking, flow** admits the attacker to distinguish between newly created objects and objects which already existed in the pre-state. This property of flow** is responsible for its compositionality:

Theorem 3 (Compositionality of flow**). *Let s_1, s_1', s_2, s_2', s_3, s_3' be states such that $s_1 \overset{\alpha_1}{\leadsto} s_2$, $s_2 \overset{\alpha_2}{\leadsto} s_3$, $s_1' \overset{\alpha_1}{\leadsto} s_2'$ and $s_2' \overset{\alpha_2}{\leadsto} s_3'$. If*

1. *flow$(s_1, \alpha_1, R_1, R_2)$,*
2. *flow$(s_2, \alpha_2, R_2, R_3)$,*
3. *agree$(R_1, s_1, s_1') \Rightarrow \{o \in Obj(R_2^{s_2}) \mid created^{s_1}(o) = tt\} \subseteq Obj(R_1^{s_1})$ and*
4. *agree$(R_2, s_2, s_2') \Rightarrow \{o \in Obj(R_3^{s_3}) \mid created^{s_2}(o) = tt\} \subseteq Obj(R_2^{s_2})$*

then
flow$(s_1, \alpha_1; \alpha_2, R_1, R_3)$ and agree$(R_1, s_1, s_1') \Rightarrow \{o \in Obj(R_3^{s_3}) \mid created^{s_1}(o) = tt\} \subseteq Obj(R_1^{s_1})$.

We omit the proofs of Lemma 7 and Thm. 3 for the sake of brevity, but they can be found in the companion technical report [7].

8   Related Work

There exists a very large body of work on language-based security. Besides the discussion below, we refer to [30] for a survey.

Security type systems are one of the most popular approaches. A prominent example in this field is the JIF system [26]. Type system approaches are efficient, but sometimes also quite imprecise. A further approach is checking the dependence graph of a program for graph-theoretical reachability properties [16]. Though this technique is substantially different from type system approaches, it is efficient and sometimes quite imprecise, too. Further approaches use abstraction and ghost code for explicit tracking of dependencies [10]. They are quite near in spirit to flow-sensitive security type systems, but have not tackled the problem of modular verification yet. All approaches mentioned so far appear to be limited to information flow between variables and it is questionable whether they can be adopted to fine-grained specifications as the one introduced in this paper.

The most popular approach in logic based information flow analysis is stating secure information flow with the help of self-composition [5,12] and using off-the-shelf software verification systems to check for it, as we do. The approach has the appealing feature that it can be arbitrarily precise as long as the used verification system has a relatively complete calculus. An important alternative in logic based information flow analysis is the usage of specialized, approximate calculi [1]. Finally, secure information flow can be formalized in higher-order logic, and higher-order theorem provers like Coq can be used for checking secure information flow [27]. This approach seems to be very expressive, but comes at the price of more and more complex interactions with the proof system.

Focusing on *object-sensitive* secure information flow, the paper closest to ours is [1]. The authors build on *region logic*, a kind of Hoare logic with concepts from separation logic, which is comparable to JavaDL. They use the same basic definition of object-sensitive secure information flow. Instead of providing verification conditions which can be discharged with a standard calculus, as we do, they introduce a specialized, more efficient calculus to show object-sensitive secure information flow. This specialized calculus uses approximate rules which avoid explicit modeling of isomorphisms, but comes with the price of imprecision. The discerning points of our work are: (1) a further investigation of the security property, allowing the restriction of isomorphisms as far as possible and thus making the explicit, non approximate modeling of isomorphisms feasible with a minimum of additional user interaction; (2) verification conditions that are discharged with an existing tool; and (3) a more flexible specification methodology.

Contributions (1) and (3) also distinguish this work from the other approaches mentioned above, including JIF, which already presented an approximative treatment of object-sensitive secure information flow for Java [26]. JIF is a practical approach to the analysis of secure information flow which covers a broad range of language features, but it has not been formally proven to enforce non-interference. Similar to JIF, [3,6] use type systems for the verification of object-oriented secure information flow. They treat a smaller set of language features, but prove that their type systems indeed enforce non-interference. A closely related approach is [9]. Here, only the information flow analysis is based on type systems; the verification task is separated from the analysis and based on program logics. Still, points (1) and (3) as well as the overall precision are discerning points of this paper. The approach in [6], already mentioned above, and the approaches [17,18] target Java Bytecode in contrast to source code, as the other approaches do. The latter is a type system approach, too, whereas the former uses abstract interpretation in combination with classical static analysis.

To the best of our knowledge, the only approach which models isomorphisms explicitly is the self-composition approach [28]. The drawback of that approach is that the specifier needs to track the isomorphism manually with the help of additional ghost code annotations. This increases the burden on the specifier, whereas our approach detects the isomorphism automatically.

Focusing on fine-grained information flow specifications, the approaches closest to ours are [4,31]. These approaches specify information flow between variables and fields only, but allow for the declassification of terms. Our approach generalizes and unifies these approaches. This generalization already proved to be useful in a recent case-study, see Sect. 3.

9   Conclusions and Future Work

Lemma 4 and Thm. 2 prove the relation between standard object-sensitive non-interference and our improved versions. These results lead to an approach to verify object-sensitive non-interference properties of Java programs by a direct translation into Dynamic Logic (Thm. 1 and Lemma 6). The approach has been implemented in the KeY tool and successfully tested on small examples. The implementation can be tested on our web page using Java Web Start. In particular, we have successfully treated the examples included in this paper, as well as the (somewhat more involved) examples by Naumann [28]. Application to a larger e-voting case study is currently underway.

In a future paper we plan to present a complementary specialized calculus for the flow predicate intended to further increase reasoning efficiency. As proved in Thm. 3, the flow** criterion is compositional and is expected to lead to a particularly efficient calculus. A specification interface to the Java Modeling Language (JML) [22] for information flow properties has been published in [31].

References

1. Amtoft, T., Bandhakavi, S., Banerjee, A.: A logic for information flow in object-oriented programs. In Proceedings POPL, pp. 91–102. ACM (2006)
2. Amtoft, T., Banerjee, A.: Information Flow Analysis in Logical Form. In: Giacobazzi, R. (ed.) SAS 2004. LNCS, vol. 3148, pp. 100–115. Springer, Heidelberg (2004)
3. Banerjee, A., Naumann, D.A.: Secure information flow and pointer confinement in a Java-like language. In: Proceedings CSFW (2002)
4. Banerjee, A., Naumann, D.A., Rosenberg, S.: Expressive declassification policies and modular static enforcement. In: IEEE Symposium on Security and Privacy. SP 2008, pp. 339–353. IEEE (2008)
5. Barthe, G., D'Argenio, P.R., Rezk, T.: Secure information flow by self-composition. CSFW 2004, pp. 100–115. IEEE CS, Washington, USA (2004)

6. Barthe, G., Pichardie, D., Rezk, T.: A certified lightweight non-interference Java bytecode verifier. Mathematical Structures in Comp. Sci., FirstView:1–50, 4 (2013)
7. Beckert, B., Bruns, D., Klebanov, V., Scheben, C., Schmitt, P.H., Ulbrich, M.: Information flow in object-oriented software : Extended version. Technical Report 2013-14, KIT (2013)
8. Beckert, B., Hähnle, R., Schmitt, P.H.: Verification of Object-Oriented Software. LNCS, vol. 4334. Springer, Heidelberg (2007)
9. Beringer, L., Hofmann, M.: Secure information flow and program logics. In: CSF, pp. 233–248 (2007)
10. Bubel, R., Hähnle, R., Weiß, B.: Abstract Interpretation of Symbolic Execution with Explicit State Updates. In: de Boer, F.S., Bonsangue, M.M., Madelaine, E. (eds.) FMCO 2008. LNCS, vol. 5751, pp. 247–277. Springer, Heidelberg (2009)
11. Cohen, E.S.: Information transmission in computational systems. In: SOSP, pp. 133–139 (1977)
12. Darvas, A., Hähnle, R., Sands, D.: A Theorem Proving Approach to Analysis of Secure Information Flow. In: Hutter, D., Ullmann, M. (eds.) SPC 2005. LNCS, vol. 3450, pp. 193–209. Springer, Heidelberg (2005)
13. Denning, D.E.: A lattice model of secure information flow. Commun. ACM **19**(5), 236–243 (1976)
14. Goguen, J.A., Meseguer, J.: Security policies and security models. In: IEEE Symposium on Security and Privacy, pp. 11–20 (1982)
15. Greiner, S., Birnstill, P., Krempel, E., Beckert, B., Beyerer, J.: Privacy preserving surveillance and the tracking paradox. In: Proceedings, Future Security Conference 2013, Berlin (2013). (To appear September 15–19, 2013)
16. Hammer, C., Krinke, J., Snelting, G.: Information flow control for Java based on path conditions in dependence graphs. In: ISSSE, pp. 87–96. IEEE (March 2006)
17. Hansen, R.R., Probst, C.W.: Non-interference and erasure policies for Java Card bytecode. In: WITS (2006)
18. Hedin, D., Sands, D.: Timing aware information flow security for a JavaCard-like bytecode. In: BYTECODE, vol. 141:1 of ENTCS, pp. 163–182 (2005)
19. Hedin, D., Sands, D.: Noninterference in the presence of non-opaque pointers. In: CSFW, pp. 217–229. IEEE Computer Society (2006)
20. Joshi, R., Leino, K.R.M.: A semantic approach to secure information flow. Science of Computer Programming **37**(1–3), 113–138 (2000)
21. Lampson, B.W.: A note on the confinement problem. Commun. ACM **16**(10), 613–615 (1973)
22. Leavens, G.T., Baker, A.L., Ruby, C.: JML: a Java Modeling Language. In: Formal Underpinnings of Java Workshop (at OOPSLA 1998) (October 1998)
23. Lindholm, T., Yellin, F.: Java Virtual Machine Specification, 2nd edn. Addison-Wesley Longman Publishing Co. Inc., Boston (1999)
24. McCarthy, J.: Towards a mathematical science of computation. Information Processing, pp. 21–28 (1962)
25. Mitchell, J.C.: Type systems for programming languages. In: Handbook of Theoretical Computer Science, Volume B: Formal Models and Sematics, pp. 365–458 (1990)
26. Myers, A.C.: JFlow: Practical mostly-static information flow control. In: POPL, pp. 228–241 (1999)
27. Nanevski, A., Banerjee, A., Garg, D.: Verification of information flow and access control policies with dependent types. In: SP, pp. 165–179 (2011)

28. Naumann, D.A.: From Coupling Relations to Mated Invariants for Checking Information Flow. In: Gollmann, D., Meier, J., Sabelfeld, A. (eds.) ESORICS 2006. LNCS, vol. 4189, pp. 279–296. Springer, Heidelberg (2006)
29. Ranise, S., Tinelli, C.: The SMT-LIB standard: Version 1.2. Tr, U. of Iowa (2006)
30. Sabelfeld, A., Myers, A.C.: Language-based information-flow security. IEEE Journal on Selected Areas in Communications 21(1), 5–19 (2003)
31. Scheben, C., Schmitt, P.H.: Verification of Information Flow Properties of JAVA Programs without Approximations. In: Beckert, B., Damiani, F., Gurov, D. (eds.) FoVeOOS 2011. LNCS, vol. 7421, pp. 232–249. Springer, Heidelberg (2012)
32. Sun, Q., Banerjee, A., Naumann, D.A.: Modular and Constraint-Based Information Flow Inference for an Object-Oriented Language. In: Giacobazzi, R. (ed.) SAS 2004. LNCS, vol. 3148, pp. 84–99. Springer, Heidelberg (2004)
33. Weiß, B.: Deductive Verification of Object-Oriented Software: Dynamic Frames, Dynamic Logic and Predicate Abstraction. PhD thesis, KIT (2011)

A Transformational Approach to Resource Analysis with Typed-Norms

Elvira Albert[1], Samir Genaim[1], and Raúl Gutiérrez[2]([✉])

[1] Complutense University of Madrid, Madrid, Spain
[2] DSIC, Universitat Politècnica de València,
Camino de Vera S/N, 46022 Valencia, Spain
`rgutierrez@dsic.upv.es`

Abstract. In order to automatically infer the resource consumption of programs, analyzers track how *data sizes* change along a program's execution. Typically, analyzers measure the sizes of data by applying *norms* which are mappings from data to natural numbers that represent the sizes of the corresponding data. When norms are defined by taking type information into account, they are named *typed-norms*. The main contribution of this paper is a transformational approach to resource analysis with typed-norms. The analysis is based on a transformation of the program into an *intermediate abstract program* in which each variable is abstracted with respect to all considered norms which are valid for its type. We also sketch a simple analysis that can be used to automatically infer the required, useful, typed-norms from programs.

1 Introduction

Automated resource analysis [17] needs to infer how the sizes of data are modified along a program's execution. Size is measured using so-called norms [5] which define how the size of a term is computed. Examples of norms are *list-length* which counts the number of elements of a list, *tree-depth* which counts the depth of a tree, *term-size* which counts the number of constructors, etc. Basically, in order to infer the resource consumption of executing a loop that traverses a data-structure, the analyzer tries to infer how the size of such data-structure decreases at each iteration w.r.t. the chosen norm. Given a tree t, using a term-size norm, we infer that a function like "def Int foo(Tree t)= case t {Leaf ↦0; Node(l,r) ↦ 1+foo(r);}" performs at most nodes(t) iterations, where function nodes returns the number of nodes in the tree. This is because size analysis infers that at each recursive call nodes(t) decreases. However, by using the tree-depth norm, we will infer that depth(t) is an upper bound on the number of iterations. The latter is obviously more precise than the former bound as depth(t)≤nodes(t).

This work was funded partially by the EU project FP7-ICT-610582 ENVISAGE: Engineering Virtualized Services (http://www.envisage-project.eu) and by the Spanish projects TIN2008-05624 and TIN2012-38137. Raúl Gutiérrez is also partially supported by a Juan de la Cierva Fellowship from the Spanish MINECO, ref. JCI-2012-13528.

© Springer International Publishing Switzerland 2014
G. Gupta and R. Peña (Eds.): LOPSTR 2013, LNCS 8901, pp. 38–53, 2014.
DOI: 10.1007/978-3-319-14125-1_3

The last two decades have witnessed a wealth of research on using norms in termination analysis, especially in the context of logic programming [5,6,9]. Early work pointed out that the choice of norm affects the precision such that the analyzer may only succeed to prove termination if a certain norm is used, while it cannot prove it with others. Later on, there has been further investigation on applying multiple norms, i.e., using two or more norms by applying them simultaneously [5]. This means that the same data in the original program is replaced by two or more abstract data each one specifying its size information w.r.t. the corresponding norm. Even a further step has been taken on using *typed-norms* which allow defining norms based on type information (namely on recursive types) [6]. Inferring norms from type information makes sense as recursive types represent recursive data-structures and thus, in termination analysis, they identify some potential sources of infinite recursion and, in resource analysis, they might influence the number of iterations that the loops perform. Besides, typed-norms allow that the same term can be measured differently depending on its type. As pointed out in [9], this is particularly useful when the same function symbol may occur in different type contexts.

In the context of resource analysis, we found early work that already pointed out that the combination of norms affects the precision of lower-bound time analysis [11]. Sized-types provide a way to consider more than one norm for each type. They have been used in the context of functional [15,16] and recently in logic programming [12]. In the former case, they are inferred by a type analysis and in the latter via abstract interpretation. In contrast, we propose a transformational approach which provides a simple and accurate way to use multiple typed-norms in resource analysis as follows: (1) we first transform the program into an *intermediate abstract program* in which each variable is abstracted with respect to all considered norms valid for its type, (2) such intermediate program is then analyzed to obtain upper and lower resource bounds automatically. Importantly, this second phase is done using existing techniques that do not need to be modified. Thus, formalizing our framework focuses only on the first step.

While allowing multiple norms might lead to more accurate bounds than adopting one norm, the efficiency of the analysis can be degraded considerably. This is because the process of finding resource bounds from abstractions that have more arguments (due to the use of multiple norms) is more costly. Thus, an essential aspect for the practical applicability of our method is to eliminate those abstractions that will not lead to further precision. As our second contribution, we outline an algorithm for the inference of typed-norms which, by inspecting the program, can detect which norms are useful to later infer the resource consumption, and discard norms that are useless for this purpose. This analysis is applied as a pre-process, such that once the relevant norms are inferred, the transformation into the abstract program is carried out w.r.t. the inferred norms.

Syntactic categories. Definitions.

T in Ground Type	$T ::= B \mid D$
B in Basic Type	$B ::= \text{Int} \mid \text{String}$
D in Data type	$Dd ::= \text{data } D = Cons[\mid \overline{Cons}];$
x in Variable	$Cons ::= Co[(\overline{T})]$
e in Expression	$F ::= \text{def } T \; fn(\overline{T\ x}) = e;$
t in Ground Term	$e ::= x \mid t \mid Co[(\overline{e})] \mid fn(\overline{e}) \mid \text{case } e \; \{\overline{br}\}$
br in Branch	$t ::= n \mid Co[(\overline{t})]$
p in Pattern	$br ::= p \Rightarrow e;$
n in Integer	$p ::= \_ \mid x \mid t \mid Co[(\overline{p})]$

Fig. 1. Syntax for the functional level. Terms $\overline{T}$ and $\overline{e}$ denote possibly empty lists over the corresponding syntactic categories, and square brackets [] optional elements.

2 The Language

We present the simple functional language on which our framework is defined. It corresponds to the functional sublanguage of ABS [10], a modeling language for concurrent distributed systems which has been used to implement two industrial case studies (both of them of more than 1,000 lines of code). The functional sublanguage of ABS is used to define and manipulate the data structures used in the program, while the imperative sublanguage is used to handle its concurrency and distribution aspects. The reason why we chose ABS is double: first, because the funcional part of the language is appropriate to present our results in a clear and simple manner; and second, because our final goal is to integrate typed-norms in the complexity analysis of concurrent and distributed systems modeled in ABS. Sec. 2.1 defines the syntax of our functional language, and Sec. 2.2 introduces the intermediate form to which the programs are translated to define the analysis later.

2.1 A Simple Functional Language

The language defines data types and functions, as shown in Fig. 1. Ground types T consist of basic types B as well as names D for data types. In data type declarations Dd, a data type D has at least one constructor $Cons$, which has a name Co and a list of ground types T for its arguments. Function declarations F consist of a return type T, a function name fn, a list of variable declarations $\overline{x}$ of types $\overline{T}$, and an expression e. *Expressions* e include variables x, (ground) terms t, constructor expressions $Co(\overline{e})$, function expressions $fn(\overline{e})$ and case expressions case $e \; \{\overline{br}\}$. Ground terms t are integer numbers and constructors applied to ground terms $Co(\overline{t})$. Case expressions have a list of branches $p \Rightarrow e$, where p is a pattern. The branches are evaluated in the listed order. Patterns include wild cards $\_$, variables x, terms t, and constructor patterns $Co(\overline{p})$. Abusing notation, fn in e can be a function name or a built-in function $(+, -, >, =, \geq)$. We assume that the considered programs are well-typed and unambiguous.

```
 1 module Library;                        14 def Int is_coauthor(Author a,Authors as)
 2 type Author = String ;                 15  = case as {
 3 type Title = String ;                   16    Nil => 0;
 4 data Authors = Nil                       17    Cons(a,as') => 1;
 5   | Cons(Author,Authors);                18    Cons(a',as')=> is_coauthor(a, as'); };
 6 data Titles = Nil                        19
 7   | Cons(Title,Titles);                 20 def Titles written_by(String a,Books bs)
 8 data Book=Pair(Title,Authors);          21  = case bs {
 9 data Books = EmptyMap                    22    EmptyMap => Nil;
10   | InsertAssoc(Book,Books);            23    InsertAssoc(b,bs')
11 data Ref = Pair(Author,Titles);         24     => case b {
12 data Refs = EmptyMap                     25       Pair(t,as)
13   | InsertAssoc(Ref,Refs);              26        => case is_coauthor(a, as) {
                                            27          1 => Cons(t,written_by(a,bs'));
                                            28          0 => written_by(a,bs'); }; }; };

29  def Refs sort_books_by_author(Authors as,Books bs)
30  = case as {
31     Nil => EmptyMap;
32     Cons(a,as')
33      => InsertAssoc(Pair(a,written_by(a,bs)),sort_books_by_author(as', bs)); };
```

Fig. 2. Motivating example (data type declarations and three functions)

Example 1. Our running example is showed in Fig. 2. It defines a function sort_books_by_author (and several auxiliary functions) for sorting books by author given a list of authors and a list of books.

2.2 Intermediate Form

From now on, we develop our analysis on a typed intermediate representation (IR) similar to those defined in [2,8,13,14]. The translation from our simple functional language to the IR is straightforward and follows exactly the same steps as the one formalized in [2]. Essentially, the IR of each function is obtained by translating each basic block in its control flow graph (CFG) into a procedure, defined by means of *rules* that adhere to the following grammar:

$r ::= m(\bar{x}, y) \mapsto g, b_1, \ldots, b_n.$
$b ::= x{:=}t \mid m(\bar{x}, y)$
$g ::= true \mid g \wedge g \mid e\ op\ e \mid match(x, t) \mid nonmatch(x, t)$
$t ::= e \mid Co(\bar{t})$
$e ::= x \mid n \mid e{+}e \mid e{-}e$

where $op \in \{>, =, \geq\}$, $m(\bar{x}, \bar{y})$ is the *head* of the rule, g specifies the conditions for the rule to be applicable and $b_1, \ldots, b_n$ is the rule's *body*. Calls are of the form $m(\bar{x}, y)$ where the variables $\bar{x}$ are the properly typed formal parameters and the variable y is the properly typed return value. Guards $match(x, t)$ and $nonmatch(x, t)$ simulate case-expressions and x and t are of the same type. We assume $x \notin vars(t)$. Terms are constructed using $Co(\bar{t})$, where Co is a data symbol and $\bar{t}$ are the arguments (e.g., $\mathsf{Cons}(x, y)$), variables x, integer numbers n and arithmetic expressions ($e + e$ and $e - e$). A function is thus defined by a (global) set of rules. The dynamics of the data-structures are preserved by using the guard *match*, which fixes the shape of the input variables in the rules.

Example 2. Fig. 3 shows the IR of function is_coauthor. For each function definition, we have a rule with the same number of arguments plus a new argument at the end that represents the output of the function call. The case expression is split into three new rules, one rule for each possible matching alternative.

def Int is_coauthor(Author a, Authors as)
   = **case** as {
     Nil => 0;
     $\mathsf{Cons}(a,as')$ => 1;
     $\mathsf{Cons}(a',as')$
     => is_coauthor(a, as');
   };
	is_coauthor(a,as,y) $\mapsto$ case$_0$(a,as,y).
	   case$_0$(a,as,y) $\mapsto$ match(as,Nil),
	         $y := 0$.
	   case$_0$(a,as,y) $\mapsto$ nonmatch(as,Nil),
	         match(as,$\mathsf{Cons}(a,as')$),
	         $y := 1$.
	   case$_0$(a,as,y) $\mapsto$ nonmatch(as,Nil),
	         nonmatch(as,$\mathsf{Cons}(a,as')$),
	         match(as,$\mathsf{Cons}(a',as')$),
	         is_coauthor(a,as',y).

Fig. 3. IR of function is_coauthor from the example in Fig. 2

3   Size Abstraction Using Typed-Norms

The cost analysis framework that we rely on [2] is performed in two steps: (1) the program is first transformed into an abstract version that is used to track how the sizes of the different data-structures change, when moving from one control point to another; and (2) the abstract program is then analyzed to infer lower and upper bounds on the resource consumption. As the second step remains unchanged, we focus only on the first step.

    Abstract programs are obtained from the source program (in the intermediate form) as follows: (1) the program variables are replaced by numerical variables that represent their corresponding sizes; and (2) the instructions are replaced by linear constraints, over the new variables, to simulate the effect of their execution on the sizes of the corresponding data-structures. When data refer to numerical values, their sizes are defined as their values, and when they refer to data-structures then size functions, commonly known as *norms*, are used to measure their sizes. Note that our goal is not to obtain the real size of data-structures, but to use the data-size information to obtain a more accurate complexity of the recursions in the program.

3.1 Preliminaries on Typed-Norms

Among all norms in the literature, the *term-size* norm is probably the most well-known one. It has been introduced, and intensively used, in the context of termination analysis of logic programs. Intuitively, it counts the number of data constructors in a given data-structure, and can be defined as follows:

$$\|t\|_{ts} = \begin{cases} 1 + \sum_{i=1}^{n} \|t_i\|_{ts} & \text{if } t = Co(t_1, \ldots, t_n) \\ 1 & \text{otherwise} \end{cases} \tag{1}$$

The main shortcoming of the term-size norm is that it considers all data types equal, which leads to imprecision when used in the context of cost analysis.

Example 3. The recursive function written_by in the example traverses Authors and Books recursive data-structures. Using term-size norm, a static analysis obtains that the complexity is $O(n^2)$, because each recursion in the data-structure is abstracted to n. However, it is more accurate if we can say that the complexity is $O(bs \times as)$ where bs refers to the number of books and as the maximum length of the lists Authors for each of the books in bs, because recursions are applied to different data-types.

To overcome the imprecision issues discussed above we use typed-norms, which are designed to distinguish data constructors according to their types. For example, they can measure the length of a list, and the size of its elements separately. Such norms have been used before in the context of termination analysis (see [6] and its references), and can be defined as follows:

$$\|t\|_\sigma = \begin{cases} t & \sigma = \text{Int and } t \text{ is an integer} \\ length(t) & \sigma = \text{String and } t \text{ is a string} \\ 1 + \sum_{i=1}^{n} \|t_i\|_\sigma & \text{if } t = Co(t_1, \ldots, t_n) \text{ and } type(t) = \sigma \\ \sum_{i=1}^{n} \|t_i\|_\sigma & \text{if } t = Co(t_1, \ldots, t_n) \text{ and } type(t) \neq \sigma \end{cases} \tag{2}$$

Intuitively, $\|t\|_\sigma$ counts the number of data-constructs of type σ in t. Basic types are treated in a special way: integers keep their values, and strings are abstracted to their lengths. This means that $\|t\|_{\text{Int}}$ equals to the sum of all integer values in the data-structure t. We modify the above typed-norm scheme to the following one

$$\|t\|_\sigma = \begin{cases} t & \sigma = \text{Int and } t \text{ is an integer} \\ length(t) & \sigma = \text{String and } t \text{ is an string} \\ 1 + \sum_{i=1}^{n} \|t_i\|_\sigma & \text{if } t = Co(t_1, \ldots, t_n) \text{ and } type(t) = \sigma \\ \max_{i=1}^{n} \|t_i\|_\sigma & \text{if } t = Co(t_1, \ldots, t_n) \text{ and } type(t) \neq \sigma \end{cases} \tag{3}$$

The difference from (2) is that, instead of summing the sizes of the inner elements, it just keeps the maximal one. For instance, consider the recursive function written_by. By using (3), we will be able to infer that the cost is bounded by $O(bs \times as)$ where bs denotes the length of the recursive data-structure Books and as is the maximal length of the recursive data-structure Authors for each book.

This is because when abstracting the list using the Authors norm, the fourth case applies and the maximum value of all elements of the list is taken as worst case cost. Using (2), we add the length of Authors as many times as Books we have (at most bs books). Thus, obtaining the less accurate bound $O(bs^2 \times as)$. We argue that scheme (3) is more suitable than (2) for the cost analysis framework we rely on. This is because this framework is based on compositional reasoning that assumes worst-case for each iteration (i.e., when processing the inner elements of a data-structure), and then multiplies it by the number of iterations (which usually depends on the size of the skeleton). Note that one could also define in an analogous way a norm that estimates the minimum value, by replacing max with min in (3). This is in particular useful for inferring lower-bounds [12]. A variation of (3) is implicitly used in works on sized types [12,15] (see Sec. 6 for more details).

3.2   Our Transformational Approach

Next we describe our abstraction procedure based on typed-norms. Our approach allows maintaining several abstractions even for the same variable at the same time as in [6]. Thus, it allows estimating the size of a variable using different measures. This is important since two different parts of the program might traverse two different parts of the same data-structure. Having both measures allows us to provide tighter bounds. Note that although we are interested in using typed-norms following scheme (3), our techniques are also valid for scheme (2).

b	b^α
$g_1 \wedge g_2$	$g_1^\alpha \wedge g_2^\alpha$
$match(x, t)$	$\wedge \{X_\sigma = \|t\|_\sigma \mid \sigma \in \texttt{typed\_norms}(x)\}$
$nonmatch(x, t)$	$true$
$e_1 \ op \ e_2$	$(e_1 \ op \ e_2)[y/Y_{\texttt{Int}}]$ if $\texttt{Int} \in \texttt{typed\_norms}(x)$; otherwise $true$
$p(\bar{x}, \bar{y})$	$p(X, Y)$
$x := t$	$\wedge \{X_\sigma = \|t\|_\sigma \mid \sigma \in \texttt{typed\_norms}(x)\}$
$true$	$true$

Fig. 4. Size abstraction for the instructions

We first introduce some concepts. Given two types σ_1 and σ_2, we write $\sigma_1 \preceq \sigma_2$ if the definition of type σ_2 uses (either directly or transitively) type σ_1. If $\sigma \preceq \sigma$ we say that the type is recursive. For simplicity, we assume that recursive types are in direct recursive form (thus, its form can be checked by just inspecting its definition). We use $type(x)$ to refer to the type of x, and $\texttt{typed\_norms}(x)$ to refer to the set of types w.r.t. which we want to measure the size of x. In Sec. 4 we explain how to automatically infer $\texttt{typed\_norms}(x)$. For $\texttt{typed\_norms}$ to be *valid*, we require that $\sigma' \preceq \sigma$ if $type(x) = \sigma$ and $\sigma' \in \texttt{typed\_norms}(x)$. For instance, $\texttt{typed\_norms}(x) = \{\mathsf{Authors}, \mathsf{String}\}$ is a valid typed-norm for x with $type(x) = \mathsf{Authors}$. Given a type $\sigma \in \texttt{typed\_norms}(x)$, we let X_σ be an integer valued variable representing the size of (the value of) x w.r.t the typed-norm $\|\cdot\|_\sigma$. If $\sigma \neq \mathsf{Int}$, then we implicitly assume $X_\sigma \geq 0$. For a sequence of variables $\bar{x}$, we let $\bar{X}$ be a sequence that results from replacing each x_i by $X_{\sigma_1}, \ldots, X_{\sigma_n}$, where $\texttt{typed\_norms}(x_i) = \{\sigma_1, \ldots, \sigma_n\}$. Given an arithmetic expression e, we abstract e as $e[y/Y_{\mathsf{Int}}]$, where we use $e[y/Y_{\mathsf{Int}}]$ to denote the expression that results from replacing each variable y in e by Y_{Int}.

Given a typed-norm as in scheme (2) or (3), its *symbolic* version is an extension to handle terms that include variables, e.g., $Cons(x, xs)$ where x and xs are variables. It is obtained from the corresponding typed-norms definition by adding the following extra cases: when t is a variable of type σ_1, then $\|t\|_\sigma = T_\sigma$ if $\sigma \preceq \sigma_1$ and $\|t\|_\sigma = 0$ otherwise. In what follows, we abuse notation and use $\|t\|_\sigma$ to refer to this symbolic version of typed-norm.

For the sake of simplifying the presentation, we assume that the input program is in *single static assignment* form. A *size abstraction* is a conjunction of linear constraints that describe the effect of the corresponding instruction. Given an instruction b, its abstract version b^α is defined as in Fig. 4. Let us explain the abstraction for the different instructions: conjunctions are abstracted by recursively abstracting each of their conjuncts; a match guard on x adds as many constraints as typed-norms apply to the variable x, each constraint assigns to the abstract variable the abstraction of the matched term w.r.t. the considered norm; as we do not keep inequality constraints, *nonmatch* guards are abstracted to *true*; in the expressions involving arithmetic operations, each variable y is replaced by an abstract variable Y_{Int}; the arguments in the calls are replaced by their corresponding abstract names; assignments are abstracted analogously to match guards.

Definition 1. *Given a program P, its size abstraction P^α is a program obtained by replacing each rule $p(\bar{x}, \bar{y}) \mapsto g, b_1, \ldots, b_n \in P$ by $p(\bar{X}, \bar{Y}) \mapsto g^\alpha, b_1^\alpha, \ldots, b_n^\alpha$.*

When using the typed-norm scheme (3), then P^α might include constraints of the form $X_\sigma = E$ where E is an arithmetic expression that involves max. Such non-linear constraints can be approximated by linear ones as follows: replace the sub-expression $\max(B_1, \ldots, B_n)$ by a new auxiliary variable A, and add the constraints $A \geq B_1 \wedge \cdots \wedge A \geq B_n$; this might be applied repeatedly in case of nested or multiple occurrences of max. When the max has zero operands, it can be safely replaced by 0. Note also that if non-linear arithmetic is allowed in our language, then P^α might include non-linear constraints. These can also be approximated by linear ones as in [4].

Example 4. Fig. 5 shows, in the right column, the abstraction of the instructions which appear in the corresponding left column for function is_coauthor. We use underlining to denote abstractions that are useless, as it will be explained in the next section. The $\texttt{typed\_norms}$ that we use in is_coauthor is: $\texttt{typed\_norms}(x) = \{\mathsf{String}\}$ if $type(x) = \mathsf{String}$; $\texttt{typed\_norms}(x) = \{\mathsf{String}, \mathsf{Authors}\}$ if $type(x) = \mathsf{Authors}$; and $\texttt{typed\_norms}(x) = \{\mathsf{Int}\}$ if $type(x) = \mathsf{Int}$. Observe that the first

is_coauthor(a,as,y)$\mapsto$ $\quad$ case$_0$(a,as,y). case$_0$(a,as,y)$\mapsto$ $\quad$ match(as,Nil), $y := 0$. case$_0$(a,as,y)$\mapsto$ $\quad$ nonmatch(as,Nil), $\quad$ match(as,Cons(a,as')), $\quad y := 1$. case$_0$(a,as,y)$\mapsto$ $\quad$ nonmatch(as,Nil),nonmatch(as,Cons(a,as')), $\quad$ match(as,Cons(a',as')), $\quad$ is_coauthor(a,as',y).	is_coauthor($\underline{a_1,as_1,as_2,y_1}$) $\mapsto$ $\quad$ case$_0$($\underline{a_1,as_1,as_2,y_1}$). case$_0$($\underline{a_1,as_1},as_2,y_1$)$\mapsto$ $\quad$ \{$as_1 = 0, as_2 = 1$\}, \{$y_1 = 0$\}. case$_0$($\underline{a_1,as_1},as_2,y_1$)$\mapsto$ $\quad$ \{\}, $\quad$ \{$as_1 \geq a_1, a_1 \geq 0, as_1 \geq as_1'$, $\quad\quad as_1' \geq 0, as_2 = as_2' + 1, as_2' \geq 1$\}, $\quad$ \{$y_1 = 1$\}. case$_0$($\underline{a_1,as_1},as_2,y_1$) $\mapsto$ $\quad$ \{\}, \{\}, $\quad$ \{$as_1 \geq a_1', a_1' \geq 0, as_1 \geq as_1'$, $\quad\quad as_1' \geq 0, as_2 = as_2' + 1, as_2' \geq 1$\}, $\quad$ is_coauthor($\underline{a_1,as_1',as_2',y_1}$).

Fig. 5. Abstraction of function is_coauthor

argument a of is_coauthor is abstracted by the variable a_1 using the type String and the second argument as is abstracted in variables as_1 and as_2, one for each element of typed_norms(as). It is interesting to see that the abstraction of the guard $match(as, Cons(a',as'))$ on the third case$_0$ rule uses as_1 to denote the maximum length of a String in the recursive data-structure as, so we have to add the constraints $as_1 \geq a_1'$ (a_1' represents the abstraction of the first argument of $Cons$) and $as_1 \geq as_1'$. Note that if we use (2) in Sec. 3.1 then as_1 corresponds to the length of the concatenation of every String in as, i.e., $as_1 = a_1' + as_1'$. Since a_1' and as_1' represent String lengths, their value cannot be lower than 0 and we add constraints for that. Also, as_2' represents the length of Authors (and Nil corresponds to size 1), then as_2' must be at least of size 1. In order to assess the impact of our approach, we show in Fig. 6 the exact upper bounds obtained from an abstraction using only the term-size norm (left) and the abstraction using typed-norms (right) for our three functions. The upper bounds are given as functions of the sizes of the input parameters w.r.t. the different abstractions (hence the output parameter is not included). As explained in Ex. 3, the upper bounds obtained for written_by are more accurate using typed-norms. The largest gain is obtained for sort_books_by_authors as it uses the upper bounds of the two other functions, namely we achieve $O(n \times m \times l)$, where n represents the number of authors in as, m represents the number of books in bs and l represents the maximum length of Authors for each book in bs.

$\quad$ Intuitively, the analyzer obtains this upper bound following this reasoning. As function sort_books_by_authors has a recursive call that decreases the number of authors of as, we have that the maximum number of recursive calls is bound by n (number of authors in as), thus its cost is $O(n * cost\_body_1)$ where $cost\_body_1$ is the cost of each application of the body of the function. Now, in order to compute $cost\_body_1$, we have to analyze the cost of function written_by as it is called in the body. In this case, we also have recursive calls that decrease the size of the second argument bs (i.e., the number of books denoted as m). By applying

term-size norm	typed-norms
is_coauthor(a, as) $= 4 + 5 \times (\frac{as}{2} - \frac{1}{2})$	is_coauthor(a_1, as_1, as_2) $= 4 + 5 \times (as_2 - 1)$
written_by(a, bs) $= 3 + (\frac{bs}{4} - \frac{1}{4}) \times (14 + 5 \times (\frac{bs}{2} - \frac{5}{2}))$	written_by(a_1, bs_1, bs_2, bs_3) $= 3 + (bs_2 - 1) \times (14 + 5 \times (bs_3 - 1))$
sort_books_by_author(as, bs) $= 3 + (\frac{as}{2} - \frac{1}{2}) \times (10 + (\frac{bs}{4} - \frac{1}{4})$ $\times (14 + 5 \times (\frac{bs}{2} - \frac{5}{2})))$	sort_books_by_author$(as_1, as_2, bs_1, bs_2, bs_3)$ $= 3 + (as_2 - 1) \times (10 + (bs_3 - 1)$ $\times (14 + 5 \times (bs_2 - 1)))$

Fig. 6. Upper bounds comparison term-size vs. typed-norms (a_1, as_1 and bs_1 represent String-norms, as_2 and bs_2 represent Authors-norms and bs_3 represents Books-norm)

a similar reasoning, the cost of written_by is bound by $O(m * cost\_body_2)$. Again, we need to compute the cost of the call to is_coauthor, as it determines the cost of the body of written_by. Finally, we have a recursive call in is_coauthor that decreases the size of l (maximum size of Authors). By replacing each $cost\_body$ by the computed cost, we get the cubic cost above as upper bound. By using term-size, we obtain $O(n * m^2)$ where n is the size of as and m the size of bs. The difference is that the whole data structure is abstracted by m, thus the cost of method is_coauthor is bound by the whole m, instead of by the length of the author's lists (denoted l above) which are a subterm of m. This might lead to an important loss of precision when the data structure m is large.

4 Inference of Typed-Norms

In Sec. 3, we have assumed that each variable x is assigned a set of types, given by typed_norms(x), whose size we want to track. In principle, one could abstract each variable w.r.t. all norms valid for its type. However, this would threaten the efficiency of the analysis, as the complexity of the solving procedure for finding resource bounds from abstractions exponentially grows with the number of variables. In this section we develop an analysis that eliminates useless abstractions in two dimensions: (1) As it was observed in [3], one can remove variables that do not affect the cost. In particular, the cost of a given program (mainly) depends on the number of recursions performed, which in turn is controlled by the corresponding guards (conditions to stop the recursion). This means that any variable that does not affect, directly or indirectly, the value of a guard, can be completely ignored. (2) We push this observation further, and besides eliminating *useless* variables (and their abstractions), we also eliminate useless (typed) size information for those variables that are useful and thus have not been eliminated in (1). In some sense we eliminate *useless* types, and thus typed-norms, from each variable.

We say that a guard instruction g is *cost-significant* if it appears in a guard. In practice, we identify such instructions by examining the (recursive) strongly connected components of the corresponding control flow graph. The variables that are involved in the guards are the source for the size information that we want to track. For example, if a cost-significant guard is of the form $match(x,t)$, and $type(x) = \sigma$ where σ is a *recursive* type, then $\|.\|_\sigma$ is a norm that we should use for x (because the corresponding recursion might be traversing this part of x). Our analysis is done in two steps: (1) first the cost-significant guards are used to initialize typed_norms(x) for the variables involved in these guards, and (2) this information is propagated to other variables in the program by means of backwards data-flow analysis. Below we sketch these two steps.

Initialization. This step starts by setting typed_norms(x) to $\emptyset$ for each variable x in the program. Then, it identifies the set of cost-significant guards, and uses each such guard to modify related typed_norms(x) as follows:

- If the guard is $match(x,t)$, variable x has a type σ, and σ is a recursive type, then σ is added to typed_norms(x).
- If the guard is of the form $e_1 \; op \; e_2$, and variable x appears in e_1 or e_2, then Int is added to typed_norms(x).

Note that in the case of $match(x,t)$, if σ is not recursive then it is simply ignored. This is because non-recursive types cannot directly affect the number of recursions. However, they might have some inner recursive types that do, those will be propagated to x (from other guards) in the second step.

Propagation. The initial information computed in the first step must be propagated backwards to other variables in the program. Intuitively, the propagation step works as follows: suppose we have an instruction $x := Cons(y, ys)$, and we know that $\sigma \in$ typed_norms(x) (after the instruction). This means that we want to track the size of x w.r.t. the type σ, but to do so precisely we must track this information in all parts of x, i.e., in y and ys, thus we add σ to typed_norms(y) and typed_norms(ys), if they are valid norms for the corresponding types. The propagation rules for the different instructions are defined as follows:

- For $match(x,t)$ and $nonmatch(x,t)$, if $y \in vars(t)$, and $\sigma \in$ typed_norms(y), then we add σ to typed_norms(x).
- For $x := t$, if $\sigma \in$ typed_norms(x) we add σ to typed_norms(y) for each variable $y \in vars(t)$ as far as $type(y) \preceq \sigma$.
- For $m(x_1, \ldots, x_n, y)$, if there is a rule $m(w_1, \ldots, w_n, z) \mapsto g, b_1, \ldots, b_m$ and $\sigma \in$ typed_norms(w_i) we add σ to typed_norms(x_i), for each $1 \leq i \leq n$.
- For $m(x_1, \ldots, x_n, y)$, if there is a rule $m(w_1, \ldots, w_n, z) \mapsto g, b_1, \ldots, b_m$ and $\sigma \in$ typed_norms(y) we add σ to typed_norms(z).
- For any pair of rules $m(x_1, \ldots, x_n, y) \mapsto g, b_1, \ldots, b_m$ and $m(w_1, \ldots, w_n, z) \mapsto g', b'_1, \ldots, b'_k$, if $\sigma \in$ typed_norms(x_i) then $\sigma \in$ typed_norms(w_i), and if $\sigma \in$ typed_norms(y) then $\sigma \in$ typed_norms(z) (this forces rules with the same name and number of arguments to be abstracted to rules with the same name and same number of abstracted arguments).

Function	Initialization	Propagation
is_coauthor	$\{\}_a,\{\}_{as},\{\}_y$	$\{\}_a,\{\text{Authors}\}_{as},\{\}_y$
case$_0$ (1st rule)	$\{\}_a,\{\text{Authors}\}_{as},\{\}_y$	$\{\}_a,\{\text{Authors}\}_{as},\{\}_y$
case$_0$ (2nd rule)	$\{\}_a,\{\text{Authors}\}_{as},\{\}_y,\{\}_{as'}$	$\{\}_a,\{\text{Authors}\}_{as},\{\}_y,\{\}_{as'}$
case$_0$ (3rd rule)	$\{\}_a,\{\text{Authors}\}_{as},\{\}_y,\{\}_{a'},\{\}_{as'}$	$\{\}_a,\{\text{Authors}\}_{as},\{\}_y,\{\}_{a'},\{\text{Authors}\}_{as'}$

Fig. 7. Inference on is_coauthor

- There are some built-in functions that are treated as built-in instructions, e.g., $length(s,x)$ which binds x to the length of the string s. In such case, if Int $\in$ typed_norms(x) then we add String to typed_norms(s).
- All other instructions do not modify any information.

The propagation step is applied iteratively, using standard backwards data-flow analysis, until a fix-point is reached, i.e., the values of all typed_norms(x) become stable. Note that this data-flow analysis also propagates information between the rules (no special treatment is required). Termination is guaranteed because the number of typed-norms is finite.

Example 5. Fig. 7 shows the obtained typed norms on each variable after initialization and propagation on is_coauthor and case$_0$ rules. We use $\{\}_x$ notation to represent typed_norms(x) in a compact way. The algorithm works in the following way:

- Initialization sets typed_norms$(as) = \{\text{Author}\}$ and typed_norms$(x) = \emptyset$ for any other variable x in the program because all the guards in the program are of the form $match(as,t)$.
- Then, $\{\text{Author}\}$ is propagated in the following way:
  1. The second argument of case$_0$ propagates $\{\text{Authors}\}$ to is_coauthor rule, making typed_norms$(as) = \{\text{Author}\}$ on is_coauthor.
  2. The second argument of is_coauthor propagates $\{\text{Authors}\}$ to the third case$_0$ rule, making typed_norms$(as') = \{\text{Author}\}$ on the third case$_0$.
  3. Guard $match(as, Cons(a', as'))$ on the third case$_0$ rule adds $\{\text{Authors}\}$ to typed_norms(as), but typed_norms(as) already contains $\{\text{Authors}\}$, and the process stops.

When a variable has an empty set of candidate norms, it means that it is not relevant to obtain the cost expression. In our example, String-norm and Int-norm are useless to obtain an upper bound. The result of applying our inference of typed-norms on the running program is the abstraction in Fig. 5 removing all underlined variables and associated constraints.

5 Experimental Evaluation

We have implemented the resource analysis detailed in this paper in the static analyzer for ABS programs SACO (http://costa.ls.fi.upm.es/saco). Our analysis is currently being integrated in the web interface of SACO and will be available by selecting the `typed-norms` option within the settings section soon. Our experiments aim at evaluating both the accuracy and efficiency of our analysis. Experimental evaluation has been carried out on the functional modules of the *Replication System* case study (an industrial case study whose source code is available from the SACO website). A total of 88 functions are used in the replication system. We have used three different configurations for the analysis with norms: (1) term-size, (2) typed-norms considering all possible norms, and (3) significant typed-norms obtained by the inference algorithm as described in Sec. 4. An upper bound was obtained on 61 out of the 88 functions in configuration (1) and in 62 out of the 88 functions on configurations (2) and (3). A notable result of our experiment is that for one function (*'itemMapToSchedule'*) an upper bound has been obtained using configurations (2) and (3) but cannot be obtained in (1) since it requires a more refined abstraction than term-size.

As regards accuracy, in Fig. 8 we compare the quality of the upper bounds obtained using term-size and typed-norms (note that in (2) and (3) we infer the same upper bounds). Since the term-size norm measures the size of the input in a different way from the typed-norms, a fair comparison of the results can be done by actually evaluating the corresponding upper bounds on some (random) concrete input. We used quickCheck [7] to generate 10 random concrete inputs for each upper bound, so for each case we obtain 10 different quotients.

For each random input, the diagram in Fig. 8 shows the quotient between the value of the upper bound obtained using term size, and the value of the upper bound using typed-norms. The x-axis corresponds to the benchmark number, and to improve readability we have sorted the benchmarks according to the corresponding values in the y-axis. We have ignored constant upper bounds since they correspond to functions without any recursion (i.e., the term-size norm and typed-norm should give the same answer), and thus remained with 32 non-constant upper bounds (the horizontal axe of the diagram corresponds to these 32 upper bounds). Values below 1 mean the analysis based on the typed-norms is more precise than the term-size one (the smaller the value, the bigger is the improvement), which is the case in all 32 cases.

We have also compared the performance of the different configurations. The run-time of each configuration (for all benchmarks together, using the average of 5 runs) is depicted in Table 1. We divide the total time into 3 parts: T_{sa} is the time for processing the input program in order to define the typed-norms, for configuration (3) this also includes the typed-norms inference, and for configuration (1) this step does not exist and thus it costs 0; T_{ac} is the time for generating the abstract program; and T_{ub} is the time for solving the abstract program into an upper bound. As expected, using all typed-norms introduces a significant overhead in configuration (2) when compared to (1). Importantly, by using the typed-norms inference we reduce the number of typed-norms significantly and thus the overhead becomes reasonable in configuration (3) when compared to (1). The experiments have been performed on an Intel Core 2 Duo at 2.4GHz with 8GB of RAM, running OS X 10.9.

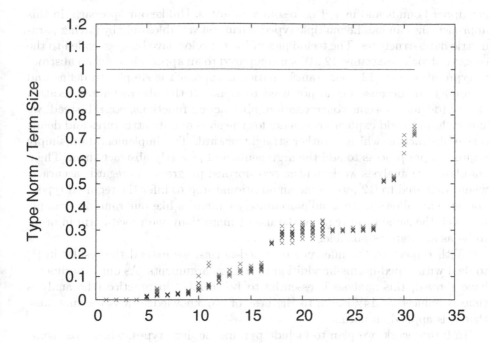

Fig. 8. Upper Bound Comparison

Table 1. Run-Time statistics (in ms.) on 61 functions: (1) using term-size; (2) using all type-norms; and (3) using only significant typed-norms obtained as in Sec. 4

Configuration	T_{sa}	Avg. T_{sa}	T_{ac}	Avg. T_{ac}	T_{ub}	Avg. T_{ub}	$T_{sa}+T_{ac}+T_{ub}$
(1)	0	0	120	2	3911	65	4031
(2)	1633	27	631	11	14230	234	16494
(3)	2161	36	255	5	4488	74	6904

6   Conclusions, Related and Future Work

We have presented a novel transformational approach to resource analysis with typed-norms which has the advantage that its formalization can be done by only adapting the first phase of cost analysis in which the program is transformed into an intermediate abstract program. Besides its simple formal development, the implementation has been easily integrated into the previous system as a pre-phase to the existing analysis.

Our work is inspired by [9] where the authors introduce the notion of typed-based norm in the context of termination analysis, and show how types can be very useful for finding suitable norms even for untyped languages like Prolog. They also illustrate that typed-based norms sometimes must be combined to get a termination proof. In [15], Vasconcelos introduces an enriched typing to get upper bounds and uses it on resource analysis. Unlike our approach, in this approach one can handle multiple typed-norms on variables only by having parametric data-structures. The techniques of Vasconcelos have been extended to the context of logic programs [12]. When compared to an approach based on abstract interpretation like [12], our transformational approach is simpler to define and to implement because we do not need to re-do all the abstract interpretation theory (defining specific concretization, abstraction functions, etc.). Instead, we simply have to add explicit arguments for the sizes of data structures and define a size abstraction which is rather straightforward. The implementation simply requires a pre-process to add the arguments and properly abstract them. Then, standard size analysis works on the transformed program. As regards accuracy, when compared to [12], we define an additional step to infer the required typed-norms. This allows us to handle accurately examples like our running example in which the same term requires the use of more than one typed-norm in order to be as accurate as possible.

With respect to the inference of typed-norms, we extend the results in [1] to deal with typed-norms in addition to useless arguments. As our experiments have showed, this analysis is essential to be scalable in practice (the analysis time is reduced 58.14%) and, to the best of our knowledge, it is the first time that it is applied on norms.

In future work, we plan to include parametric data types, which pose some challenges in the definition of the framework. Also, we want to enrich types with positions so that we can measure differently the same type when it appears in different type contexts. E.g., the type **data** $t = Pair(\mathtt{Int}, \mathtt{Int})$ is enriched to **data** $t = Pair(\mathtt{Int}^1, \mathtt{Int}^2)$, and thus we will have the two different norms $\|.\|_{\mathtt{Int}^1}$ and $\|.\|_{\mathtt{Int}^2}$.

References

1. Albert, E., Arenas, P., Genaim, S., Gómez-Zamalloa, M., Puebla, G.: Cost Analysis of Concurrent OO Programs. In: Yang, H. (ed.) APLAS 2011. LNCS, vol. 7078, pp. 238–254. Springer, Heidelberg (2011)
2. Albert, E., Arenas, P., Genaim, S., Puebla, G., Zanardini, D.: Cost Analysis of Java Bytecode. In: De Nicola, R. (ed.) ESOP 2007. LNCS, vol. 4421, pp. 157–172. Springer, Heidelberg (2007)
3. Albert, E., Arenas, P., Genaim, S., Puebla, G., Zanardini, D.: Removing Useless Variables in Cost Analysis of Java Bytecode. In: Proc. of SAC 2008, pp. 368–375. ACM (2008)
4. Alonso, D., Arenas, P., Genaim, S.: Handling Non-linear Operations in the Value Analysis of COSTA. In: Proc. of BYTECODE 2011. ENTCS, vol. 279, pp. 3–17. Elsevier (2011)
5. Bossi, A., Cocco, N., Fabris, M.: Proving Termination of Logic Programs by Exploiting Term Properties. In: Proc. of TAPSOFT 1991. LNCS, vol. 494, pp. 153–180. Springer (1991)
6. Bruynooghe, M., Codish, M., Gallagher, J., Genaim, S., Vanhoof, W.: Termination Analysis of Logic Programs through Combination of Type-Based norms. TOPLAS 29(2), Art. 10 (2007)

7. Claessen, K., Hughes, J.: QuickCheck: A Lightweight Tool for Random Testing of Haskell Programs. In: Proc. of ICFP 2000, pp. 268–279. ACM (2000)
8. Fähndrich, M.: Static Verification for Code Contracts. In: Cousot, R., Martel, M. (eds.) SAS 2010. LNCS, vol. 6337, pp. 2–5. Springer, Heidelberg (2010)
9. Genaim, S., Codish, M., Gallagher, J.P., Lagoon, V.: Combining Norms to Prove Termination. In: Cortesi, A. (ed.) VMCAI 2002. LNCS, vol. 2294, pp. 123–138. Springer, Heidelberg (2002)
10. Johnsen, E.B., Hähnle, R., Schäfer, J., Schlatte, R., Steffen, M.: ABS: A Core Language for Abstract Behavioral Specification. In: Aichernig, B.K., de Boer, F.S., Bonsangue, M.M. (eds.) Formal Methods for Components and Objects. LNCS, vol. 6957, pp. 142–164. Springer, Heidelberg (2011)
11. King, A., Shen, K., Benoy, F.: Lower-bound Time-complexity Analysis of Logic Programs. In: Proc. of ILPS 1997, pp. 261–275. MIT Press (1997)
12. Serrano, A., Lopez-Garcia, P., Bueno, F., Hermenegildo, M.: Sized Type Analysis for Logic Programs. In: Tech. Comms. of ICLP 2013. Cambridge U. Press (2013) (to appear)
13. Spoto, F., Mesnard, F., Payet, É.: A Termination Analyser for Java Bytecode based on Path-Length. TOPLAS 32(3), Art. 8 (2010)
14. Vallée-Rai, R., Hendren, L., Sundaresan, V., Lam, P., Gagnon, E., Co, P.: Soot - a Java Optimization Framework. In: Proc. of CASCON 1999. pp. 125–135. IBM (1999)
15. Vasconcelos, P.: Space Cost Analysis using Sized Types. Ph.D. thesis, School of CS, University of St. Andrews (2008)
16. Vasconcelos, P.B., Hammond, K.: Inferring Cost Equations for Recursive, Polymorphic and Higher-Order Functional Programs. In: Trinder, P., Michaelson, G.J., Peña, R. (eds.) IFL 2003. LNCS, vol. 3145, pp. 86–101. Springer, Heidelberg (2004)
17. Wegbreit, B.: Mechanical Program Analysis. Commun. ACM 18(9), 528–539 (1975)

A Finite Representation of the Narrowing Space

Naoki Nishida[1](✉) and Germán Vidal[2]

[1] Graduate School of Information Science, Nagoya University,
Furo-cho, Chikusa-ku, Nagoya 4648603, Japan
nishida@is.nagoya-u.ac.jp
[2] MiST, DSIC, Universitat Politècnica de València,
Camino de Vera, s/n, 46022 Valencia, Spain
gvidal@dsic.upv.es

Abstract. Narrowing basically extends rewriting by allowing free variables in terms and by replacing matching with unification. As a consequence, the search space of narrowing becomes usually infinite, as in logic programming. In this paper, we introduce the use of some operators that allow one to always produce a finite data structure that still represents all the narrowing derivations. Furthermore, we extract from this data structure a novel, compact equational representation of the (possibly infinite) answers computed by narrowing for a given initial term. Both the finite data structure and the equational representation of the computed answers might be useful in a number of areas, like program comprehension, static analysis, program transformation, etc.

1 Introduction

The narrowing relation [28], originally introduced in the context of theorem proving, was later adopted as the operational semantics of so called functional logic programming languages (like Curry [15]). Basically, narrowing extends term rewriting by allowing terms with variables and by replacing matching with unification. Therefore, narrowing has many similarities with the SLD resolution principle of logic programming. Indeed, both narrowing and SLD resolution usually produce an infinite search space, i.e., an infinite tree-like structure where several branches are created every time a function call matches with the left-hand side of more than one program rule. Currently, narrowing is regaining popularity in a number of areas other than functional logic programming, like protocol verification [10,18], model checking [8,11], partial evaluation [1,27], refining methods for proving the termination of rewriting [5,6], etc. In many—if not all—of these applications, producing a finite representation—usually in the form of a finite graph—of the narrowing space is essential.

This work has been partially supported by the Spanish *Ministerio de Economía y Competitividad (Secretaría de Estado de Investigación, Desarrollo e Innovación)* under grant TIN2013-44742-C4-1-R and by the *Generalitat Valenciana* under grant PROMETEO/2011/052.

© Springer International Publishing Switzerland 2014
G. Gupta and R. Peña (Eds.): LOPSTR 2013, LNCS 8901, pp. 54–71, 2014.
DOI: 10.1007/978-3-319-14125-1_4

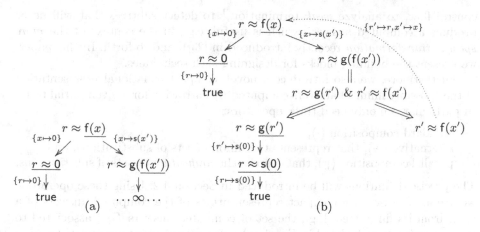

Fig. 1. Building a finite representation of the narrowing space for $f(x)$

The generation of a finite representation of the narrowing space has been tackled, e.g., by partial evaluation techniques (see, e.g., [1]). Here, so called *subsumption* and *abstraction* operators are introduced in order to stop potentially infinite derivations. However, no previous work has formally considered how the use of these operators can be used to construct a finite tree that still represents all possible derivations. In this work, we present a new approach to produce a finite data structure that still represents all the narrowing derivations for any given term. For this purpose, we introduce two basic operators: splitting and flattening. *Splitting* a conjunction like e_1 & e_2 implies the parallel evaluation of the conjuncts e_1 and e_2. On the other hand, *flattening* an equation e returns a conjunction of the form $x \approx e|_p$ & $e[x]_p$, where the subterm $e|_p$ of e is replaced by a fresh variable x not in e and a new equation is added. These two operations suffice to always produce a finite representation of the narrowing space.

Example 1. Consider the simple program (a term rewriting system) { $f(0) \rightarrow$ 0, $f(s(x)) \rightarrow g(f(x))$, $g(s(0)) \rightarrow s(0)$ }. where natural numbers are built using the constructors 0 and $s(\_)$. Given the initial equation $r \approx f(x)$, the narrowing space using an *innermost* strategy is infinite, as shown in Fig. 1 (a), where the terms selected to be narrowed are underlined. Even by using some sort of memoization (as in [4]), where variants of a previously narrowed term are not unfolded, we still get an infinite narrowing space. In contrast, by using flattening (depicted with a double line) and splitting (depicted with a double arrow), we can obtain a finite tree that still represents all the possible narrowing derivations, as shown in Fig. 1 (b), where dotted arrows are used to point to a previous variant of a term or equation. We consider these dotted arrows *implicit* to keep the data structure a tree (i.e., they are only used to identify the occurrence of a variant of the given node, and could be replaced by just adding the information locally to this node).

Designing a technique for producing finite trees can be useful in many different areas. For instance, one can use them to better understand the program's control flow, to analyze *weak* termination,[1] to detect subtrees that will never produce a computed answer (which is useful, e.g., in the context of the *more specific transformation* recently introduced in [23]), and so forth. In this paper, we present the building blocks for designing such techniques.

Furthermore, we also introduce a novel, compact equational representation of the (possibly infinite) answers computed by narrowing for a given initial term. In particular, we only need three operators:

- standard composition ($\cdot$),
- alternative ($+$), that represents the union of sets of substitutions, and
- parallel composition ($\Uparrow$), that denotes the *unification* on sets of substitutions.

The precise definitions will be introduced in Section 4.2. Using these operators, we are able to produce compact representations of the computed answers of a term from its finite tree. E.g., the set of computed answers $\Gamma_{f(x)}$ associated to the narrowing tree depicted in Fig. 1 (a) can be succinctly represented by

$$\Gamma_{f(x)} = \{x \mapsto 0, r \mapsto 0\} + \{x \mapsto s(x')\} \cdot (\{r \mapsto s(0), r' \mapsto s(0)\} \Uparrow \{r' \mapsto r, x' \mapsto x\} \cdot \Gamma_{f(x)})$$

which is extracted from the finite tree in Fig. 1 (b). Interestingly, one can easily see that there is no solution to

$$\{r \mapsto s(0), r' \mapsto s(0)\} \Uparrow \{r' \mapsto r, x' \mapsto x\} \cdot \Gamma_{f(x)}$$

since $\{r \mapsto s(0), r' \mapsto s(0)\}$ maps r' to $s(0)$ while $\{r' \mapsto r, x' \mapsto x\} \cdot \Gamma_{f(x)}$ can only bind r' to 0 (because the only non-recursive solution of $\Gamma_{f(x)}$ binds r to 0), and $s(0)$ and 0 clearly do not unify. Therefore, one can conclude that the only solution is $\{x \mapsto 0, r \mapsto 0\}$ despite the fact that the original narrowing tree is infinite. In this case, this was already obvious from the inspection of the narrowing tree. In general, however, our equational representation may be useful to analyze the computed answers of more complex programs.

This paper is organized as follows. In Section 2, we briefly review some notions and notations of term rewriting and narrowing. Section 3 presents some results on the compositionality of narrowing, introduces the flattening operator and proves its correctness. Section 4 then presents our method to produce finite trees by using subsumption, constructor decomposition, flattening, and splitting. We also introduce an equational representation for the computed answers in this section. Finally, Section 6 concludes and points out some directions for future research. Proofs of technical results can be found in the appendix available from http://users.dsic.upv.es/~gvidal/german/lopstr13a_lncs/paper.pdf.

2   Preliminaries

We assume familiarity with basic concepts of term rewriting and narrowing. We refer the reader to, e.g., [7], [25], and [14] for further details.

Terms and Substitutions. A *signature* $\mathcal{F}$ is a set of function symbols. Given a set of variables $\mathcal{V}$ with $\mathcal{F} \cap \mathcal{V} = \varnothing$, we denote the domain of *terms* by $\mathcal{T}(\mathcal{F}, \mathcal{V})$.

---

[1] A TRS is weakly terminating if any term has at least one normal form [13].

We assume that $\mathcal{F}$ always contains at least one constant $f/0$. We use $f, g, \ldots$ to denote functions and $x, y, \ldots$ to denote variables. A *position* p in a term t is represented by a finite sequence of natural numbers, where ϵ denotes the root position. The set of positions of a term t is denoted by $\mathcal{P}os(t)$. We let $t|_p$ denote the *subterm* of t at position p and $t[s]_p$ the result of *replacing the subterm* $t|_p$ by the term s. $\mathcal{V}ar(t)$ denotes the set of variables appearing in t. A term t is *ground* if $\mathcal{V}ar(t) = \varnothing$.

A *substitution* $\sigma : \mathcal{V} \mapsto \mathcal{T}(\mathcal{F}, \mathcal{V})$ is a mapping from variables to terms such that $\mathcal{D}om(\sigma) = \{x \in \mathcal{V} \mid x \neq \sigma(x)\}$ is its domain. Substitutions are extended to morphisms from $\mathcal{T}(\mathcal{F}, \mathcal{V})$ to $\mathcal{T}(\mathcal{F}, \mathcal{V})$ in the natural way. We denote the application of a substitution σ to a term t by $t\sigma$ rather than $\sigma(t)$. The identity substitution is denoted by id. A *variable renaming* is a substitution that is a bijection on $\mathcal{V}$. A substitution σ is *more general* than a substitution θ, denoted by $\sigma \leqslant \theta$, if there is a substitution δ such that $\delta \cdot \sigma = \theta$, where "$\cdot$" denotes the composition of substitutions (i.e., $\sigma \cdot \theta(x) = (x\theta)\sigma = x\theta\sigma$). A substitution σ is *idempotent* if $\sigma \cdot \sigma = \sigma$. The *restriction* $\theta\restriction_V$ of a substitution θ to a set of variables V is defined as follows: $x\theta\restriction_V = x\theta$ if $x \in V$ and $x\theta\restriction_V = x$ otherwise. We say that $\theta = \sigma \; [V]$ if $\theta\restriction_V = \sigma\restriction_V$.

A term t_2 is an *instance* of a term t_1 (or, equivalently, t_1 is *more general* than t_2), in symbols $t_1 \leqslant t_2$, if there is a substitution σ with $t_2 = t_1\sigma$. Two terms t_1 and t_2 are *variants* (or equal up to variable renaming) if $t_1 = t_2\rho$ for some variable renaming ρ. A *unifier* of two terms t_1 and t_2 is a substitution σ with $t_1\sigma = t_2\sigma$. This notion is naturally extended to a set of equations: σ is a unifier of a set of equations $\{s_1 = t_1, \ldots, s_n = t_n\}$ if $s_i\sigma = t_i\sigma$ for $i = 1, \ldots, n$; furthermore, σ is the *most general unifier* of $\{s_1 = t_1, \ldots, s_n = t_n\}$, denoted by $\mathsf{mgu}(\{s_1 = t_1, \ldots, s_n = t_n\})$ if, for every other unifier θ of $\{s_1 = t_1, \ldots, s_n = t_n\}$, we have that $\sigma \leqslant \theta$.

TRSs and Rewriting. A set of rewrite rules $l \rightarrow r$ such that l is a non-variable term and r is a term whose variables appear in l is called a *term rewriting system* (TRS for short); terms l and r are called the left-hand side (lhs) and the right-hand side (rhs) of the rule, respectively. We restrict ourselves to finite signatures and TRSs. Given a TRS $\mathcal{R}$ over a signature $\mathcal{F}$, the *defined* symbols $\mathcal{D}_\mathcal{R}$ are the root symbols of the lhs's of the rules and the *constructors* are $\mathcal{C}_\mathcal{R} = \mathcal{F} \setminus \mathcal{D}_\mathcal{R}$. *Constructor terms* of $\mathcal{R}$ are terms over $\mathcal{C}_\mathcal{R}$ and $\mathcal{V}$, i.e., $\mathcal{T}(\mathcal{C}_\mathcal{R}, \mathcal{V})$. We omit $\mathcal{R}$ from $\mathcal{D}_\mathcal{R}$ and $\mathcal{C}_\mathcal{R}$ if it is clear from the context. A substitution σ is a *constructor substitution* (of $\mathcal{R}$) if $x\sigma \in \mathcal{T}(\mathcal{C}_\mathcal{R}, \mathcal{V})$ for all variables x. A TRS $\mathcal{R}$ is a *constructor system* if the lhs's of its rules have the form $f(s_1, \ldots, s_n)$ where s_i are constructor terms, i.e., $s_i \in \mathcal{T}(\mathcal{C}, \mathcal{V})$, for all $i = 1, \ldots, n$.

For a TRS $\mathcal{R}$, we define the associated rewrite relation $\rightarrow_\mathcal{R}$ as the smallest binary relation satisfying the following: given terms $s, t \in \mathcal{T}(\mathcal{F}, \mathcal{V})$, we have $s \rightarrow_\mathcal{R} t$ iff there exist a position p in s, a rewrite rule $l \rightarrow r \in \mathcal{R}$ and a substitution σ with $s|_p = l\sigma$ and $t = s[r\sigma]_p$; the rewrite step is usually denoted by $s \rightarrow_{p, l \rightarrow r} t$ to make explicit the position and rule used in this step. The instantiated lhs $l\sigma$ is called a *redex*. A term t is called *irreducible* or in *normal form* w.r.t. a TRS $\mathcal{R}$ if there is no term s with $t \rightarrow_\mathcal{R} s$. A *derivation* is a (possibly empty) sequence of rewrite steps. Given a binary relation $\to$, we denote by $\to^*$ its reflexive and transitive closure. Thus $t \to_{\mathcal{R}}^* s$ means that t can be reduced to s in $\mathcal{R}$ in zero or more steps.

Narrowing. The *narrowing* relation [28] mainly extends term rewriting by replacing pattern matching with unification, so that terms containing logic variables can also be reduced by non-deterministically instantiating these variables. Formally, given a TRS $\mathcal{R}$ and two terms $s, t \in \mathcal{T}(\mathcal{F}, \mathcal{V})$, we have that $s \leadsto_{\mathcal{R}} t$ is a *narrowing step* iff there exist a non-variable position p of s, a variant $l \to r$ of a rule in $\mathcal{R}$, and a substitution $\sigma = \mathsf{mgu}(\{s|_p = l\})$,[2] such that $t = (s[r]_p)\sigma$. We usually write $s \leadsto_{p, l \to r, \theta} t$ (or simply $s \leadsto_\theta t$) to make explicit the position, rule, and substitution of the narrowing step.

A *narrowing derivation* $t_0 \leadsto_\sigma^* t_n$ denotes a sequence of narrowing steps $t_0 \leadsto_{\sigma_1} \cdots \leadsto_{\sigma_n} t_n$ with $\sigma = \sigma_n \cdot \cdots \cdot \sigma_1$ (if $n = 0$ then $\sigma = id$). Given a narrowing derivation $s \leadsto_\sigma^* t$ with t a constructor term, we say that σ is a *computed answer* for s.

Innermost Narrowing. In this paper, we consider a particular narrowing strategy called *innermost narrowing* (see, e.g., [12]). Innermost narrowing only reduces subterms of the form $f(t_1, \ldots, t_n)$, with f a defined function symbol and $t_1, \ldots, t_n$ constructor terms; if there are several such subterms, we consider in this paper that the leftmost one is selected. Innermost narrowing steps are denoted using arrows of the form "$\overset{i}{\leadsto}$". A well-known result for innermost narrowing states its completeness for (confluent and terminating) constructor TRSs that are *completely defined* (CD) (or *sufficiently complete*): TRSs in which no function symbol occurs in any ground term in normal form (i.e., functions are always reducible on all ground terms). The CD condition is common when using types and each function is defined for all constructors of its argument types. It is easy to extend innermost narrowing to incompletely defined functions, by just adding a so called innermost *reflection* rule which skips an innermost function call that cannot be reduced [17], given rise to so called innermost *basic* narrowing. For the sake of simplicity, here we assume that the CD condition holds for all functions so that innermost narrowing suffices to compute all answers.

Example 2. Consider the TRS $\mathcal{R} = \{ \ (R_1) \ \mathsf{add}(0, y) \to y, \ (R_2) \ \mathsf{add}(\mathsf{s}(x), y) \to \mathsf{s}(\mathsf{add}(x, y)) \ \}$ defining the addition $\mathsf{add}/2$ on natural numbers built from $0/0$ and $\mathsf{s}/1$. Given the term $\mathsf{add}(x, \mathsf{s}(0))$, we have infinitely many innermost narrowing derivations starting from $\mathsf{add}(x, \mathsf{s}(0))$, e.g.,

$$\mathsf{add}(x, \mathsf{s}(0)) \overset{i}{\leadsto}_{\epsilon, R_1, \{x \mapsto 0\}} \quad \mathsf{s}(0)$$
$$\mathsf{add}(x, \mathsf{s}(0)) \overset{i}{\leadsto}_{\epsilon, R_2, \{x \mapsto \mathsf{s}(y_1)\}} \mathsf{s}(\mathsf{add}(y_1, \mathsf{s}(0))) \overset{i}{\leadsto}_{1, R_1, \{y_1 \mapsto 0\}} \mathsf{s}(\mathsf{s}(0))$$
$$\ldots$$

with computed answers $\{x \mapsto 0\}$, $\{x \mapsto \mathsf{s}(0)\}$, etc.

---

[2] We consider the so called *most general* narrowing, i.e., the mgu of the selected subterm and the lhs of a rule—rather than an arbitrary unifier—is computed at each narrowing step.

3   Compositionality and Flattening

The compositionality property can be simply formalized at the level of equations, i.e., we say that narrowing is compositional when the computed answers of e_1 & e_2 can be obtained from the computed answers of e_1 and e_2, where "&" denotes the Boolean conjunction operator. As for the flattening operation, given an equation $x \approx f(g(y))$,[3] its flattening w.r.t. the position 2.1 (i.e., w.r.t. $g(y)$ since $x \approx f(g(y))|_{2.1} = g(y)$) returns $x' \approx g(y)$ & $x \approx f(x')$, where x' is a fresh variable. Therefore, flattening can be used to *distribute* the narrowing tasks among different equations.

Intuitively speaking, compositionality holds for any narrowing strategy that fulfills the following conditions:

- Independence of the context. This is the case, for instance, of unrestricted narrowing, basic narrowing, innermost narrowing, etc. Lazy or needed narrowing, in contrast, are not independent of the context because, given an expression $s[t]_p$, we cannot determine whether t should be narrowed (and to what extent) without looking at the context $s[\ ]_p$.
- Terms introduced by instantiation should not be narrowable. This is the case, for instance, of basic narrowing, innermost narrowing, lazy and needed narrowing (for left-linear constructor systems), etc. This is not the case of unrestricted narrowing though.

In the following, we will focus on (unconditional) innermost narrowing (though other narrowing strategies would also be equally appropriate, e.g., basic narrowing or innermost basic narrowing). Furthermore, some strategies not fulfilling the above conditions, like lazy and needed narrowing, can also be proved compositional by restricting the narrowing derivations to head normal form (so that they become essentially independent of the context).

In this paper, we consider the usual definitions for syntactic equality and conjunction: $\mathcal{R}_{eq} = \{x \approx x \to \text{true}\}$, $\mathcal{R}_{\&} = \{\text{true } \& \ x \to x, \text{ false } \& \ x \to \text{false}\}$. Hence, we have that $s \approx t$ holds if s and t are syntactically equal. Also, when using innermost narrowing, we can only reduce $s \approx t$ using the rule $x \approx x \to \text{true}$ if both s and t are constructor terms. Narrowing deals with equations and conjunctions as ordinary terms. We often call such terms *equational* terms to make it explicit that they contain occurrences of "$\approx$" and/or "&". In the following, we assume that every TRS implicitly includes the rules of $\mathcal{R}_{eq} \cup \mathcal{R}_{\&}$.

Here, we only aim at preserving the answers computed in *successful* derivations, i.e., derivations ending with a constructor term (true, when the initial term is an equation or a conjunction of equations).

Definition 3 (Success Set). *Let $\mathcal{R}$ be a TRS and let t be a term. We define the* success set *$\mathcal{S}_{\mathcal{R}}(t)$ of t in $\mathcal{R}$ as follows:*

$$\mathcal{S}_{\mathcal{R}}(t) = \{\sigma|_{\mathcal{V}ar(t)} \mid t \overset{i}{\leadsto}{}^*_\sigma c \text{ in } \mathcal{R} \text{ and } c \in \mathcal{T}(\mathcal{C}, \mathcal{V}) \text{ is a constructor term}\}$$

---

[3] Here, "$\approx$" is a binary symbol to denote syntactic equality on terms, see below.

Observe that function $\mathcal{S}$ does not return the computed normal forms. Nevertheless, we can still get the computed normal form as follows: given a term t, we consider an initial equation of the form $x \approx t$, where x is a fresh variable not occurring in t; therefore, x will be bound to the normal form of t in any successful derivation (i.e., any derivation that ends with true).

Let us now recall the definition of *parallel composition* of substitutions, denoted by $\Uparrow$ in [16,26]. Informally speaking, this operation corresponds to the notion of unification generalized to substitutions. Here, $\widehat{\theta}$ denotes the *equational representation* of a substitution θ, i.e., if $\theta = \{x_1 \mapsto t_1, \ldots, x_n \mapsto t_n\}$ then $\widehat{\theta} = \{x_1 = t_1, \ldots, x_n = t_n\}$.

Definition 4 (Parallel Composition [26]). *Let θ_1 and θ_2 be two idempotent substitutions. Then, we define $\Uparrow$ as follows:*

$$\theta_1 \Uparrow \theta_2 = \begin{cases} \mathsf{mgu}(\widehat{\theta_1} \cup \widehat{\theta_2}) & \text{if } \widehat{\theta_1} \cup \widehat{\theta_2} \text{ has a solution (a unifier)} \\ \mathsf{fail} & \text{otherwise} \end{cases}$$

Parallel composition is extended to sets of substitutions in the natural way:

$$\Theta_1 \Uparrow \Theta_2 = \{\theta_1 \Uparrow \theta_2 \mid \theta_1 \in \Theta_1, \ \theta_2 \in \Theta_2, \ \theta_1 \Uparrow \theta_2 \neq \mathsf{fail}\}$$

Now, we state the main compositional result for innermost narrowing:

Theorem 5. *Let $\mathcal{R}$ be a constructor CD TRS. Let $e_1 \ \& \ e_2$ be an equational term. Then, we have $\mathcal{S}_{\mathcal{R}}(e_1 \ \& \ e_2) = \mathcal{S}_{\mathcal{R}}(e_1) \Uparrow \mathcal{S}_{\mathcal{R}}(e_2)$ up to variable renaming.*

As a useful consequence of the above compositionality result, we can state the following corollary:

Corollary 6. *Let $\mathcal{R}$ be a constructor CD TRS. Let $e_1 \ \& \ e_2$ be an equational term. Then, we have $\mathcal{S}_{\mathcal{R}}(e_1 \ \& \ e_2) = \mathcal{S}_{\mathcal{R}}(e_2 \ \& \ e_1)$ up to variable renaming.*

In practice, this result implies that innermost narrowing can select the equations to be narrowed in any order (and not necessarily in a left-to-right order) while preserving the computed answers. This is equivalent to the *independence of the selection rule* of logic programming.

Now, we recall the flattening transformation (called *unfolding* in [24]) that will become useful in the next section, and prove its correctness.

Definition 7 (Flattening). *Let e be an equational term and $p \in \mathcal{P}os(e)$ be a position of e such that $e|_p$ is not a variable, and the root of $e|_p$ is neither $\approx$ nor $\&$. Then, the flattening of e w.r.t. p is given by $x \approx e|_p \ \& \ e[x]_p$.*

We say that a flattening is trivial when e has an equation $y \approx t$ and flattening just replaces it with $x \approx t \ \& \ y \approx x$ (so that just another level of indirection is created). In the following, we assume that all flattenings are non-trivial.

The following property states the correctness of the flattening operation:

Theorem 8. *Let $\mathcal{R}$ be a constructor CD TRS. Let e be an equational term and e' be a non-trivial flattening of e w.r.t. some position p. Then, we have $\mathcal{S}_{\mathcal{R}}(e) = \mathcal{S}_{\mathcal{R}}(e') \ [\mathcal{V}ar(e)]$ up to variable renaming.*

4   A Finite Representation of the Narrowing Space

First, we introduce a framework to obtain a finite representation of a (possibly infinite) narrowing space. Then, we also present a method to extract an equational representation of the success set of a given term.

4.1   Constructing Finite Narrowing Trees

We produce finite trees representing all the (possibly infinite) narrowing derivations of a term as follows. Basically, we proceed as in the construction of a standard narrowing tree, but we also introduce some new operators in order to ensure that the tree can be kept finite.

Definition 9 (Extended Narrowing Tree). *Let $\mathcal{R}$ be a TRS and t be a term. An* extended narrowing tree *for t in $\mathcal{R}$ is a directed rooted node- and edge-labeled graph τ built as follows:*

- *the root node of τ is labeled with $x \approx t$, where x is a fresh variable not occurring in t;*
- *a leaf is either a node labeled with* true *(a success node) or a node containing defined functions that cannot be further narrowed, which is labeled with* fail *to make it explicit that it represents a failing derivation;*
- *subsumption: if a node is labeled with a non-constructor term e that is a variant of a previous node e' in the same root-to-leaf derivation, i.e., $e\vartheta = e'$, it is also considered a leaf, and we add an implicit edge between these nodes labeled with ϑ;* [4]
- *otherwise, given a node labeled with e, we expand it (do not care nondeterministically) using one of the following rules:*
   narrowing: *if e is narrowable, we have an output edge labeled with σ from node e to node e' for each innermost narrowing step $e \overset{i}{\leadsto}_\sigma e'$;*
   constructor decomposition: *if $e \equiv (y \approx c(t_1, \ldots, t_n) \mathbin{\&} e')$ $(c \in \mathcal{C})$, we add an edge to a node $y_1 \approx t_1 \mathbin{\&} \ldots \mathbin{\&} y_n \approx t_n \mathbin{\&} e'$, with $y_1, \ldots, y_n$ fresh variables, and the edge is labeled with $\{y \mapsto c(y_1, \ldots, y_n)\}$;*
   splitting: *if $e \equiv (e_1 \mathbin{\&} \cdots \mathbin{\&} e_{n-1} \mathbin{\&} e_n)$, we add output edges from e to new nodes labeled with $e_1, \ldots, e_{n-1}$, and e_n;*
   flattening: *we add an output edge from node e to a node $y \approx e|_p \mathbin{\&} e[y]_p$, where y is a fresh variable not occurring anywhere in the tree.*

The operations considered in the previous definition can also be found in the literature (perhaps with slightly different definitions). For instance, flattening is introduced in [24] (where it is called *unfolding*); subsumption is used in many different contexts (e.g., [1,4]); (constructor) decomposition rules are used in different narrowing calculi (see, e.g., [19]); finally, splitting is considered when proving compositionality results (e.g., [3]) and in the partial evaluation of logic programs [9].

In the following, we will use these graphical conventions when depicting the steps of an extended narrowing tree:

---

[4] We consider these edges *implicit* to keep the data structure a tree.

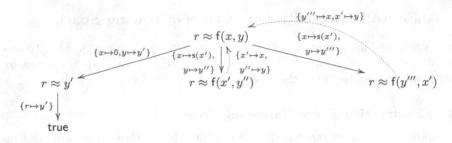

Fig. 2. Finite narrowing tree for $f(x, y)$

- narrowing and constructor decomposition: (labeled) solid arrow ($\longrightarrow$);
- subsumption: (labeled) dotted arrow ($\cdots\cdots\succ$);
- flattening: double line ($=\!=$);
- splitting: double arrow ($\Longrightarrow$).

By abuse of notation, we often use in the text $e \longrightarrow_\sigma^* e'$ to denote a path in the tree, no matter the type of rules applied from node e to node e'—except subsumption—where σ is the composition of the substitutions in the labeled edges along this path (if any, and id otherwise).

Let us now illustrate the construction of finite extended narrowing trees with some examples (where no fixed strategy is considered). Note that rule variables are always renamed with fresh names; this is mandatory to produce correct equations in the next section.

Example 10. Consider the (non-confluent) TRS $\mathcal{R} = \{\ f(0, y) \to y,\ f(s(x), y) \to f(x, y),\ f(s(x), y) \to f(y, x)\ \}$. Given the initial term $f(x, y)$, the narrowing space is clearly infinite because of the recursive calls to f. Here, a couple of subsumption steps suffice to get a finite extended narrowing tree, as shown in Fig. 2.

Example 11. Consider the TRS $\mathcal{R} = \{\ f(0, y) \to y,\ f(s(x), y) \to c(f(x, y), f(y, x))\}$ and the initial term $f(x, y)$. In this case, subsumption does not suffice and constructor decomposition and splitting becomes necessary, as shown in Fig. 3. This is a simple pattern that could be routinely applied to all constructor-rooted terms in order to get a finite representation of the narrowing space.

Observe that the constructor decomposition step is not really needed and could be mimicked by performing two flattening steps and, then, reducing the last equation as follows:

$$
\begin{aligned}
r &\approx c(f(x', y''), f(y'', x')) \\
&=\!=\qquad\qquad r' \approx f(x', y'')\ \&\ r \approx c(r', f(y'', x')) \\
&=\!=\qquad\qquad r' \approx f(x', y'')\ \&\ r'' \approx f(y'', x')\ \&\ r \approx c(r', r'') \\
&\longrightarrow_{\{r \mapsto c(r', r'')\}} r' \approx f(x', y'')\ \&\ r'' \approx f(y'', x')\ \&\ \text{true}
\end{aligned}
$$

However, we prefer to keep the constructor decomposition steps for simplicity.

Example 12. Finally, consider the TRS $\mathcal{R} = \{\ 0 + y \to y,\ s(x) + y \to s(x + y),\ 0 * y \to 0,\ s(x) * y \to y + (x * y)\ \}$. Given the initial term $x * y$, both flattening and splitting are necessary to produce a finite extended narrowing tree, as shown in Fig. 4.

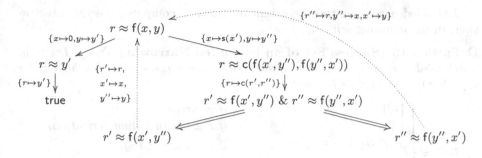

Fig. 3. Finite narrowing tree for $f(x, y)$

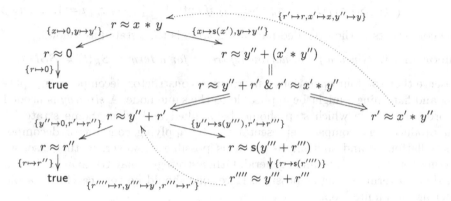

Fig. 4. Finite narrowing tree for $x * y$

In the following, we use the following notation. Given an extended narrowing tree τ, we let $\mathsf{root}(\tau)$ denote the root of τ. We also let $\tau \equiv (t \to_\sigma \tau')$ denote the fact that τ is rooted by term t and has a (possibly labeled) output edge to a subtree τ'. Moreover, we use the auxiliary function $\mathsf{out}(\tau)$ that returns the output edges from $\mathsf{root}(\tau)$ (if any). E.g., let τ be the extended narrowing tree of Fig. 4; here, we have $\mathsf{out}(\tau) = \{r \approx x * y \to_{\{x \mapsto 0, y \mapsto y'\}} \tau_1, \ r \approx x * y \to_{\{x \mapsto s(x'), y \mapsto y''\}} \tau_2\}$, where τ_1 and τ_2 are the subtrees rooted by $r \approx 0$ and $r \approx y'' + (x' * y'')$, respectively. Finally, we let $\mathsf{subtrees}(\tau)$ denote the set of subtrees of a tree τ that are obtained by partitioning τ into those subtrees that are rooted by a term with an incoming subsumption edge. E.g., for the tree τ of Fig. 4, $\mathsf{subtrees}(\tau)$ returns two subtrees, one rooted by $r \approx x * y$ and another one rooted by $r \approx y'' + r'$.

The relevance of the notion of extended narrowing tree is that, thanks to the use of the rules of flattening, constructor decomposition,[5] and splitting, one can always produce a tree with finitely many non-variant nodes. We do not provide a formal proof of this claim, but it is an easy consequence of the fact that using flattening—which involves replacing a subterm by a fresh variable—and splitting one can keep the set of non-variant terms finite.

---

[5] The rule of constructor decomposition is mainly introduced for simplicity, but could be replaced by a sequence of flattening steps.

Extended narrowing trees represent all possible computed answer substitutions in the following sense:

Definition 13 (Success Set of an Extended Narrowing Tree). *Let τ_0 be a extended narrowing tree for a term t. Then, the success set of a subtree τ for τ_0, $SS(\tau)$, is defined as follows:*[6]

$$
SS(\tau) = \begin{cases}
\{id\} & \text{if } \tau \equiv \mathsf{true} \\
\{\,\} & \text{if } \tau \equiv \mathsf{fail} \text{ (a failing derivation)} \\
\sigma \cdot SS(\tau') & \text{if } \tau \equiv (\, t \dashrightarrow_\sigma \tau'\,) \\
SS(\tau') & \text{if } \tau \equiv (e = \tau') \\
SS(\tau_1) \Uparrow \cdots \Uparrow SS(\tau_n) & \text{if } \mathsf{out}(\tau) = \{e \Rightarrow \tau_i \mid i = 1, \dots, n\} \\
\sigma_1 \cdot SS(\tau_1) \cup \cdots \cup \sigma_n \cdot SS(\tau_n) & \text{if } \mathsf{out}(\tau) = \{e \to_\sigma \tau_i \mid i = 1, \dots, n\}
\end{cases}
$$

The correctness of the extended narrowing trees is then stated as follows:

Theorem 14. *Given a finite narrowing tree τ for a term t, $S_{\mathcal{R}}(t) = SS(\tau)$.*

Observe that the four operations—narrowing, constructor decomposition, splitting and flattening—might be applicable to the same node. A *strategy* is needed in order to decide which step should be applied and when. Some strategies can produce very compact representations by applying constructor decomposition/flattening and splitting as much as possible. However, in this case, we also get less accurate results in general. Other strategies may try to avoid breaking down a term as long as possible. Here, one should be very careful to avoid entering an infinite loop.

For instance, a simple strategy that always guarantees the construction of a finite extended narrowing tree may proceed as follows. Basically, every time a node e is narrowed at some position p with $e|_p$ rooted by a defined function symbol: $e \overset{i}{\leadsto}_\sigma e[r]_p \sigma'$ with $\sigma' = \sigma \upharpoonright_{\mathsf{Var}(e)}$, we apply a flattening step:

$$
e[r]_p \sigma' \;=\; x \approx r \,\&\, e[x]_p \sigma'
$$

followed by these splitting steps:
$$
x \approx r \,\&\, e[x]_p \sigma' \Longrightarrow
\begin{cases}
x \approx r \\
e[x]_p \\
\widehat{\sigma'}
\end{cases}
$$

By abuse of notation, for $\sigma' = \{x_1 \mapsto t_n, \dots, x_n \mapsto t_n\}$, we use $\widehat{\sigma'}$ to denote the equational term $x_1 \approx t_1 \,\&\, \cdots \,\&\, x_n \approx t_n$. Roughly speaking, the construction of the extended narrowing tree will be finite since i) the number of nodes of the form $x \approx r$, with r an rhs of the TRS, is finite modulo variable renaming; ii) the new node $e[x]_p$ contains strictly less defined function symbols than e; and iii) $\widehat{\sigma'}$ only contains constructor symbols, $\approx$, and $\&$.

More refined strategies involve the use of appropriate orders on terms so that flattening and/or splitting steps are only applied when there is a risk of non-termination. We refer the interested reader to [1,2], where terminating strategies

---

[6] Observe that a failing derivation returns an empty set. Here, we assume that both $\sigma \cdot \{\,\} = \{\,\}$ and $\{\,\} \Uparrow \Theta = \Theta \Uparrow \{\,\} = \{\,\}$.

for narrowing-driven partial evaluation are introduced. Similar strategies could be defined using the operations of Definition 9.

4.2 Success Set Equations

In this section, we introduce an equational notation for representing the success set of a term, that we call its *success set equations*. Here, we consider the following three operators:

- Composition ($\cdot$). For simplicity, besides the standard composition of substitutions, we also consider its extension to sets of substitutions as follows. Given a set of substitutions Θ and a substitution σ, we let $\sigma \cdot \Theta = \{\sigma \cdot \theta \mid \theta \in \Theta\}$ and $\Theta \cdot \sigma = \{\theta \cdot \sigma \mid \theta \in \Theta\}$.
- Alternative ($+$). In our context, an expression like $ss_1 + ss_2$ denotes the union of the success sets denoted by ss_1 and ss_2. Again, for simplicity, we let a substitution denote a singleton set with this substitution.
- Parallel composition ($\Uparrow$). This is the standard parallel composition operator introduced in Definition 4.

As for the operator precedence, we assume that composition has a higher priority than parallel composition, which has a higher priority than alternative.

Now, we introduce a technique to extract the success set equations of a term from a given (finite) extended narrowing tree. Loosely speaking, substitutions along derivations with narrowing steps are just composed; the success sets of the different branches issuing from a term are put together using the alternative operator; flattening and constructor decomposition steps are ignored; splitting steps involve computing the parallel composition of the success sets of the different branches; finally, for subsumption steps, we compose the current set with the substitution labeling the step and, then, with the success set of the previous variant term.

Definition 15 (Success Set Equations). *Let τ be a finite extended narrowing tree for a term t. Let $\mathcal{T} = \mathsf{subtrees}(\tau)$. Then, we produce a success set equation $\Gamma_t = \mathcal{SF}(\tau')$ for each tree in $\tau' \in \mathcal{T}$ with $\mathsf{root}(\tau') = t$, where the auxiliary function $\mathcal{SF}$ is defined as follows:*

$$
\mathcal{SF}(\tau) = \begin{cases}
id & \text{if } \tau \equiv \mathsf{true} \\
fail & \text{if } \tau \equiv \mathsf{fail} \text{ (a failing derivation)} \\
\sigma \cdot \Gamma_{t'} & \text{if } \tau \equiv (t \dashrightarrow_{\sigma} \tau'),\ t' = \mathsf{root}(\tau') \\
\mathcal{SF}(\tau') & \text{if } \tau \equiv (e \Longrightarrow \tau') \\
\mathcal{SF}(\tau_1) \Uparrow \cdots \Uparrow \mathcal{SF}(\tau_n) & \text{if } \mathsf{out}(\tau) = \{e \Rightarrow \tau_i \mid i = 1, \ldots, n\} \\
\sigma_1 \cdot \mathcal{SF}(\tau_1) + \cdots + \sigma_n \cdot \mathcal{SF}(\tau_n) & \text{if } \mathsf{out}(\tau) = \{e \rightarrow_{\sigma} \tau_i \mid i = 1, \ldots, n\}
\end{cases}
$$

For clarity, when no confusion can arise, we often label function Γ with term t rather than with the equation $x \approx t$.

Example 16. Given the extended narrowing tree of Fig. 2, we produce the following success set equation:

$$\begin{aligned}
\Gamma_{\mathsf{f}(x,y)} = \ & \{x \mapsto 0, y \mapsto y', r \mapsto y'\} \\
& + \{x \mapsto \mathsf{s}(x'), y \mapsto y''\} \cdot (\{x' \mapsto x, y'' \mapsto y\} \cdot \Gamma_{\mathsf{f}(x,y)}) \\
& + \{x \mapsto \mathsf{s}(x'), y \mapsto y'''\} \cdot (\{y''' \mapsto x, x' \mapsto y\} \cdot \Gamma_{\mathsf{f}(x,y)})
\end{aligned}$$

Informally speaking, the (infinite) solutions of this equation can be enumerated iteratively as follows. One starts with $\Gamma^0_{\mathsf{f}(x,y)} = \{\ \}$. Then, we compute the next iteration $i > 0$ as follows:

$$\begin{aligned}
\Gamma^i_{\mathsf{f}(x,y)} = \ & \{x \mapsto 0, y \mapsto y', r \mapsto y'\} \\
& + \{x \mapsto \mathsf{s}(x'), y \mapsto y''\} \cdot (\{x' \mapsto x, y'' \mapsto y\} \cdot \Gamma^{i-1}_{\mathsf{f}(x,y)}) \\
& + \{x \mapsto \mathsf{s}(x'), y \mapsto y'''\} \cdot (\{y''' \mapsto x, x' \mapsto y\} \cdot \Gamma^{i-1}_{\mathsf{f}(x,y)})
\end{aligned}$$

Therefore, we have the following infinite sequence:[7]

$$\begin{aligned}
\Gamma^1_{\mathsf{f}(x,y)} = \ & \{\{x \mapsto 0, y \mapsto y'\}\} \\
\Gamma^2_{\mathsf{f}(x,y)} = \ & \Gamma^1_{\mathsf{f}(x,y)} \cup \{\{x \mapsto \mathsf{s}(0), y \mapsto y'\}, \{x \mapsto \mathsf{s}(y'), y \mapsto 0\}\} \\
\Gamma^3_{\mathsf{f}(x,y)} = \ & \Gamma^2_{\mathsf{f}(x,y)} \cup \{\{x \mapsto \mathsf{s}(\mathsf{s}(0)), y \mapsto y'\}, \{x \mapsto \mathsf{s}(\mathsf{s}(y')), y \mapsto 0\}, \\
& \{x \mapsto \mathsf{s}(y'), y \mapsto \mathsf{s}(0)\}, \{x \mapsto \mathsf{s}(0), y \mapsto \mathsf{s}(y')\}\}
\end{aligned}$$

$\dots$

In the following, we denote by $\mathsf{sols}(\Gamma_t)$ the (possibly infinite) set of solutions of the success set equation Γ_t for some term t. Let us consider a set of success set equations $\Gamma_{t_1} = r_1, \dots, \Gamma_{t_n} = r_n$ associated to the narrowing derivations starting from term t_1. A procedure to enumerate the substitutions in $\mathsf{sols}(\Gamma_{t_1})$ can proceed as follows:

1. Initialization. $\Gamma^0_{t_1} = \cdots = \Gamma^0_{t_n} = \{\ \}$.
2. Iterative process. for all $i > 0$, we compute the following sets:

$$\Gamma^i_{t_1} = r_1[\Gamma_t \mapsto \Gamma^{i-1}_t] \quad \dots \quad \Gamma^i_{t_n} = r_n[\Gamma_t \mapsto \Gamma^{i-1}_t]$$

where $r_j[\Gamma_t \mapsto \Gamma^{i-1}_t]$ denotes the expression that results from r_j by replacing every occurrence of Γ_t by Γ^{i-1}_t, with $j = 1, \dots, n$ and $t \in \{t_1, \dots, t_n\}$.

Then, we have $\mathsf{sols}(\Gamma_{t_1}) = \bigcup_{i>0} \Gamma^i_{t_1}$, where the $\Gamma^i_{t_1}$ are computed as above.

We do not formally prove the correctness of the above procedure for computing $\mathsf{sols}(\Gamma_t)$, but it is rather straightforward.

Example 17. Given the extended narrowing tree shown in Fig. 3, we produce the following success set equation:

$$\begin{aligned}
\Gamma_{\mathsf{f}(x,y)} = \ & \{x \mapsto 0, y \mapsto y', r \mapsto y'\} \\
& + \{x \mapsto \mathsf{s}(x'), y \mapsto y'', r \mapsto \mathsf{c}(r', r'')\} \cdot (\{r' \mapsto r, x' \mapsto x, y'' \mapsto y\} \cdot \Gamma_{\mathsf{f}(x,y)} \\
& \qquad\qquad\qquad \Uparrow \{r'' \mapsto r, y'' \mapsto x, x' \mapsto y\} \cdot \Gamma_{\mathsf{f}(x,y)})
\end{aligned}$$

---

[7] We restrict substitutions to $\mathcal{V}ar(\mathsf{f}(x,y))$ for conciseness.

Computing the success set is slightly more difficult now since it involves parallel compositions. The sequence of success sets is as follows:

$$\Gamma^0_{f(x,y)} = \{\,\}$$
$$\Gamma^1_{f(x,y)} = \{\{x \mapsto 0, y \mapsto y', r \mapsto y'\}\}$$
$$\Gamma^2_{f(x,y)} = \Gamma^1_{f(x,y)} \cup \{\{x \mapsto s(x'), y \mapsto y'', r \mapsto c(r', r'')\}$$
$$\cdot(\{r' \mapsto y', x' \mapsto 0, y'' \mapsto y', x \mapsto 0, y \mapsto y', r \mapsto y'\}$$
$$\Uparrow \{r'' \mapsto y', y'' \mapsto 0, x' \mapsto y', x \mapsto 0, y \mapsto y', r \mapsto y'\})\}$$
$$= \Gamma^1_{f(x,y)} \cup \{\{x \mapsto s(x'), y \mapsto y'', r \mapsto c(r', r'')\}$$
$$\cdot\{r' \mapsto 0, r'' \mapsto 0, x' \mapsto 0, y'' \mapsto 0, x \mapsto 0, y \mapsto 0, r \mapsto 0\}\}$$
$$= \Gamma^1_{f(x,y)} \cup \{\{x \mapsto s(0), y \mapsto 0, r \mapsto c(0,0)\}\}$$
$$\dots$$

Example 18. Given the extended narrowing tree shown in Fig. 1, we produce the following success set equation:

$$\Gamma_{f(x)} = \{x \mapsto 0, r \mapsto 0\} + \{x \mapsto s(x')\} \cdot (\{r' \mapsto s(0), r \mapsto r'\} \Uparrow \{r' \mapsto r, x' \mapsto x\} \cdot \Gamma_{f(x)})$$

The sequence of success sets is as follows:

$$\Gamma^0_{f(x)} = \{\,\}$$
$$\Gamma^1_{f(x)} = \{\{x \mapsto 0, r \mapsto 0\}\}$$
$$\Gamma^2_{f(x)} = \Gamma^1_{f(x)} \cup \{\{x \mapsto s(x')\} \cdot (\{r' \mapsto s(0), r \mapsto r'\} \Uparrow \{r' \mapsto 0, x' \mapsto 0, x \mapsto 0, r \mapsto 0\})\}$$
$$= \Gamma^1_{f(x)}$$

Thus, the success set equation denote the singleton set $\{\{x \mapsto 0, r \mapsto 0\}\}$.

Example 19. Given the extended narrowing tree shown in Fig. 4, we produce the following success set equations:

$$\Gamma_{x*y} = \{x \mapsto 0, y \mapsto y', r \mapsto 0\}$$
$$+ \{x \mapsto s(x'), y \mapsto y''\} \cdot (\Gamma_{y''+r'} \Uparrow \{r' \mapsto r, x' \mapsto x, y'' \mapsto y\} \cdot \Gamma_{x*y})$$
$$\Gamma_{y''+r'} = \{y'' \mapsto 0, r' \mapsto r'', r \mapsto r''\}$$
$$+ \{y'' \mapsto s(y'), r' \mapsto r, r \mapsto s(r), r'''' \mapsto r, y''' \mapsto y', r''' \mapsto r'\} \cdot \Gamma_{y''+r'}$$

The success set is the obvious one for addition and multiplication.

The correctness of success set equations can be stated as follows:

Theorem 20. *Let $\mathcal{R}$ be a constructor CD TRS and let t be a term. Let τ be a finite extended narrowing tree for t in $\mathcal{R}$ rooted with $x \approx t$, and let $\Gamma_{x \approx t}$ be its associated success set equation. Then, we have $S_{\mathcal{R}}(x \approx t) = \text{sols}(\Gamma_{x \approx t})$ up to variable renaming.*

5   Related Work

There are basically two closely related lines of research. On the one hand, we have a work by Antoy and Ariola [4] that aims at finding a finite representation of the (possibly infinite) narrowing space. In contrast to our approach, however, they only consider subsumption. Therefore, there is no guarantee that the representation of the narrowing space is going to be finite. They also propose a finite representation inspired by regular expressions to denote a (possibly infinite) enumeration of computed answers. This is somehow similar to our success set equations; nevertheless, our equations are more complex since they may also include parallel compositions.

On the other hand, there are a number of papers on the so called *narrowing-driven* partial evaluation (see [1] and references herein) that also require the construction of a finite representation of the narrowing space. In contrast to [4], other operators like generalization (i.e., replacing some subterms by fresh variables) and splitting are used to ensure that the representation of the narrowing space is finite. However, no single narrowing tree is constructed, but a sequence of (possibly incomplete) narrowing trees, which are then used to extract the residual program (a sequence of *resultants* associated to each root-to-leaf narrowing derivation). The correctness of the transformation is proved for some narrowing strategies (under the *closedness* condition of the narrowing trees). However, no general properties are proved for the different operators.

Our approach can be seen as a combination of the above lines of research. We aim at constructing finite representations of the narrowing space, as in [4], but we also allow the use of powerful operators like flattening and splitting, similarly to the works on narrowing-driven partial evaluation.

6   Conclusion and Future Work

In this work, we have introduced a framework that provides the building blocks that are required to produce a finite representation of the (possibly infinite) narrowing space. For this purpose, we have considered three simple operations: constructor decomposition, flattening and splitting, and have proved its correctness. Then, we have introduced the notion of *extended* narrowing tree, where the above operations can be applied to make the tree finite. Finally, we have introduced a compact equational representation of the success set that follows the structure of a finite extended narrowing tree.

Let us note that our approach could easily be transferred to other logic-based programming languages like Prolog. For instance, the splitting operation is well-known in this context and allows one to partition a query Q into a number of queries $Q_1, \ldots, Q_n$ such that $Q = Q_1, \ldots, Q_n$ (see, e.g., [9] for a precise definition, where the reordering of atoms in a query is also allowed). As for flattening, it can be seen as a simplified version of our notion since predicate symbols cannot be nested. For instance, the flattening of a query $p(X), q(f(Y), Z)$ w.r.t. the position of $f(Y)$ would be $p(X), W = f(Y), q(W, Z)$, where Z is a fresh variable and "$=$" is the syntactic equality defined by the clause $X = X \leftarrow$. Thus, it should not be difficult to adapt the notions of extended narrowing tree and success set equations to logic programming.

Among the possible applications, one can consider the use of extended narrowing trees and success set equations to better understand the program's control

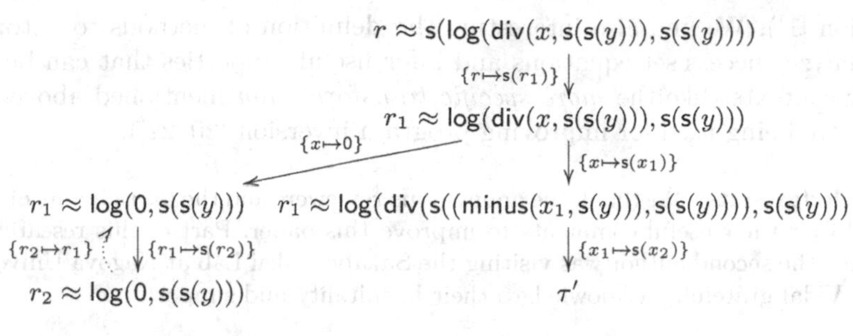

$$r \approx s(\log(\mathsf{div}(x, s(s(y))), s(s(y))))$$

$$\downarrow {\scriptstyle \{r \mapsto s(r_1)\}}$$

$$r_1 \approx \log(\mathsf{div}(x, s(s(y))), s(s(y)))$$

$$\swarrow {\scriptstyle \{x \mapsto 0\}} \qquad\qquad \downarrow {\scriptstyle \{x \mapsto s(x_1)\}}$$

$$r_1 \approx \log(0, s(s(y))) \qquad r_1 \approx \log(\mathsf{div}(s((\mathsf{minus}(x_1, s(y))), s(s(y)))), s(s(y)))$$

$${\scriptstyle \{r_2 \mapsto r_1\}} \nwarrow \; \downarrow {\scriptstyle \{r_1 \mapsto s(r_2)\}} \qquad\qquad \downarrow {\scriptstyle \{x_1 \mapsto s(x_2)\}}$$

$$r_2 \approx \log(0, s(s(y))) \qquad\qquad \tau'$$

Fig. 5. Finite narrowing tree for $s(\log(\mathsf{div}(x, s(s(y))), s(s(y))))$

flow, to analyze *weak* termination [13], to detect subtrees that will never produce a computed answer as in Example 1 (which could be useful, e.g., in the context of the *more specific transformation* (MSV) recently introduced in [23]), and so forth. For instance, let us consider the following TRS:

$$\log(s(0), s(s(y))) \to 0$$
$$\log(x, s(s(y))) \to s(\log(\mathsf{div}(x, s(s(y))), s(s(y))))$$
$$\mathsf{div}(0, s(y)) \to 0$$
$$\mathsf{div}(s(x), s(y)) \to s(\mathsf{div}(\mathsf{minus}(x, y), s(y)))$$
$$\mathsf{minus}(x, 0) \to x$$
$$\mathsf{minus}(s(x), s(y)) \to \mathsf{minus}(x, y)$$

which is obtained by applying the inverse transformation of [21]. Now, we aim at producing a non-overlapping definition of function $\log$. Unfortunately, by applying the original MSV transformation [23] to the body of the second rule, we construct an incomplete narrowing tree for $r \approx s(\log(\mathsf{div}(x, s(s(y))), s(s(y))))$ that still produces overlapping (partial) computed answers. In this context, we can construct the finite extended narrowing tree shown in Fig. 5 instead.

In the extended narrowing tree, one can easily see that the leftmost subtree rooted by $r_1 \approx \log(0, s(s(y)))$ cannot produce any computed answer since there is no leaf. Therefore, it is still safe if the MSV transformation ignores the substitution $\{x \mapsto 0\}$ of the leftmost subtree. Thus we know that the variable x of the rule $\log(x, s(s(y))) \to s(\log(\mathsf{div}(x, s(s(y))), s(s(y))))$ needs to be bound only to $s(s(x_2))$, so that the following non-overlapping definition of $\log$ is obtained:[8]

$$\log(s(0), s(s(y))) \to 0$$
$$\log(s(s(x_2)), s(s(y))) \to s(\log(\mathsf{div}(s(s(x_2)), s(s(y))), s(s(y))))$$

This work opens many possibilities for future work. In particular, we would like to design fully automatic strategies for producing finite extended narrowing trees (e.g., following the methods used in the context of narrowing-driven partial

---

[8] Actually, since the initial TRS is not completely-defined, a reflection rule for innermost narrowing is required, as discussed in Section 2. However, since the result would be the same, we prefer to ignore this rule here and keep the example simpler.

evaluation [1]). We find also interesting the definition of methods to automatically analyze success set equations and infer useful properties that can be used in other contexts (like the *more specific transformation* mentioned above, that is currently being used for improving program inversion [20–22]).

Acknowledgments. We thank the anonymous reviewers and the participants of LOP-STR 2013 for their useful comments to improve this paper. Part of this research was done while the second author was visiting the Sakabe/Sakai Lab at Nagoya University. Germán Vidal gratefully acknowledges their hospitality and support.

References

1. Albert, E., Vidal, G.: The Narrowing-Driven Approach to Functional Logic Program Specialization. New Generation Computing **20**(1), 3–26 (2002)
2. Alpuente, M., Falaschi, M., Vidal, G.: Partial Evaluation of Functional Logic Programs. ACM Transactions on Programming Languages and Systems **20**(4), 768–844 (1998)
3. Alpuente, M., Falaschi, M., Vidal, G.: Compositional Analysis for Equational Horn Programs. In: Rodríguez-Artalejo, M., Levi, G. (eds.) ALP 1994. LNCS, vol. 850, pp. 77–94. Springer, Heidelberg (1994)
4. Antoy, S., Ariola, Z.: Narrowing the Narrowing Space. In: Hartel, P.H., Kuchen, H. (eds.) PLILP 1997. LNCS, vol. 1292, pp. 1–15. Springer, Heidelberg (1997)
5. Arts, T., Giesl, J.: Termination of term rewriting using dependency pairs. Theoretical Computer Science **236**(1–2), 133–178 (2000)
6. Arts, T., Zantema, H.: Termination of Logic Programs Using Semantic Unification. In: Proietti, M. (ed.) LOPSTR 1995. LNCS, vol. 1048, pp. 219–233. Springer, Heidelberg (1996)
7. Baader, F., Nipkow, T.: Term Rewriting and All That. Cambridge University Press (1998)
8. Bae, K., Escobar, S., Meseguer, J.: Abstract Logical Model Checking of Infinite-State Systems Using Narrowing. In: Proceedings of the 24th International Conference on Rewriting Techniques and Applications. LIPIcs, vol. 21, pp. 81–96. Schloss Dagstuhl - Leibniz-Zentrum für Informatik (2013)
9. De Schreye, D., Glück, R., Jørgensen, J., Leuschel, M., Martens, B., Sørensen, M.: Conjunctive partial deduction: foundations, control, algorihtms, and experiments. Journal of Logic Programming **41**(2&3), 231–277 (1999)
10. Escobar, S., Meadows, C., Meseguer, J.: A rewriting-based inference system for the NRL Protocol Analyzer and its meta-logical properties. Theoretical Computer Science **367**(1–2), 162–202 (2006)
11. Escobar, S., Meseguer, J.: Symbolic Model Checking of Infinite-State Systems Using Narrowing. In: Baader, F. (ed.) RTA 2007. LNCS, vol. 4533, pp. 153–168. Springer, Heidelberg (2007)
12. Fribourg, L.: SLOG: A Logic Programming Language Interpreter Based on Clausal Superposition and Rewriting. In: Proceedings of the Symposium on Logic Programming, pp. 172–185. IEEE Press (1985)
13. Gnaedig, I., Kirchner, H.: Proving weak properties of rewriting. Theoretical Computer Science **412**(34), 4405–4438 (2011)
14. Hanus, M.: The integration of functions into logic programming: From theory to practice. Journal of Logic Programming **19&20**, 583–628 (1994)

15. Hanus, M. (ed.): Curry: An integrated functional logic language (vers. 0.8.3) (2012). http://www.curry-language.org
16. Hermenegildo, M., Rossi, F.: On the Correctness and Efficiency of Independent And-Parallelism in Logic Programs. In: Lusk, E., Overbeck, R. (eds.) Proceedings of the 1989 North American Conf. on Logic Programming, pp. 369–389. The MIT Press, Cambridge (1989)
17. Hölldobler, S. (ed.): Foundations of Equational Logic Programming. LNCS, vol. 353. Springer, Heidelberg (1989)
18. Meseguer, J., Thati, P.: Symbolic Reachability Analysis Using Narrowing and its Application to Verification of Cryptographic Protocols. Electronic Notes in Theoretical Computer Science **117**, 153–182 (2005)
19. Middeldorp, A., Okui, S.: A Deterministic Lazy Narrowing Calculus. Journal of Symbolic Computation **25**(6), 733–757 (1998)
20. Nishida, N., Sakai, M., Sakabe, T.: Generation of Inverse Computation Programs of Constructor Term Rewriting Systems. IEICE Transactions on Information and Systems **J88–D–I**(8), 1171–1183 (2005) (in Japanese)
21. Nishida, N., Sakai, M., Sakabe, T.: Partial Inversion of Constructor Term Rewriting Systems. In: Giesl, J. (ed.) RTA 2005. LNCS, vol. 3467, pp. 264–278. Springer, Heidelberg (2005)
22. Nishida, N., Vidal, G.: Program inversion for tail recursive functions. In: Schmidt-Schauß, M. (ed.) Proceedings of the 22nd International Conference on Rewriting Techniques and Applications. LIPIcs, vol. 10, pp. 283–298. Schloss Dagstuhl - Leibniz-Zentrum für Informatik (2011)
23. Nishida, N., Vidal, G.: Computing More Specific Versions of Conditional Rewriting Systems. In: Albert, E. (ed.) LOPSTR 2012. LNCS, vol. 7844, pp. 137–154. Springer, Heidelberg (2013)
24. Nutt, W., Réty, P., Smolka, G.: Basic Narrowing Revisited. Journal of Symbolic Computation **7**(3/4), 295–317 (1989)
25. Ohlebusch, E.: Advanced Topics in Term Rewriting. Springer, London, UK (2002)
26. Palamidessi, C.: Algebraic Properties of Idempotent Substitutions. In: Paterson, M. (ed.) ICALP 1990. LNCS, vol. 443, pp. 386–399. Springer, Heidelberg (1990)
27. Ramos, J.G., Silva, J., Vidal, G.: Fast Narrowing-Driven Partial Evaluation for Inductively Sequential Systems. In: Danvy, O., Pierce, B.C. (eds.) Proceedings of the 10th ACM SIGPLAN International Conference on Functional Programming, pp. 228–239. ACM Press (2005)
28. Slagle, J.R.: Automated theorem-proving for theories with simplifiers, commutativity and associativity. Journal of the ACM **21**(4), 622–642 (1974)

Energy Consumption Analysis of Programs Based on XMOS ISA-Level Models

Umer Liqat[1], Steve Kerrison[2], Alejandro Serrano[1], Kyriakos Georgiou[2],
Pedro Lopez-Garcia[1,3]([⊠]), Neville Grech[2],
Manuel V. Hermenegildo[1,4], and Kerstin Eder[2]

[1] IMDEA Software Institute, Madrid, Spain
{umer.liqat,alejandro.serrano,pedro.lopez,
manuel.hermenegildo}@imdea.org
[2] University of Bristol, Bristol, UK
{steve.kerrison,kyriakos.georgiou,n.grech,
kerstin.eder}@bristol.ac.uk
[3] Spanish Council for Scientific Research (CSIC), Madrid, Spain
[4] Universidad Politécnica de Madrid (UPM), Madrid, Spain

Abstract. Energy consumption analysis of embedded programs requires the analysis of low-level program representations. This is challenging because the gap between the high-level program structure and the low-level energy models needs to be bridged. Here, we describe techniques for recreating the structure of low-level programs and transforming these into Horn clauses in order to make use of a generic resource analysis framework (CiaoPP). Our analysis, which makes use of an energy model we produce for the underlying hardware, characterises the energy consumption of the program, and returns energy formulae parametrised by the size of the input data. We have performed an initial experimental assessment and obtained encouraging results when comparing the statically inferred formulae to direct energy measurements from the hardware running a set of benchmarks. Static energy estimation has applications in program optimisation and enables more energy-awareness in software development.

Keywords: Energy consumption analysis · Energy models · Resource usage analysis · Static analysis

1 Introduction

Energy consumption and the environmental impact of computing technologies are a major focus. Despite advances in power-efficient hardware, more energy savings can be achieved by improving the way current software technologies make use of such hardware. Many optimization techniques that can be used for producing energy-efficient software need estimations of the energy consumption of software segments prior to their execution, in order to make decisions about the optimal way of executing them. These a priori estimations are also very useful to software engineers to better understand the effect of their designs on the

© Springer International Publishing Switzerland 2014
G. Gupta and R. Peña (Eds.): LOPSTR 2013, LNCS 8901, pp. 72–90, 2014.
DOI: 10.1007/978-3-319-14125-1_5

energy consumption early on during the software development process, and make more informed design decisions (e.g., using the appropriate data structures), even when there are parts not developed yet.

In this paper we combine static analysis and *low level* energy modelling techniques to implement a tool capable of estimating the energy consumption of an embedded program (and its constituent parts, such as procedures and functions) as a function on several parameters of the input data (e.g., sizes), and the hardware platform where they are executed (e.g., clock frequency and voltage). We show the feasibility of our proposal with a concrete case study: analysis of ISA (Instruction Set Architecture) code compiled from XC [24]. XC is a high-level C-based programming language that includes extensions for concurrency, communication, input/output operations, and real-time behaviour. XC libraries share a common API with standard C libraries and therefore C code can commingle with XC code in a single application.

Since energy consumption analysis depends on the underlying hardware, the analyser requires information expressing the effect of the execution of a software segment (e.g., an assembly instruction) on the hardware. Such information is represented using *models*. In our approach these models express information using assertions. These are propagated during the static analysis process in order to infer information for higher-level entities such as functions. For instance, using assertions we abstract the operations in the language in terms of their effect on the size of the runtime data and the energy exerted. Energy models at lower levels (e.g., at the ISA level) are more precise than at higher levels (e.g., XC source code), since the closer to the hardware, the easier it is to determine the effect of the execution of the program on the hardware. For this reason, we have produced models for the ISA level, which we use when analysing ISA code generated by the XCC compiler.

Our approach leverages the CiaoPP tool [6], the preprocessor of the Ciao programming environment [7]. CiaoPP includes a generic, parametric analysis framework for resource usage that can be instantiated to infer bounds on resources of interest (energy consumption in our case), for different languages [14]. In CiaoPP, a resource is a user-defined *counter* representing a (numerical) non-functional global property, such as execution time, execution steps, number of bits sent or received by an application over a socket, etc. The CiaoPP resource analysis can infer upper and lower bounds on the usage made of such resources by programs by working on an intermediate block-based representation, the Horn clause (HC) IR. In this representation, each block is written as a *Horn clause*, i.e., a head followed by a sequence of primitive operations or calls to other blocks. Assertions describe the resources to be analyzed. We propose a transformation of the ISA program into this HC IR (containing Horn clauses and assertions), which allows us to analyse the transformed program with CiaoPP. The control and data flow encoded through the procedural interpretation of these Horn-clause programs, coupled with the resource-related information contained in the assertions (such as the energy consumption models at the ISA level), allow the

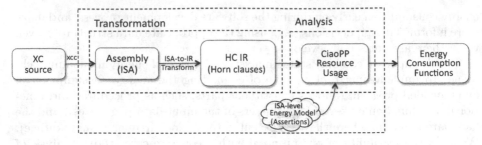

Fig. 1. Overview of the analysis framework for XC programs

```
int fact(int N) {
  if (N <= 0) return 1;
  return N * fact(N - 1);
}
```

Fig. 2. An XC source (factorial) function

resource analysis to infer static bounds on the energy consumption of the blocks that are directly applicable to the original ISA programs.

Figure 1 shows the main steps of our approach for energy consumption analysis, which starts with an XC program (e.g., the `fact` function in Figure 2). The ISA program corresponding to it is generated using the XC compiler tool XCC (left hand side of Figure 3). The resulting ISA program is passed to a translator which generates the associated Horn clauses (right hand side of Figure 3). Such program, together with the information contained in the energy models at the ISA level (represented using the mentioned assertion language), is passed to the resource analysis which outputs the energy consumption for all procedures in the HC IR program. In our example, the resource analysis infers an estimation of the energy consumed by a call to `fact` as $(26.0\ N + 19.4)$ nano-Joules. This is parametric with N, the input argument to `fact`.

In this work we have successfully bridged the gap between researchers closer to the hardware area, needed to produce the low level energy models, and others from software, with expertise in static analysis techniques and tools. In this multidisciplinary research, we have faced some challenges and produced some original contributions that we describe in this paper and summarise as follows:

1. Development of an energy model for a multi-threaded architecture (XMOS XS1-L), that can be applied at instruction set simulation level or higher, with specialisation for high-level, single-threaded benchmarks.
2. Design and implementation of a translation from ISA programs into a Horn-clause representation (HC IR).
3. Instantiation of the CiaoPP general resource analysis framework to infer energy consumption using the low-level energy consumption model.

```
 1   <fact>:
 2   001:  entsp  0x2
 3   002:  stw    r0,  sp[0x1]
 4   003:  ldw    r1,  sp[0x1]
 5   004:  ldc    r0,  0x0
 6   005:  lss    r0,  r0,  r1
 7   006:  bf     r0,  <008>


11   007:  bu     <010>
12   010:  ldw    r0,  sp[0x1]
13   011:  sub    r0,  r0,  0x1
14   012:  bl     <fact>

16   013:  ldw    r1,  sp[0x1]
17   014:  mul    r0,  r1,  r0
18   015:  retsp  0x2


21   008:  mkmsk  r0,  0x1
22   009:  retsp  0x2
```

```
 1   fact(R0,R0_3):-
 2       entsp(0x2),
 3       stw(R0,SpOx1),
 4       ldw(R1,SpOx1),
 5       ldc(R0_1,b0x0),
 6       lss(R0_2,bR0_1,R1),
7a       bf(R0_2,0x8),
7b       fact_aux(R0_2,SpOx1,R0_3,
             R1_1).

10   fact_aux(1,SpOx1,R0_4,R1):-
11       bu(0x0A),
12       ldw(R0_1,SpOx1),
13       sub(R0_2,R0_1,0x1),
14a      bl(fact),
14b      fact(R0_2,R0_3),
16       ldw(R1,SpOx1),
17       mul(R0_4,R1,R0_3),
18       retsp(0x2).

20   fact_aux(0,SpOx1,R0,R1):-
21       mkmsk(R0,0x1),
22       retsp(0x2).
```

Fig. 3. An ISA (factorial) program (left) and its Horn-clause representation (right)

4. Overall design and implementation of a fully automatic system that statically estimates the energy consumption of functions and procedures written in a high-level, C-based programming language, giving the results as functions on input data sizes.

5. Experimental assessment of the developed energy usage static analyser.

Point 4 above may look simple at first sight, given that we have taken advantage of a number of existing tools, mainly the CiaoPP general resource analyser. However, in practice the implementation has required the development of a significant number of new modules and functionalities, as well as interfaces between these existing tools, all of which posed substantial design and implementation challenges and problems that we have successfully solved.

In the rest of the paper, energy characterisation and modelling for our case study architecture (XMOS XS1-L) is explained in Section 2. Then, Section 3 describes the translation from ISA programs into Horn clauses and Section 4 the instantiation of the CiaoPP general resource usage analysis framework. In Section 5, we have performed an experimental assessment of our approach, showing that the estimation of energy consumption is reasonably accurate. Section 6 comments on related work. Finally, Section 7 summarises our conclusions and comments on ongoing and future work.

2   Energy Characterization and Modelling

The assertion-based model uses power consumption data collected during hardware measurement. We have developed an ISA-level model that provides software energy consumption estimates based on Instruction Set Simulation (ISS) statistics. The hardware, the measurement process, as well as the construction of the ISS-driven model, are detailed in [10], with the key components relevant to this paper explained in the rest of this section.

The practicality and accuracy of our approach to energy consumption analysis relies on a good characterisation of energy consumption and generating good energy consumption models. A trade-off needs to be found between the simplicity of the models, which improves the efficiency of the analysis, and the accuracy of the models, which improves the accuracy of the global analysis. Although we analyse single-threaded code, the energy profiling must consider the hardware multi-threading of the architecture, which has an energy impact even when only a single thread is executed.

Further, the nature of the architecture requires specific approaches in order to gather energy profiling data, but these same characteristics preclude certain energy effects from static analysis. For example, the effects of interleaving instructions or re-use of operands from the previous instruction become less relevant in a hardware multi-threaded pipeline, and impossible to determine statically. Although manifested in a specific way in this particular processor architecture, such traits also exist in other processors, such as super-scalar designs. In this paper we describe an initial proposal that offers a good compromise between the above issues, and also eliminates factors that are determined to be insignificant.

2.1   Energy Profiling Framework and Strategy

An energy profiling framework, xmprofile, is used to generate sequences of instructions under various constraints in order to profile the energy characteristics of the hardware. This data is essential for the accurate application of models at any analysis level. The hardware used is shown in Figure 4. A master processor issues test programs to and measures the power used by a slave processor, the Device Under Test (DUT).

Currently, a subset of the ISA, including arithmetic operations, logic operations, and condition tests, has been characterised. Other instructions are at the moment approximated using a single average value, based on typical observed behaviour.

2.2   ISA-Level Model

An ISA-level model, xmmodel, gives an energy estimate for a program based on ISS output. Data from the measurement framework feeds this model.

Our model is based on that devised by Tiwari [22]. Tiwari's approach is shown in Equation (1). The energy of an ISA program, E_p, is characterised as the sum

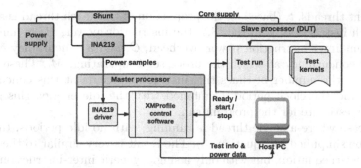

Fig. 4. Overview of test harness hardware and software structure, with a slave processor executing test kernels and a master processor collecting power samples

of base energy cost, B_i, for all ISA instructions, i, multiplied by the number of executions of each instruction, N_i. An inter-instruction overhead energy, $O_{i,j}$, is then accounted for by enumerating for all instruction combinations i, j and their frequency, $N_{i,j}$. Finally, additional contributions to program energy can be accounted for by k external effects, E_k, which may include externally modelled behaviours such as cache memory.

$$E_p = \sum_{i \in \text{ISA}} (B_i \times N_i) + \sum_{i,j \in \text{ISA}} (O_{i,j} \times N_{i,j}) + \sum_{k \in \text{ext}} E_k \qquad (1)$$

The XS1 architecture is hardware multi-threaded. This necessitates a fundamental revision of the model equation. In addition, for performance reasons, the ISS collects instruction statistics rather than a full trace. This reduces the execution time by an order of magnitude, such that it is approximately 100 times slower than the hardware when simulation is run on a modern computer.

Equation (2) describes the energy of a program, E_p, using a similar method to Equation (1), but with several key differences. Time is an explicit component, multiplied by power terms in order to calculate energy. This separation enables future exploration of idle periods, external event timing, and variable operating frequencies. Inter-instruction overhead is represented as a single component, rather than considering it for all possible pairs of instructions, on account of a statistics-based approach rather than cycle-by-cycle instruction tracing. Finally, the level of concurrency must be accounted for, something that was not necessary for the architecture targeted by Equation (1). The concurrency level is the number of threads that are active at a given time. In the case of the XS1-L, the concurrency level represents how full the pipeline is and therefore how much activity is generated within it as each stage switches between instructions from the active threads.

$$E_p = P_{\text{base}} N_{\text{idle}} T_{\text{clk}} + \sum_{t=1}^{N_t} \sum_{i \in \text{ISA}} ((M_t P_i O + P_{\text{base}}) N_{i,t} T_{\text{clk}}) \qquad (2)$$

The base power, P_{base}, is present in both active and idle periods. The number of idle periods, N_{idle}, is counted and multiplied by the clock period, T_{clk}, to account for the energy consumed when no threads are active. For each number of concurrent threads, t, (based on the proportion of time each thread is active), and for each instruction, i, in the ISA, the instruction power, P_i, is multiplied by a constant inter-instruction power overhead, O, and a concurrency cost for the level of concurrency at which the processor is operating, M_t. These are all multiplied by the number of times this instruction occurs at this concurrency level, $N_{i,t}$, and the clock period. Combined with the idle energy, this gives a total energy estimate for the program run.

In the case where a single thread is running, with no idle periods, then the above can be simplified to Equation (3). The result is very similar to the single-threaded Tiwari equation, but with only a single, generic inter-instruction power overhead component, O, and with no external "k" components as the memory of the XS1-L is single-cycle with no cache, with no other effects that need to be considered at this point. There is only ever one active thread, so we use the concurrency cost for one thread, M_1. Again, in Equation (3), time is an explicit component. The overhead, O, is a constant because the inter-instruction effect cannot be known statically in the XS1 architecture, and during profiling the variation in inter-instruction effect was shown to be an order of magnitude less than the instruction cost and would average out over program runs.

$$E_\mathrm{p} = \sum_{i \in \mathrm{ISA}} \left((M_1 P_i O + P_\mathrm{base}) \times (N_i T_\mathrm{clk}) \right) \tag{3}$$

Our ISS-based model, using the same energy data as the static analysis, will be used as an additional comparison point between actual hardware energy measurements and the static analysis results.

3 Transforming ISA Programs into Horn Clauses

In this section we describe the transformation from ISA programs into Horn clauses (HC IR) mentioned in Section 1, which is used for analysis. Such representation consists of a sequence of *blocks* (as in the right hand side of Figure 3). Each block is represented as a *Horn clause*:

$$< block\_id > (< params >) :- \ S_1, \ \dots \ , S_n.$$

which has an entry point, that we call the *head* of the block (to the left of the $:-$ symbol), including a number of parameters $< params >$, and a sequence of steps (the *body*, to the right of the $:-$ symbol), each of which is either, (the representation of) an ISA *instruction*, or a *call* to another (or the same) block. The analyser deals with the HC IR always in the same way, independently of its origin. The transformation ensures that the program information relevant to resource usage is preserved, so that the energy consumption functions of the HC IR programs inferred by the resource analysis are applicable to the original ISA programs.

ISA programs are expressed using the XS1 instruction set [13]. The transformation framework currently works on a subset of this instruction set. The ISA program is parsed and a control flow analysis is carried out, yielding an inter-procedural control flow graph (CFG). This process starts by identifying control transfer instructions such as branch or call instructions. Basic blocks are then constructed, which are annotated with input/output arguments and transformed into Static Single Assignment (SSA) form. Finally, the target HC IR (i.e., Horn clauses) is emitted.

A basic block over a CFG is a maximal sequence of distinct instructions, S_1 through S_n, such that all instructions $S_k, 1 < k < n$ have exactly one in-edge and one out-edge (excluding call-return edges), S_1 has one out-edge, and S_n has one in-edge. A basic block therefore has exactly one entry point at S_1 and one exit point at S_n. All call instructions are assumed to eventually return. Using the basic block definition a block control flow graph is constructed by the analyser, where each node represents a block. Edges between the blocks are derived from calls/jumps between blocks. This process involves iterating through the CFG of the ISA program and marking block boundaries, which are instructions that either begin or end a basic block.

Inferring Block Input/Output Parameters. In order to treat each block as a Horn clause, the block's input and output arguments need to be inferred. For the entry block, the input and output arguments are derived from the original function's signature. We define the functions $params_{in}$ and $params_{out}$, which infer input and output parameters of a block respectively. These perform a backwards analysis of the program, and are recomputed until a least fixpoint is reached on these functions.

$$params_{out}(b) = kill(b) \cup \bigcup_{b' \in next(b)} params_{out}(b')$$
$$params_{in}(b) = gen(b) \cup \bigcup_{b' \in next(b)} params_{in}(b')$$

where $next(b)$ denotes the set of immediate target blocks that can be reached from block b with a call or jump, while $gen(k)$ and $kill(k)$ are the read and written variables in a block respectively, which we define as:

$$kill(b) = \bigcup_{k=1}^{n} def(k), \qquad gen(b) = \bigcup_{k=1}^{n} \{v \mid v \in ref(k) \wedge \forall(j < k).v \notin def(j)\}$$

and $def(k)$ and $ref(k)$ denote the variables written or referred to at a node in the block respectively.

Our approach here is closely related to that of the live variable analysis (LVA) [18] used in compilers, and in dead code elimination in particular. A variable is live at a program point if it may get referenced later in the program (which is decided by considering the whole CFG of the program). In LVA, for each program point, a set of live variables is computed using functions similar to our $kill$ and gen functions with data flow equations. In our approach however, instead of computing liveness information for each program point, we compute a least fixpoint of our $params_{out}$ and $params_{in}$ functions over the program's block control flow graph. This is an efficient solution that safely over-approximates the set of input/output arguments to each block, so that the extra arguments inferred for block heads due to such over-approximation do not affect the energy consumption estimations, since they are not used in the analysis of procedures corresponding to the original XC code.

Resolving Branching to Multiple Blocks. In the XS1 instruction set, conditional branch instructions (e.g., bt, bf) jump to one of the two target blocks based on the value of the branching variable. For example, in Figure 3, at line 7 the bf instruction (branch if fail) will jump to address 008 if $r0 = 0$, otherwise to address 007. In the HC IR this branch needs to be a call to one of the two blocks.

We use a similar approach to the one described in [14] to resolve branches to multiple blocks. The multiple target blocks of a jump instruction are assigned the same head, which essentially are clauses of the same HC IR predicate. This is achieved by merging the heads of the target clauses so that each clause has the same head. The algorithm is trivial, since we have already inferred the input/output parameters to each block's head. The input/output parameters to the new head of the clauses are the union of the input/output parameters of all the clauses along with the branching variable. This enables preservation of the branching semantics of the original ISA program in the HC IR form.

For example in Figure 3, the bf instruction at line 7 of the ISA program is changed to a dummy literal at line 7a in the HC IR, plus a predicate call to fact_aux on line 7b. The predicate fact_aux has two clauses, each representing one of the target blocks of the bf instruction. The dummy literal for the bf instruction is created so that the resource usage analysis can take it into account when inferring energy usage functions.

Static Single Assignment form (SSA). The last step is to convert the block representation into static single assignment (SSA) form, where each variable is assigned exactly once and multiple assignments to the same variable create new versions of that variable.

In compilers, the SSA form is generated at the function level (e.g., at LLVM [11] level) where a function might consist of multiple basic blocks. However, we follow the approach of generating the SSA form at the block level, and therefore we do not need to generate ϕ nodes. A ϕ node is an instruction used to select a version of the variable depending on the predecessor of the current block. Since each block is already annotated with input/output arguments, any predecessor block will pass the appropriate values as input parameters when making a call to the target block.

In Figure 3, the HC IR (right hand side) is already in SSA form, where each variable is defined exactly once and stack references are transformed to local variables. Each instruction is transformed into a HC IR literal with input/output variables.

Analysis on low level (ISA) representations, in general, suffers from the problem of extracting a precise control flow graph in the presence of indirect jumps and calls. The current implementation of our transformation is restricted to direct jumps and calls. We plan to integrate other techniques into the transformation tool to resolve such problems including recognizing code patterns used by compilers and performing static program analysis (see [26] and its references).

4   General Analysis Framework

In this section we introduce the CiaoPP general resource usage analysis framework and discuss how to instantiate it for the analysis of the HC IR programs resulting from the translation of ISA programs.

CiaoPP includes a global static analyser which is parametric with respect to resources and type of approximation (lower and upper bounds) [17]. The user can define the parameters of the analysis for a particular resource by means of assertions that associate basic cost functions with elementary operations of the base language and procedures in libraries, thus expressing how they affect the usage of a particular resource. The global static analysis can then infer bounds on the resource usage of all the procedures in the program, as functions of input data sizes.

In the rest of the section we use a running example to illustrate the main concepts and steps of the analysis framework. In particular, and for simplicity, assume that we are interested in estimating upper bounds on the energy consumed by the HC IR program in Figure 3 (right hand side) generated from its XC code in Figure 2.

4.1   Instantiating the General Framework

Defining Resources. We start by defining the identifier ("counter") associated to the energy consumption resource, through a declaration:

```
:- resource energy.
```

Expressing the Energy Model. In CiaoPP, the resource usage of primitive operations can be provided using "trust" assertions (see [7] and its references for a description of the assertion language). For example, we can write assertions for each predicate that represents an ISA instruction; these constitute the energy models. The following assertions (for the add and sub instructions) are part of the simple energy model that we used in the static analysis, which assigns a constant energy consumption to these ISA instructions (values 1215439 and 1210759 respectively):

```
:- trust pred add(X,Y,Z) + resource(avg, energy, 1215439).
:- trust pred sub(X,Y,Z) + resource(avg, energy, 1210759).
```

Note that the first argument (avg) of the **resource** property (in the global computational properties field "+" of the assertions) expresses that the given energy consumption for the ISA instructions is an average value. This model is obtained using the measurement process described in Section 2, based on Equation (3), so that the energy cost for an ISA instruction i is $c_i = (M_1 \ P_i \ O +$

P_{base}) T_{clk}, expressed in the third argument of the **resource** property in femto-Joules (fJ, 10^{-15} Joules).

Assertions are also used to express other information that is instrumental in the resource usage analysis. For example, the assertion:

```
:- trust pred sub(X,Y,Z) : (var(X), int(Y), int(Z))
   => (int(X), int(Y), int(Z), size(ub,X,int(Y)-int(Z)),
        size(ub,Y,int(Y)), size(ub,Z,int(Z)))
    + (metric(X,int), metric(Y,int), metric(Z,int)).
```

indicates that if the sub(X, Y, Z) predicate (representing the "subtraction" ISA instruction) is called with X and Y bound to integer numbers and Z an unbound variable (precondition field ":"), after the successful completion of the call (postcondition field "=>"), X is an integer number whose size is the size of Y minus the size of Z. It also expresses that the size metric used for the three arguments is "int", the actual value of the integer numbers.

4.2   Performing the Analysis

Once the parameters of the general resource analysis framework have been defined, and assertions for primitives (representing the energy models) and library calls have been provided, the CiaoPP global static analysis can infer the resource usage of all the procedures/blocks in the program (as functions of input data sizes). A full description of how this is done can be found in [17].

Calling Mode Information. The resource analysis needs information referred to each argument in each predicate in the block representation (HC IR) that expresses whether it acts as an input or an output argument (its "mode"). In our approach no mode analysis is performed in order to obtain such information. The modes of the main blocks are extracted from the XC source code that the HC IR is originated from. This is possible because mode information is statically known at the XC language level and is propagated to the HC IR using (trust) assertions. There are also new intermediate predicates generated by the transformation from ISA programs into HC IR (described in Section 3), originated from conditional branching, which cannot be directly related to the XC source code. However, for such predicates information from the transformation phase, where the input/output arguments are determined for each predicate, is used, so that no mode analysis needs to be performed by CiaoPP.

Size Measure Analysis. CiaoPP uses type information to decide which metric to use to infer and express data sizes, from a set of predefined metrics (see [17] for details). As already said, our resource analysis is performed on a block-based representation (HC IR) of the ISA code generated by the XC compiler. Although XC is a typed language, most of the type information is lost in the ISA code generated by the compiler. There are a number of static and dynamic techniques developed by the reverse engineering community to reconstruct types/shape information from binaries (see [12] and its references). In our approach, we can recover and transfer types from the ISA code into some blocks (predicates) in the HC IR that are directly related to the ISA code, so that no type analysis is performed in those cases. However, we still need to perform some propagation of such types to any new intermediate blocks created by the transformation from ISA programs into Horn clauses. For example, our approach can determine that in the HC IR program in Figure 3 (right hand side) `fact` will be called with $R0$ bound to an integer and $R0\_3$ a free variable, and will succeed with $R0\_3$ bound to an integer. Also, `fact_aux` will be called with the first two arguments bound to integers, and the rest free, and, upon success, all of them will be bound to integers. Given that information, the chosen metric for all the arguments will be *int*, i.e., the integer value of the argument.

Size Analysis. It determines the relative sizes of variable bindings at different program points. For each clause, size relations are propagated to express each output data size as a function of input data sizes. For recursive functions this is done symbolically, creating a set of recurrence relations that will be solved to get a closed form function.

For our running example, the recurrence relations set up for the size of the output argument $R0\_3$ of `fact` as a function of the size of the input argument $R0$ (denoted $fact_{R0\_3}(R0)$) as well as the corresponding one for `fact_aux` are:

$$fact_{R0\_3}(R0) = fact\_aux_{R0\_4}(0 \le R0, R0)$$
$$fact\_aux_{R0\_4}(B, R0) = \begin{cases} R0 * fact_{R0\_3}(R0 - 1) & \text{if } B \text{ is } \mathbf{true} \text{ (i.e., } 0 \le R0) \\ 1 & \text{if } B \text{ is } \mathbf{false} \text{ (i.e., } 0 > R0) \end{cases}$$

These inferred recurrence relations/equations are then fed into a computer algebra system (e.g., CiaoPP's internal solver or an external solver such as Mathematica, used for the results presented in this paper) that gives the following closed form function for it: $fact_{R0\_3}(R0) = R0!$

Resource Usage Analysis. It uses the size information inferred by the size analysis to set up recurrence equations representing the resource usage of predicates (blocks), and computes bounds to their solutions. Remember that c_i represents the energy cost of each instruction, taken from the energy model. Let b_e denote the energy consumption function for a predicate (block) b. Then, the inferred equations for `fact` are:

$$fact_e(R0) = fact\_aux_e(0 \le R0, R0) + c_{entsp} + c_{stw} + c_{ldw} + c_{ldc} + c_{lss} + c_{bf}$$
$$fact\_aux_e(B, R0) = \begin{cases} fact_e(R0 - 1) + c_{bu} + 2\,c_{ldw} + c_{sub} + \\ \qquad\qquad + c_{bl} + c_{mul} + c_{retsp} & \text{if } B \text{ is } \mathbf{true} \\ c_{mkmsk} + c_{retsp} & \text{if } B \text{ is } \mathbf{false} \end{cases}$$

If we assume (for simplicity of exposition) that each instruction has unitary cost, i.e., $c_i = 1$ for all i, we obtain (using the mentioned computer algebra system) the energy consumed by `fact` as a function of its input data size ($R0$): $fact_e(R0) = 13\,R0 + 8$.

Table 1. Description of benchmark functions used in experiments and their corresponding energy functions

Function name	Description	Energy function
fact(N)	Calculates $N!$	$26.0\ N + 19.4$
fibonacci(N)	Nth Fibonacci no.	$30.1 + 35.6\ \phi^N + 11.0\ (1 - \phi)^N$
sqr(N)	Computes N^2	$103.0\ N^2 + 205.8\ N + 188.32$
poweroftwo(N)	Calculates 2^N	$62.4 \cdot 2^N - 312.3$
power(base,exp)	Calculates $base^{exp}$	$6.3\ (\log_2 exp + 1) + 6.5$

Note that our approach based on setting up recurrence equations and solving them using a computer algebra system allows inferring different types of (resource usage) functions, such as polynomial, factorial, exponential, logarithmic, and summatory.

Note also that using average values in the model implies that the energy function for the whole program inferred by the upper-bound resource analysis is an approximation of the actual upper bound that can possibly be below it. To ensure that the analysis infers a strict upper bound, we would need to use strict upper bounds as well in the energy models. However, with the current models such bounds would be very conservative, causing a loss in accuracy that would make the analysis not useful in practice. Thus, the current approach is a practical compromise.

5   Benchmarks, Results and Evaluation

The aim of the experimental evaluation is to perform a first comparison of actual hardware energy measurements, in terms of accuracy, with the values obtained from both the low-level Instruction Set Simulation (ISS) model and the Static Resource Analysis (SRA) implemented within the CiaoPP framework, to obtain an early estimation of the feasibility of the approach. To this end, we describe a selection of currently analysable benchmarks, the method by which data was collected, and an evaluation of the analysis framework accuracy vs. the low-level ISS model and hardware measurements.

Benchmarks. For this type of evaluation we use as benchmarks mainly small mathematical functions. The structure of these programs is either iterative or recursive, with their cost depending on the function argument. For such programs state of the art solvers can easily provide the cost functions, by solving the system of recurrence relations provided by the SRA framework. Table 1 shows the benchmarks used in this comparison, their execution behaviour in relation to each function's parameters, and the cost function inferred.

Table 2. Actual and estimated energy consumption for the `fact(N)` function over a range of N

SRA cost function(nJ)	N	HW measured energy (nJ)	Model energy (nJ)		Error vs. HW	
			ISS	SRA	ISS	SRA
	1	53.1	62.8	45.3	1.18	0.85
	2	78.0	83.8	71.3	1.07	0.91
	4	127.7	125.7	123.1	0.98	0.96
$26.0\ N + 19.4$	8	227.1	209.6	226.8	0.92	1.00
	16	426.0	377.4	434.2	0.89	1.02
	32	823.8	713.4	849.0	0.87	1.03
	64	1690.5	1387.0	1678.4	0.82	0.99

Experimental Method. Hardware energy readings were obtained by repeatedly executing a benchmark function over a 0.5 second period, T, collecting a set of power samples, P, whilst counting the number of executions, N_{fn}. From this, the energy of a single function call, $E_{\text{fn}} = \frac{\text{mean}(P) \times T}{N_{\text{fn}}}$ is calculated. This was performed using a similar method to the collection of energy model data described in Section 2, but was performed on separate hardware so as to decouple modelling from testing.

ISS modelling involved simulating the same function a smaller number of times than on the hardware in order to keep simulation time adequately low. The instruction statistics were then processed in order to produce an energy figure, and then that figure divided by N_{fn} was used during ISS in order to extract the energy of a single call. The ISS modelling framework currently has a less efficient test loop than the hardware, potentially reducing accuracy for very short function calls. Similarly, if too few function calls are made during the simulation due to a long-executing function, overrun in the test time may skew low-level energy figures.

Static resource usage analysis was performed by evaluating the produced cost function for a given benchmark with respect to the input arguments, immediately providing the energy cost of a single function call.

Results. Table 2 provides an example of test data for the `fact` (factorial) function. The hardware (HW), low-level Instruction Set Simulation model (ISS), and Static Resource Analysis (SRA) model energy figures are compared. The relative errors of ISS and SRA are compared with respect to the HW energy and normalised as such. The cost function provided for this particular example demonstrates the relationship between the input parameter, N, and the SRA estimate of such a call. This, together with data for a number of further benchmarks are presented in graph form in Figure 5.

In Figure 5, hardware measured energy is compared directly to ISS and SRA energy predictions for the set of four benchmarks. The relative errors are also plotted. In all cases, the ISS model is seen to improve in accuracy as the input parameter N increases, in line with the expected inaccuracies arising from inefficiencies in the modelling loop used in simulation, as described in the previous

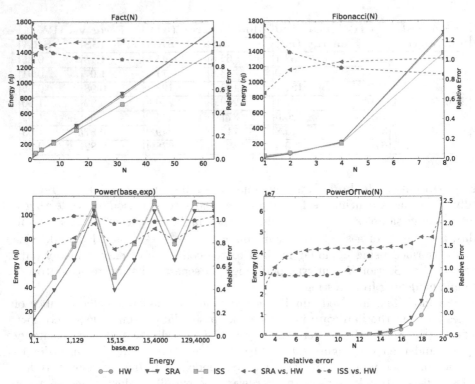

Fig. 5. Hardware energy, estimations and relative errors for (starting top-left, moving clock-wise) `fact`, `fibonacci`, `poweroftwo` and `power`

subsection. In the case of the `poweroftwo` function, time limitations prevent the ISS model from approximating the function above $N = 13$, approaching which the error begins to increase markedly. The `power` function behaves in a similar way and demonstrates the relationship between multiple input arguments.

The CiaoPP SRA model does not suffer the same deficiencies, although it does incur a greater underestimation of energy for small values of N. The HW measurements unavoidably contain some loop code beyond the target function being examined and small N values will increase the effects of this in the measurement. ISS in fact models this inefficiency directly, whereas SRA does not, hence the roughly symmetrical relative errors for the two models, particularly in the `fact` and `fibonacci` cases.

Both approaches are reliant on the same underlying instruction energy figures. Given that some instructions are not directly profiled and, instead, given an average value, accuracy is reduced when the distribution of instructions in a given program is such that the number of profiled instructions is low.

Overall, these results demonstrate both models' capabilities to estimate energy, with encouraging accuracy that can be improved upon. Further, the SRA approach is less restrictive, particularly in situations where simulation time might be prohibitively long.

6 Related Work

Static cost analysis techniques based on setting up and solving recurrence equations date back to Wegbreit's seminal paper [25], and have been developed significantly in subsequent work [1–3,16,17,19,21,23]. This approach was first applied to energy consumption in [15], which inferred statically upper-bounds on the energy consumption of Java programs as functions of input data sizes. As herein, this work used the generic framework of [6,17], specializing it for Java bytecode [14,16] by translating the Jimple (a typed three-address code) representation of Java bytecode into the Horn clause-based IR of the analyzer [14]. However, we employ transformations at lower level (XS1-ISA), irrespective of source language in general, where much of the program structure and typing information is trimmed away. Our transformation employs analysis techniques to reverse engineer ISA programs and reconstruct the control flow graph so that the equivalent HC IR safely approximates the semantics of the original ISA program. In addition, [15] did not compare the results with actual, measured energy consumptions and used a comparatively simple energy model.

Other approaches to cost analysis, such as, e.g., those based on the potential method [8], are limited to polynomial bounds, and do not allow inferring non-polynomial energy functions, as in the recurrence equation method. A number of static analyses are aimed at inferring worst case execution time (WCET, see, e.g., [4] and its references) and related techniques have been applied in [9] to derive a worst-case energy analysis. However, WCET methods typically do not infer cost functions on input data sizes but rather absolute maximum values, and they generally require manual annotation of loops with an upper bound on the number of iterations.

Other transformation-based approaches have been proposed in order to analyse low level microprocessor code [5] and Java source and bytecode [1] (outside the context of energy analysis).

Instruction Set Simulation can be used to estimate the energy of a program running on a suitably profiled hardware platform. Simple models for single-threaded architectures have been demonstrated [22]. These have then been expanded upon, leading to models capable of modelling more complex hardware such as that used in this paper, which comprises a multi-threaded architecture [10].

7 Conclusions and Future Work

In this paper we introduce an approach for estimating the energy consumption of programs compiled for the XS1 architecture, based on a Horn clause transformation and the use of ISA level models that we have produced. We have shown the feasibility of the approach with a prototype implementation within the CiaoPP system, which has been successful in statically finding a good approximation of the energy consumed by a set of selected programs in our experiments.

The XS1 architecture is inherently multi-threaded, and the simulation-based model is able to provide energy estimates for this. Statically analysing multiple concurrent threads adds a significant new dimension of complexity to the modelling exercise. This is a goal of further work in order to provide meaningful analysis for contemporary multi-threaded programs running on this architecture.

We also plan to produce and deal with energy models that take into account the switching cost among pairs of ISA instructions (i.e., the energy consumed by bit flipping), since our analysis framework allows it. The improvement in accuracy from this approach can vary between architectures, for example research such as [20], shows that a simple model can be sufficient in some cases, due to bit flipping effects averaging out over time. Thus, the impact in the context of any target architectures must therefore be considered in this future work, in order to establish whether the increased complexity of analysis delivers a worthwhile gain in accuracy.

We also intend to improve upon the energy measurements of commonly used instructions, which involves more complex techniques such as linear regression. This technique can also be used to construct energy models of intermediate compiler representations such as LLVM IR [11], which would enable us to apply our analysis techniques to more structured program representations. Another method for analysing LLVM IR would involve mapping low-level program instruction segments to LLVM IR segments and reusing the ISA-level energy models.

Acknowledgments. The research leading to these results has received funding from the European Union 7th Framework Programme under grant agreement 318337, ENTRA - Whole-Systems Energy Transparency, Spanish MINECO TIN'12-39391 *StrongSoft* and TIN'08-05624 *DOVES* projects, and Madrid TIC-1465 *PROMETIDOS-CM* project. We also thank John Gallagher for useful and fruitful discussions and feedback in general, and in particular for his help on the implementation of a translation for removing mutual recursions in Horn clause programs, which is performed prior to setting up recurrence equations.

References

1. Albert, E., Arenas, P., Genaim, S., Puebla, G., Zanardini, D.: Cost Analysis of Java Bytecode. In: De Nicola, R. (ed.) ESOP 2007. LNCS, vol. 4421, pp. 157–172. Springer, Heidelberg (2007)
2. Debray, S.K., Lin, N.-W., Hermenegildo, M.: Task Granularity Analysis in Logic Programs. In: Proc. of the 1990 ACM Conf. on Programming Language Design and Implementation, pp. 174–188. ACM Press (June 1990)
3. Debray, S.K., López-García, P., Hermenegildo, M., Lin, N.-W.: Lower Bound Cost Estimation for Logic Programs. In: 1997 International Logic Programming Symposium, pp. 291–305. MIT Press, Cambridge (October 1997)
4. Wilhelm, R., et al.: The worst-case execution-time problem - overview of methods and survey of tools. ACM Trans. Embedded Comput. Syst. **7**(3) (2008)

5. Henriksen, K.S., Gallagher, J.P.: Abstract interpretation of PIC programs through logic programming. In: Sixth IEEE International Workshop on Source Code Analysis and Manipulation (SCAM 2006), pp. 184–196. IEEE Computer Society (2006)
6. Hermenegildo, M., Puebla, G., Bueno, F., López-García, P.: Integrated Program Debugging, Verification, and Optimization Using Abstract Interpretation (and The Ciao System Preprocessor). Science of Computer Programming **58**(1–2), 115–140 (2005)
7. Hermenegildo, M.V., Bueno, F., Carro, M., López, P., Mera, E., Morales, J.F., Puebla, G.: An Overview of Ciao and its Design Philosophy. Theory and Practice of Logic Programming **12**(1–2), 219–252 (2012)
8. Hoffmann, J., Aehlig, K., Hofmann, M.: Multivariate amortized resource analysis. ACM Trans. Program. Lang. Syst. **34**(3), 14 (2012)
9. Jayaseelan, R., Mitra, T., Li, X.: Estimating the worst-case energy consumption of embedded software. In: IEEE Real Time Technology and Applications Symposium, pp. 81–90. IEEE Computer Society (2006)
10. Kerrison, S., Eder, K.: Energy modelling and optimisation of software for a hardware multi-threaded embedded microprocessor. ACM Transactions on Embedded Computing Systems (TECS) (to appear, 2015)
11. Lattner, C., Adve, V.S.: LLVM: A compilation framework for lifelong program analysis and transformation. In: Proc. of the 2004 International Symposium on Code Generation and Optimization (CGO), pp. 75–88. IEEE Computer Society (2004)
12. Lee, J.H., Avgerinos, T., Brumley, D.: TIE: Principled Reverse Engineering of Types in Binary Programs. In: Proceedings of the Network and Distributed System Security Symposium, NDSS 2011. The Internet Society (February 2011)
13. May, D.: The XMOS XS1 architecture (2013). http://www.xmos.com/published/xmos-xs1-architecture
14. Méndez-Lojo, M., Navas, J., Hermenegildo, M.V.: A flexible, (C)LP-based approach to the analysis of object-oriented programs. In: King, A. (ed.) LOPSTR 2007. LNCS, vol. 4915, pp. 154–168. Springer, Heidelberg (2008)
15. Navas, J., Méndez-Lojo, M., Hermenegildo, M.: Safe Upper-bounds Inference of Energy Consumption for Java Bytecode Applications. In: The Sixth NASA Langley Formal Methods Workshop (LFM 2008) (April 2008) (Extended Abstract)
16. Navas, J., Méndez-Lojo, M., Hermenegildo, M.: User-Definable Resource Usage Bounds Analysis for Java Bytecode. In: Proceedings of BYTECODE. Electronic Notes in Theoretical Computer Science, vol. 253, pp. 65–82. Elsevier - North Holland (March 2009)
17. Navas, J., Mera, E., López-García, P., Hermenegildo, M.V.: User-Definable Resource Bounds Analysis for Logic Programs. In: Dahl, V., Niemelä, I. (eds.) ICLP 2007. LNCS, vol. 4670, pp. 348–363. Springer, Heidelberg (2007)
18. Nielson, F., Nielson, H.R., Hankin, C.: Principles of Program Analysis. Springer (1999)
19. Rosendahl, M.: Automatic Complexity Analysis. In: 4th ACM Conference on Functional Programming Languages and Computer Architecture (FPCA 1989). ACM Press (1989)
20. Russell, J.T., Jacome, M.F.: Software power estimation and optimization for high performance, 32-bit embedded processors. In: ICCD, pp. 328–333 (1998)
21. Serrano, A., Lopez-Garcia, P., Hermenegildo, M.: Resource Usage Analysis of Logic Programs via Abstract Interpretation Using Sized Types. In: Theory and Practice of Logic Programming, 30th Int'l. Conference on Logic Programming (ICLP 2014) Special Issue, vol. 14(4–5), pp. 739–754 (2014)

22. Tiwari, V., Malik, S., Wolfe, A., Lee, M.T.C. Instruction level power analysis and optimization of software. In: Proceedings of VLSI Design, pp. 326–328 (1996)
23. Vasconcelos, P.B., Hammond, K.: Inferring Cost Equations for Recursive, Polymorphic and Higher-Order Functional Programs. In: Trinder, P., Michaelson, G.J., Peña, R. (eds.) IFL 2003. LNCS, vol. 3145, pp. 86–101. Springer, Heidelberg (2005)
24. Watt, D.: Programming XC on XMOS Devices. XMOS Limited (2009)
25. Wegbreit, B.: Mechanical program analysis. Commun. ACM **18**(9), 528–539 (1975)
26. Xu, L., Sun, F., Su, Z.: Constructing Precise Control Flow Graphs from Binaries. University of California, Davis, Tech. Rep (2009)

From Outermost Reduction Semantics
to Abstract Machine

Olivier Danvy$^{(\boxtimes)}$ and Jacob Johannsen

Department of Computer Science, Aarhus University, Aarhus, Denmark
{danvy,cnn}@cs.au.dk

Abstract. Reduction semantics is a popular format for small-step operational semantics of deterministic programming languages with computational effects. Each reduction semantics gives rise to a reduction-based normalization function where the reduction sequence is enumerated. Refocusing is a practical way to transform a reduction-based normalization function into a reduction-free one where the reduction sequence is not enumerated. This reduction-free normalization function takes the form of an abstract machine that navigates from one redex site to the next without systematically detouring via the root of the term to enumerate the reduction sequence, in contrast to the reduction-based normalization function.

We have discovered that refocusing does not apply as readily for reduction semantics that use an outermost reduction strategy and have overlapping rules where a contractum can be a proper subpart of a redex. In this article, we consider such an outermost reduction semantics with backward-overlapping rules, and we investigate how to apply refocusing to still obtain a reduction-free normalization function in the form of an abstract machine.

1 Introduction

A Structural Operational Semantics [27] is a small-step semantics where reduction steps are specified with a relation. For a deterministic programming language, this relation is a function, and evaluation is defined as iterating this one-step reduction function until a normal form is found, if there is one. This way of evaluating a term is said to be "reduction-based" because it enumerates each reduct in the reduction sequence, reduction step by reduction step. A reduction step from a term t_i to the reduct t_{i+1} is carried out by locating a redex r_i in t_i, contracting r_i into a contractum c_i, and then constructing t_{i+1} as an instance of t_i where c_i replaces r_i. In a Structural Operational Semantics, the context of every redex is represented logically as a proof tree.

A Reduction Semantics [13] is a small-step semantics where the context of every redex is represented syntactically as a term with a hole. To reduce the term t_i to the reduct t_{i+1}, t_i is decomposed into a redex r_i and a reduction context $C_i[\ ]$, r_i is contracted into a contractum c_i, and $C_i[\ ]$ is recomposed with c_i to form t_{i+1}. Graphically:

$$t_i = C_i[r_i] \rightarrow C_i[c_i] = t_{i+1}$$

© Springer International Publishing Switzerland 2014
G. Gupta and R. Peña (Eds.): LOPSTR 2013, LNCS 8901, pp. 91–108, 2014.
DOI: 10.1007/978-3-319-14125-1_6

A reduction step is therefore carried out by rewriting a redex into a contractum according to the reduction rules, with a rewriting strategy that matches the reduction order and is reflected in the structure of the reduction context. If the reduction strategy is deterministic, it can be implemented with a function. Applying this decomposition function to a term which is not in normal form gives a reduction context and a potential redex.

Reduction is stuck for terms that are in normal form (i.e., where no potential redex occurs according to the reduction strategy), or if a potential redex is found which is not an actual one (e.g., if an operand has a type that the semantics deems incorrect).

For a deterministic programming language, the reduction strategy is deterministic, and so it yields a unique next potential redex to be contracted, if there is one. Furthermore, for any actual redex, only one reduction rule can apply. Therefore, there are no critical pairs and rewriting is confluent.

The format of reduction semantics lends itself well to ensure properties such as type safety [32], thanks to the subject reduction property from type theory. It also makes it possible to account for control operators and first-class continuations by making the reduction context part of the reduction rules [3,13]. Today reduction semantics are in common use in the area of programming languages [14,25].

1.1   Reduction-Based vs. Reduction-Free Evaluation

Evaluating a term is carried out by enumerating its reduction sequence, reduction step after reduction step:

$$\ldots \to \overbrace{C_{i-1}[c_{i-1}] = C_i[r_i]}^{t_i} \to \overbrace{C_i[c_i] = C_{i+1}[r_{i+1}]}^{t_{i+1}} \to \overbrace{C_{i+1}[c_{i+1}] = C_{i+2}[r_{i+2}]}^{t_{i+2}} \to \ldots$$

This reduction-based enumeration requires all of the successive reducts to be constructed, which is inefficient. So in practice, alternative, reduction-free evaluation functions are sought, often in the form of an abstract machine, and many such abstract machines are described in the literature.

Over the last decade, the first author and his students have been putting forward a methodology for systematically constructing such abstract machines [4,5,9]: instead of recomposing the reduction context with the contractum to obtain the next reduct in the reduction sequence and then decomposing this reduct into the next potential redex and its reduction context, we simply continue the decomposition of the contractum in its reduction context, as depicted with a squiggly arrow:

$$\ldots \to C_{i-1}[c_{i-1}] \rightsquigarrow C_i[r_i] \to C_i[c_i] \rightsquigarrow C_{i+1}[r_{i+1}] \to C_{i+1}[c_{i+1}] \rightsquigarrow C_{i+2}[r_{i+2}] \to \ldots$$

This shortcut works for deterministic reduction strategies where after recomposition, decomposition always comes back to the contractum and its reduction context before continuing [9]. In particular, it always works for innermost reduction, and has given rise to a 'syntactic correspondence' between reduction semantics and abstract machines [2,3].

This syntactic correspondence has proved successful to reconstruct many pre-existing abstract machines as well as to construct new ones [1, 6, 16, 29], even in the presence of control operators [3, 7]. For a class of examples, it applies to all the reduction semantics of Felleisen et al.'s latest textbook [14]. More generally, it concretizes Plotkin's connection between calculi and programming languages [26] in that it mechanizes the connection between reduction order (in the small-step world) and evaluation order (in the big-step world), and between not getting stuck (in the small-step world) and not going wrong (in the big-step world).

That said, we have discovered that for reduction semantics that use an outermost strategy and have backward-overlapping rules [11, 17, 18], refocusing does not apply as readily: indeed after recomposition, decomposition does not always come back to the contractum and its reduction context – it might stop before, having found a potential redex that was in part constructed by the previous contraction. The goal of our work here is to study reduction semantics that use an outermost strategy ("outermost reduction semantics") and that have backward-overlapping rules, and to investigate how to apply refocusing to still obtain an abstract machine implementing a reduction-free normalization function.

1.2 Overview

We first illustrate reduction semantics for arithmetic expressions with an innermost reduction strategy (Section 2), where all the elements of our domain of discourse are touched upon: BNF of terms; reduction rules and contraction function; reduction strategy and BNF of reduction contexts; recomposition of a context with a term; decomposition of a term either into a normal form or into a potential redex and a reduction context; left inverseness of recomposition with respect to decomposition; one-step reduction as decomposition, contraction, and recomposition; reduction-based evaluation as the iteration of one-step reduction; refocusing; and reduction-free evaluation. We then turn to the issue of overlapping rules (Section 3). With respect to refocusing, the only problematic combination of overlaps and strategies is backward-overlapping rules and outermost strategy (Section 4). To solve the problem, we suggest to backtrack after contracting a redex, which enables refocusing (Section 5). For symmetry, we also consider foretracking (Section 6). We then review related work (Section 7).

2 A Simple Example with an Innermost Strategy

We consider a simple language of arithmetic expressions with a zero-ary constructor 0, a unary constructor S, and a binary constructor A. The goal is to normalize a given term into a normal form using only the constructors 0 and S.

Terms: The BNF of terms reads as follows:

$$t ::= 0 \mid S(t) \mid A(t, t)$$

Terms in normal form: The BNF of terms in normal form reads as follows:

$$t^{\mathrm{nf}} ::= 0 \mid S(t^{\mathrm{nf}})$$

Reduction rules: The BNF of potential redexes reads as follows:

$$pr ::= A(0, t_2) \mid A(S(t_1^{\mathrm{nf}}), t_2)$$

The reduction rules read as follows:

$$A(0, t_2) \mapsto t_2$$
$$A(S(t_1^{\mathrm{nf}}), t_2) \mapsto S(A(t_1^{\mathrm{nf}}, t_2))$$

Note the occurrence of t_1^{nf}, which is in normal form, in the left-hand side of the second reduction rule: it is characteristic of innermost reduction.

All potential redexes are actual ones, i.e., no terms are stuck. We can thus implement contraction as a total function:

$$\frac{pr \mapsto c}{contract(pr) = c}$$

Reduction strategy: We are looking for the leftmost-innermost redex. This reduction strategy is materialized with the following grammar of reduction contexts:

$$C[\,] ::= \Box[\,] \mid C[S[\,]] \mid C[A([\,], t)]$$

We obtained this grammar by CPS-transforming a search function implementing the innermost reduction strategy and then defunctionalizing its continuation [10].

Lemma 1 (Unique Decomposition). *Any term not in normal form can be decomposed into exactly one reduction context and one potential redex.*

Recomposition: As usual, a reduction context is iteratively recomposed with a term using a left fold, as specified by the following abstract-machine transitions:

$$\langle \Box[t] \rangle^{\mathrm{rec}} \uparrow t$$
$$\langle C[S[t]] \rangle^{\mathrm{rec}} \uparrow \langle C[S(t)] \rangle^{\mathrm{rec}}$$
$$\langle C[A([t_1], t_2)] \rangle^{\mathrm{rec}} \uparrow \langle C[A(t_1, t_2)] \rangle^{\mathrm{rec}}$$

This abstract machine is a deterministic finite automaton with two states: an intermediate state pairing a context and a term, and a final state holding a term. Each transition corresponds to a context constructor. There is therefore no ambiguity and no incompleteness. Recomposition is defined as the iteration of these transitions:

$$\frac{\langle C[t] \rangle^{\mathrm{rec}} \uparrow^* t'}{recompose(C, t) = t'}$$

Since a context constructor is peeled off at each iteration, making the size of the context decrease, the recomposition function is total.

Decomposition: Likewise, a term is iteratively decomposed in an innermost fashion into a potential redex and its reduction context, as specified by the following abstract-machine transitions:

$$\langle C[0]\rangle_{\text{term}}^{\text{dec}} \downarrow \langle C[0]\rangle_{\text{cont}}^{\text{dec}}$$

$$\langle C[S(t)]\rangle_{\text{term}}^{\text{dec}} \downarrow \langle C[S[t]]\rangle_{\text{term}}^{\text{dec}}$$

$$\langle C[A(t_1, t_2)]\rangle_{\text{term}}^{\text{dec}} \downarrow \langle C[A([t_1], t_2)]\rangle_{\text{term}}^{\text{dec}}$$

$$\langle \Box[t^{\text{nf}}]\rangle_{\text{cont}}^{\text{dec}} \downarrow t^{\text{nf}}$$

$$\langle C[S[t^{\text{nf}}]]\rangle_{\text{cont}}^{\text{dec}} \downarrow \langle C[S(t^{\text{nf}})]\rangle_{\text{cont}}^{\text{dec}}$$

$$\langle C[A([0], t_2)]\rangle_{\text{cont}}^{\text{dec}} \downarrow C[A(0, t_2)]$$

$$\langle C[A([S(t^{\text{nf}})], t_2)]\rangle_{\text{cont}}^{\text{dec}} \downarrow C[A(S(t^{\text{nf}}), t_2)]$$

This abstract machine is a deterministic pushdown automaton with four states where the context is the stack: two intermediate states pairing a context and a term, and two final states, one for the case where the given term is in normal form, and one for the case where it decomposes into a context and a potential redex. Each transition from the first intermediate state corresponds to a term constructor, and each transition rule from the second intermediate state corresponds to a context constructor. Each transition from the first intermediate state peels off a term constructor, and each transition from the second intermediate state peels off a context constructor. There is therefore no ambiguity and no incompleteness.

Furthermore, each transition preserves an invariant: recomposing the current context with the current term yields the original term.

Given a term to decompose, the initial machine state pairs this term with the empty context. There are two final states: one for terms in normal form (and therefore containing no redex), and one for potential redexes and their reduction context. Decomposition, which is defined as the iteration of these machine transitions, is therefore a total function:

$$\frac{\langle \Box[t]\rangle_{\text{term}}^{\text{dec}} \downarrow^* t^{\text{nf}}}{decompose(t) = t^{\text{nf}}} \qquad \frac{\langle \Box[t]\rangle_{\text{term}}^{\text{dec}} \downarrow^* C[pr]}{decompose(t) = C[pr]}$$

A notable property: Due to the invariant of the abstract machine implementing decomposition, the recomposition function is a left inverse of the decomposition function.

One-step reduction: One-step reduction is implemented as, successively, the decomposition of a given term into a potential redex and its reduction context; the contraction of this redex into a contractum; and the recomposition of the reduction context with the contractum:

$$\frac{\langle \Box[t]\rangle_{\text{term}}^{\text{dec}} \downarrow^* C[pr] \qquad pr \mapsto c \qquad \langle C[c]\rangle^{\text{rec}} \uparrow^* t'}{t \,\rangle\, t'}$$

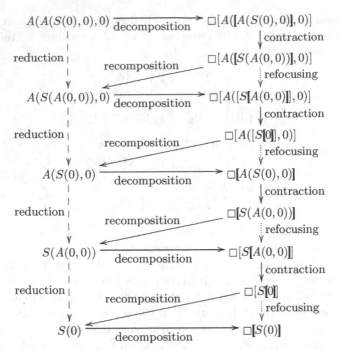

Fig. 1. Innermost reduction sequence for $A(A(S(0), 0), 0)$

Reduction-based evaluation: A term is evaluated into a normal form by iterating one-step reduction:

$$\frac{t \to^* t^{\mathrm{nf}}}{t \Rightarrow_{\mathrm{rb}} t^{\mathrm{nf}}}$$

Towards reduction-free evaluation: Between one contraction and the next, we recompose the reduction context with the contractum until the next reduct, which we decompose into the next potential redex and its reduction context. But since the reduction strategy is innermost (and deterministic), the decomposition of the next reduct will come back to the site of this contractum and this context before continuing. This offers us the opportunity to short-cut the recomposition and decomposition to this contractum and this context and thus to refocus by just continuing the decomposition *in situ*. 5 More formally, we have

$$\frac{t \downarrow^* C[pr] \qquad C[pr] \; ([\mapsto]; \downarrow^*)^* \; t^{\mathrm{nf}}}{t \Rightarrow_{\mathrm{rf}} t^{\mathrm{nf}}}$$

where $([\mapsto]; \downarrow^*)$ denotes contraction in context followed by decomposition (and was noted $\rightsquigarrow$ in Section 1.1).

An example: See Figure 1.

Reduction-free evaluation: After applying refocusing, we follow the steps of the syntactic correspondence [2,3,5], fusing the iteration and refocus functions, inlining the contract function, and compressing corridor transitions. The resulting normalizer implements a transition system described by the following abstract machine:

$$t \longmapsto \langle \square[t] \rangle_{\text{term}}$$

$$\langle C[0] \rangle_{\text{term}} \longmapsto \langle C[0] \rangle_{\text{cont}}$$
$$\langle C[S(t)] \rangle_{\text{term}} \longmapsto \langle C[S[t]] \rangle_{\text{term}}$$
$$\langle C[A(t_1, t_2)] \rangle_{\text{term}} \longmapsto \langle C[A([t_1], t_2)] \rangle_{\text{term}}$$

$$\langle \square[t^{\text{nf}}] \rangle_{\text{cont}} \longmapsto t^{\text{nf}}$$
$$\langle C[S[t^{\text{nf}}]] \rangle_{\text{cont}} \longmapsto \langle C[S(t^{\text{nf}})] \rangle_{\text{cont}}$$
$$\langle C[A([0], t_2)] \rangle_{\text{cont}} \longmapsto \langle C[t_2] \rangle_{\text{term}}$$
$$\langle C[A([S(t^{\text{nf}})], t_2)] \rangle_{\text{cont}} \longmapsto \langle C[S[A([t^{\text{nf}}], t_2)]] \rangle_{\text{cont}}$$

3   Backward-Overlapping Rules

Refocusing (i.e., the short-cutting of recomposition and decomposition after contraction) is possible when, after recomposing a reduction context with a contractum into a reduct, the subsequent decomposition of this reduct comes back to this contractum and context before continuing.

However, there are cases where decomposition of the reduct does not come back to the contractum. For example, this is the case when the reduction strategy is outermost and the contractum is a proper subpart of a potential redex: then after recomposing a reduction context with a contractum into a reduct, the subsequent decomposition of this reduct would *not* come back to this contractum and context—it would stop at the newly created potential redex, above the contractum. So when the reduction strategy is outermost and a contractum can be a subpart of a potential redex, refocusing is not possible.

A contractum can be a subpart of a potential redex when the reduction rules contain *backward overlaps*:

Definition 1 (Backward-Overlapping Rules). *Let $l_1 \to r_1$ and $l_2 \to r_2$ be two reduction rules. If l_1 decomposes into a non-empty context C and a term t that contains at least one term constructor and that unifies with r_2, then the two rules are said to be* backward-overlapping *[11, 17, 18].*

Symmetrically, if the left-hand side of one reduction rule can form a proper subpart of the right-hand side of another rule, the reduction rules are said to be *forward-overlapping.*

The combination of backward-overlapping rules and outermost reduction does not occur often in programming languages. However, it does occur in the full normalization of λ-terms using normal-order reduction, which has applications for comparing normal forms in proof assistants. Other occurrences can more readily be found outside the field of programming-language semantics, in the area of term rewriting.

We distinguish four cases of reduction strategy in combination with rule overlaps, and treat each of them in the following sections:

	innermost strategy	outermost strategy
forward-overlapping rules	Section 3.1	Section 3.2
backward-overlapping rules	Section 3.3	Section 3.4

3.1   Forward Overlaps and Innermost Strategy

In this case, a contractum may contain a potential redex. This redex will be found in due course when the contractum is decomposed. The detour via an intermediate reduct can therefore be avoided.

3.2   Forward Overlaps and Outermost Strategy

In this case, a contractum may contain a potential redex. This redex will also be found in due course when the contractum is decomposed. The detour via an intermediate reduct can therefore be avoided.

3.3   Backward Overlaps and Innermost Strategy

In this case, a contractum may be a proper subpart of a potential redex. However, it should be considered *after* the contractum has been decomposed in search for an innermost redex, which will happen in due course. The detour via an intermediate reduct can therefore be avoided.

3.4   Backward Overlaps and Outermost Strategy

In this case, a contractum may be a proper subpart of a potential redex. This potential redex should be considered *before* decomposing the contractum since it occurs further out in the term (i.e., towards its root). Avoiding the detour via an intermediate reduct would in general miss this potential redex and therefore not maintain the reduction order. Does it mean that we need to detour via every intermediate reduct to normalize a term outside-in in the presence of backward overlaps? In this worst-case scenario, reduction-free outside-in normalization would be impossible in the presence of backward overlaps.

It is our observation that this worst-case scenario can be averted: most of the detour via an intermediate reduct can be avoided if we can identify the position of the correct potential redex without detouring all the way to the root.

In the next section, we show how to systematically determine the position of the next potential redex relative to the contractum in the presence of backward overlaps. This extra piece of information makes it possible to move upwards in the term to the position of the potential redex. Most of the detour via the intermediate reduct can therefore be avoided.

4   The Simple Example with an Outermost Strategy

We now consider the same simple language of arithmetic expressions again, but this time using an outermost reduction strategy.

Terms: The BNF of terms is unchanged:

$$t ::= 0 \mid S(t) \mid A(t, t)$$

Terms in normal form: The BNF of terms in normal form is also unchanged:

$$t^{\mathrm{nf}} ::= 0 \mid S(t^{\mathrm{nf}})$$

Reduction rules: The BNF of potential redexes now reads as follows:

$$pr ::= A(0, t_2) \mid A(S(t_1), t_2)$$

The reduction rules now read as follows:

$$A(0, t_2) \mapsto t_2$$
$$A(S(t_1), t_2) \mapsto S(A(t_1, t_2))$$

Note the occurrence of t_1, which is not necessarily in normal form, in the left-hand side of the second reduction rule: it is characteristic of outermost reduction.

All potential redexes are actual ones, i.e., no terms are stuck. We can thus implement contraction as a total function:

$$\frac{pr \mapsto c}{contract(pr) = c}$$

Reduction strategy: We are looking for the leftmost-outermost redex. We materialize this reduction strategy with the same grammar of reduction contexts as in the innermost case:

$$C[\,] ::= \Box[\,] \mid C[S[\,]] \mid C[A([\,], t)]$$

As in Section 2, we obtained this grammar by CPS-transforming a search function implementing the outermost reduction strategy and then defunctionalizing its continuation.[1]

In contrast to Section 2, a term not in normal form can be decomposed into more than one reduction context and one potential redex. For example, the term $A(S(A(S(t_0), t_1)), t_2)$ can be decomposed into $\Box[A(S(A(S(t_0), t_1)), t_2)]$ and $\Box[A([S[A(S(t_0), t_1)]], t_2)]$. This non-unique decomposition puts us outside the validity conditions of refocusing [9], so we are on our own here.

---

[1] A more precise grammar for contexts exists in the outermost case. It presents the same problems for refocusing as the one used here, and the solution we present also applies to it. Being unaware of any mechanical way to derive a precise grammar for outermost reduction, we therefore present our solution using this less precise but mechanically derivable grammar.

Recomposition: It is defined as in Section 2.

Decomposition: A term is decomposed in an outermost fashion into a potential redex and its reduction context with the following abstract-machine transitions:

$$\langle C[0]\rangle_{\text{term}}^{\text{dec}} \downarrow \langle C[0]\rangle_{\text{cont}}^{\text{dec}}$$
$$\langle C[S(t)]\rangle_{\text{term}}^{\text{dec}} \downarrow \langle C[S[t]]\rangle_{\text{term}}^{\text{dec}}$$
$$\langle C[A(t_1, t_2)]\rangle_{\text{term}}^{\text{dec}} \downarrow \langle C[A(t_1, t_2)]\rangle_{\text{add}}^{\text{dec}}$$

$$\langle C[A(0, t_2)]\rangle_{\text{add}}^{\text{dec}} \downarrow C[A(0, t_2)]$$
$$\langle C[A(S(t_1), t_2)]\rangle_{\text{add}}^{\text{dec}} \downarrow C[A(S(t_1), t_2)]$$
$$\langle C[A(A(t_{11}, t_{12}), t_2)]\rangle_{\text{add}}^{\text{dec}} \downarrow \langle C[A([A(t_{11}, t_{12})], t_2)]\rangle_{\text{add}}^{\text{dec}}$$

$$\langle \square[t^{\text{nf}}]\rangle_{\text{cont}}^{\text{dec}} \downarrow t^{\text{nf}}$$
$$\langle C[S[t^{\text{nf}}]]\rangle_{\text{cont}}^{\text{dec}} \downarrow \langle C[S(t^{\text{nf}})]\rangle_{\text{cont}}^{\text{dec}}$$

As in Section 2, this abstract machine is a pushdown automaton where the context is the stack. This time, the machine has five states: two intermediate states pairing a context and a term, one intermediate state with two terms and a context (this state handles A terms – the A is shown in the transitions above, but can be left implicit in an implementation), and two final states, one for the case where the given term is in normal form, and one for the case where the term decomposes into a context and a potential redex.

Each transition rule from the first intermediate state corresponds to a term constructor. Each transition rule from the second intermediate state corresponds to a term constructor on the left-hand side of an addition. Each transition rule from the third intermediate state corresponds to a context constructor. There is no transition rule to handle A context constructors in the third state, because the machine will move to the second state if it sees a A term constructor, after which the machine is guaranteed to find a potential redex. There is therefore no ambiguity and no incompleteness.

Furthermore, each transition preserves an invariant: recomposing the current context with the current term yields the original term. Given a term to decompose, the initial machine state pairs this term with the empty context. There are two final states: one for terms in normal form (and therefore containing no redex at all), and one for potential redexes and their reduction context. Decomposition, which is defined as the iteration of these machine transitions, is therefore a total function:

$$\frac{\langle \square[t]\rangle_{\text{term}}^{\text{dec}} \downarrow^* t^{\text{nf}}}{decompose(t) = t^{\text{nf}}} \qquad \frac{\langle \square[t]\rangle_{\text{term}}^{\text{dec}} \downarrow^* C[pr]}{decompose(t) = C[pr]}$$

A notable property: Due to the invariant of the abstract machine implementing decomposition, as in Section 2, the recomposition function is still a left inverse of the decomposition function.

One-step reduction: It is defined as in Section 2.

Reduction-based evaluation: It is defined as in Section 2.

A backward overlap: The reduction rules contain a backward overlap:

$$A(0, t_2) \mapsto t_2$$
$$A(S(t_1), t_2) \mapsto S(A(t_1, t_2))$$

On the right-hand side of both reduction rules, the contractum may occur as the first subterm in the left-hand side of the second rule. Additionally, the contractum of the first rule may occur as the first subterm of the left-hand side of the first rule.

Towards reduction-free evaluation: Between one contraction and the next, we recompose the reduction context with the contractum until the next reduct, which we decompose into the next potential redex and its reduction context.

Contrary to the innermost case, we now cannot be sure that decomposition of the next reduct will come back to this contractum and context before continuing, because a contractum in the context of an addition may be a new redex.

However, we can see from the reduction rules that any new redex constructed in this way cannot be positioned any higher than one step above the contractum, so decomposition will always return at least to the site of this new redex. Hence, if we backtrack/recompose one step after each contraction, we can shortcut the recomposition and decomposition to this point, and just continue the decomposition *in situ*.

More formally, we have

$$\frac{t \downarrow^* C[pr] \qquad C[pr] \; ([\mapsto]; \uparrow; \downarrow^*)^* \; t^{nf}}{t \Rightarrow_{rf} t^{nf}}$$

where $([\mapsto]; \uparrow; \downarrow^*)$ denotes contraction under context followed by one step of backtracking/recomposition and then decomposition.

The need for backtracking is caused by the existence of backward-overlapping rules. In the present example, we only need to backtrack one step, but in general, multiple steps are needed (Section 5 explains how to determine the number of necessary backtracking steps). Our contribution here is that backtracking is sufficient to enable refocusing and therefore reduction-free evaluation.

An example: See Figure 2.

Reduction-free evaluation: After applying refocusing, we fuse the iteration and refocus functions, we inline the contract function and the backtracking function, and we compress corridor transitions. The resulting normalizer implements a transition system described by the following abstract machine:

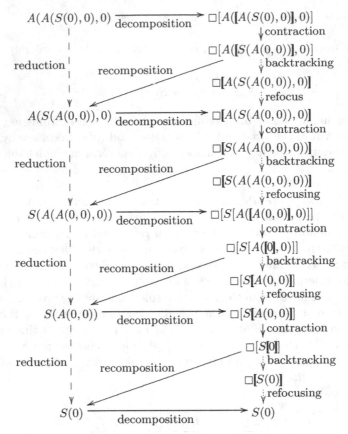

Fig. 2. Outermost reduction sequence for $A(A(S(0),0),0)$

$$t \rightarrowtail \langle \Box[t] \rangle_{\text{term}}$$

$$\langle C[0] \rangle_{\text{term}} \rightarrowtail \langle C[0] \rangle_{\text{cont}}$$
$$\langle C[S(t)] \rangle_{\text{term}} \rightarrowtail \langle C[S[t]] \rangle_{\text{term}}$$
$$\langle C[A(t_1, t_2)] \rangle_{\text{term}} \rightarrowtail \langle C[A(t_1, t_2)] \rangle_{\text{add}}$$

$$\langle \Box[A(0, t_2)] \rangle_{\text{add}} \rightarrowtail \langle \Box[t_2] \rangle_{\text{term}}$$
$$\langle C[S[A(0, t_2)]] \rangle_{\text{add}} \rightarrowtail \langle C[S[t_2]] \rangle_{\text{term}}$$
$$\langle C[A([A(0, t_2)], t_2')] \rangle_{\text{add}} \rightarrowtail \langle C[A(t_2, t_2')] \rangle_{\text{add}}$$
$$\langle \Box[A(S(t_1), t_2)] \rangle_{\text{add}} \rightarrowtail \langle \Box[S[A(t_1, t_2)]] \rangle_{\text{add}}$$
$$\langle C[S[A(S(t_1), t_2)]] \rangle_{\text{add}} \rightarrowtail \langle C[S[S[A(t_1, t_2)]]] \rangle_{\text{add}}$$
$$\langle C[A([A(S(t_1), t_2)], t_2')] \rangle_{\text{add}} \rightarrowtail \langle C[A(S(A(t_1, t_2)), t_2')] \rangle_{\text{add}}$$
$$\langle C[A(A(t_{11}, t_{12}), t_2)] \rangle_{\text{add}} \rightarrowtail \langle C[A([A(t_{11}, t_{12})], t_2)] \rangle_{\text{add}}$$

$$\langle \Box[t^{\text{nf}}] \rangle_{\text{cont}} \rightarrowtail t^{\text{nf}}$$
$$\langle C[S[t^{\text{nf}}]] \rangle_{\text{cont}} \rightarrowtail \langle C[S(t^{\text{nf}})] \rangle_{\text{cont}}$$

The effect of backtracking can be seen in the third and sixth transitions of the second intermediate state, where contraction in an addition context gives rise to a new redex above the position of the contractum. In these cases, the machine peels off a context constructor until it reaches the position of the new redex.

5 Backtracking

5.1 Identifying the Number of Backtracking Steps

In our example, it is sufficient to backtrack one step after each contraction. In general, it may be necessary to backtrack further in order to discover a new potential redex and enable refocusing.

For each contractum, the number of steps to backtrack can be determined by analyzing the reduction rules for backward overlaps, i.e., by identifying which subterms of left-hand sides the contractum unifies with. The number of steps to backtrack is the depth of the unifying subterm, i.e., the depth of the hole in the context C of Definition 1. This analysis can be performed statically because the depth of the hole is a property of the reduction rules, not of the reduction strategy. In other words, the analysis is neither performed over the constitutive elements of the normalization function (so no case-by-case semantic manipulation is required) nor during the normalization process (so no extra overhead is introduced).

Determining the existence of backward overlaps is local and mechanical, and hence, so is determining the necessary number of backtracking steps for each contractum. However, rather than determining the number of backtracking steps *for each reduction rule*, we can obtain a conservative estimate of the necessary number of backtracking steps *for all reduction rules* by

- always using the maximum depth of the left-hand sides of the reduction rules of the system; or by
- always using the maximum depth of the unifying subterms in the backward overlap analysis.

5.2 The Effect of Backtracking on the Abstract Machine

In practice, the choice of analysis (one precise number of backtracking steps for each reduction rule or one conservative number of backtracking steps for all reduction rules) has little impact on the resulting abstract machine. The reason is that any superfluous backtracking steps introduced in the abstract machine by an overly conservative analysis can be removed by the subsequent transition compressions. The contract function pattern-matches on terms, so after it is inlined, the abstract machine knows a number of term constructors of the contractum. The backtrack function pattern-matches on contexts, so after it is inlined, the abstract machine knows a number of context constructors of the immediate context. Within the window between the top-most known context constructor and the bottom-most known term constructor, transition compression makes the abstract machine move directly to the earliest position (according to the reduction strategy) at which the next redex can be found. Hence, if the context does not give rise to a redex, all the backtracking steps into that context are removed by transition compression.

Still, avoiding superfluous backtracking has two beneficial consequences; first, it simplifies transition compression because lowering the number of backtracking steps reduces the number of cases that need to be considered in the abstract machine. Second, it ensures that backtracking is only performed one new redex pattern at a time, thereby limiting the depth of pattern matching on the context.

5.3   Backward Overlaps Without the Need for Backtracking

In some cases, the combination of backward overlaps and outermost reduction can be dealt with without backtracking. Two examples in the literature illustrate cases where backtracking is not needed:

- The call-by-name λ-calculus [2,9]. In this case, the contractum that gives rise to a backward overlap is in normal form: it is a λ-abstraction that occurs on the left of an application node; this application node forms a new β-redex. Decomposing the contractum therefore does not yield a potential redex inside the contractum, and thus the decomposition process moves outwards in the term and finds the newly formed potential redex.
- Outermost tree flattening [5]. In this case, the backward overlap only occurs when contracting a redex which is not outermost, so backtracking is not needed.

6   Foretracking

Symmetrically to backtracking, one could envision foretracking as a symmetric solution to innermost reduction in reduction systems with forward overlaps, i.e., where the contraction may construct a new redex at a lower position than the contractum. However, refocusing is defined as resuming decomposition in the context of the contractum, so the newly constructed redex will be found in due time without using a separate foretracking function.

Additionally, one might envision that foretracking would result in a more efficient abstract machine, because unnecessary decomposition steps could be eliminated by a forward overlap analysis. However, the same superfluous steps are eliminated by transition compression of the abstract machine derived without foretracking.

So all in all, foretracking is not needed to go from an innermost reduction semantics to an abstract machine.

7   Related Work

Refocusing has mainly be applied for weak reduction in the λ-calculus, for normal order, applicative order, etc. For full reduction in the λ-calculus, a backward overlap exists. As analyzed in Section 3, this overlap is only problematic for outermost reduction, e.g., normal order. We are aware of two previous applications of refocusing to full normal-order reduction of the λ-calculus: one by Danvy, Millikin and Munk [8,23,24], in the mid-2000's, and a recent one by García-Pérez and Nogueira [15,16]:

– Danvy, Millikin and Munk overcome the backward overlap (without identifying it as such) by backtracking *after* applying the refocus function. In a more general setting, backtracking after refocusing would change the reduction order, so this solution does not scale. Our solution does not change the reduction order, and therefore it applies in a more general setting.
– García-Pérez and Nogueira overcome the backward overlap (without identifying it as such) by developing a notion of hybrid strategy and by integrating backtracking in the refocus function. Our solution is more minimalistic and remains mechanical: we simply analyze the reduction rules to detect backward overlaps when the reduction strategy is outermost, and in that case, we backtrack accordingly after contraction and before refocusing.

Backward and forward overlaps have been considered for some 30 years in relation to termination and confluence properties of term rewriting systems [11,12,17–19], and more recently in Jiresch's thesis [20]. Whereas term rewriting studies normalization *relations*, where any potential redex in the term may be contracted, we consider normalization operationally as *functions*, where a deterministic reduction strategy determines which potential redex to contract next.

As mentioned in Section 5, refocusing can in some cases be applied without backtracking, even if the reduction semantics contains backward overlaps. A formal definition of backward overlaps for which backtracking is needed would be similar to the definition of *narrowable terms* [31], which is a concept used in term rewriting [21,30]. However, narrowing is used to solve equations [22], and hence it is unrelated to our goal here.

8   Conclusion

We have considered refocusing for reduction semantics with an outermost reduction strategy, and we have discovered that in that case, the original conditions for refocusing [9] are not satisfied. We have then singled out backward-overlapping rules as the only stumbling block towards reduction-free normalization, and we have outlined how to overcome this stumbling block in a systematic way, by analyzing the backward overlaps in the reduction rules. In particular, we have shown how to implement the backtracking function, how to incorporate the backtracking function into the derivation, and how to statically determine the minimal number of backtracking steps, be it relative to each reduction rule or to all of them. We have also shown how to determine whether backtracking is actually necessary.

We have also analyzed all the other combinations (innermost / outermost reduction strategy and forward / backward overlaps in the reduction rules) and demonstrated how refocusing is a simple and effective way to go from reduction-based normalization in the form of a reduction semantics to reduction-free normalization in the form of an abstract machine.

Acknowledgments. We are grateful to the anonymous reviewers for their comments. Thanks are also due to Ian Zerny and Lasse R. Nielsen for their early feedback and kind encouragement.

References

1. Anton, K., Thiemann, P.: Towards deriving type systems and implementations for coroutines. In: Ueda, K. (ed.) APLAS 2010. LNCS, vol. 6461, pp. 63–79. Springer, Heidelberg (2010)
2. Biernacka, M., Danvy, O.: A concrete framework for environment machines. ACM Transactions on Computational Logic **9**(1), 1–30 (2007). Article #6. Extended version available as the research report BRICS RS-06-3
3. Biernacka, M., Danvy, O.: A syntactic correspondence between context-sensitive calculi and abstract machines. Theoretical Computer Science **375**(1–3), 76–108 (2007). Extended version available as the research report BRICS RS-06-18
4. Danvy, O.: Defunctionalized interpreters for programming languages. In: Hook, J., Thiemann, P. (eds.) Proceedings of the 2008 ACM SIGPLAN International Conference on Functional Programming (ICFP 2008). SIGPLAN Notices, 43(9), pp. 131–142. ACM Press. Invited talk, Victoria, British Columbia (September 2008)
5. Danvy, O.: From reduction-based to reduction-free normalization. In: Koopman, P., Plasmeijer, R., Swierstra, D. (eds.) AFP 2008. LNCS, vol. 5832, pp. 66–164. Springer, Heidelberg (2009)
6. Danvy, O., Johannsen, J., Zerny, I.: A walk in the semantic park. In: Khoo, S.C., Siek, J. (eds.) Proceedings of the 2011 ACM SIGPLAN Workshop on Partial Evaluation and Semantics-Based Program Manipulation (PEPM 2011), pp, pp. 1–12. ACM Press. Invited talk, Austin, Texas (2011)
7. Danvy, O., Millikin, K.: A rational deconstruction of Landin's SECD machine with the J operator. Logical Methods in Computer Science **4**(4:12), 1–67 (2008)
8. Danvy, O., Millikin, K., Munk, J.: A correspondence between full normalization by reduction and full normalization by evaluation. Manuscript (July 2007)
9. Danvy, O., Nielsen, L.R.: Refocusing in reduction semantics. Research Report BRICS RS-04-26, Department of Computer Science, Aarhus University, Aarhus, Denmark: A preliminary version was presented at the 2nd International Workshop on Rule-Based Programming (RULE 2001). Electronic Notes in Theoretical Computer Science 59.4 (November 2004)
10. Danvy, O., Zerny, I.: A prequel to reduction semantics. [33, Chapter 5]
11. Dershowitz, N.: Termination of linear rewriting systems (preliminary version). In: Even, S., Kariv, O. (eds.) Automata. Languages, and Programming, 8th Colloquium, number 115 in Lecture Notes in Computer Science, pp. 448–458. Springer-Verlag, Acre (Akko), Israel (1981)

12. Dershowitz, N.: Termination of rewriting. Journal of Symbolic Computation **3**(1/2), 69–116 (1987)
13. Felleisen, M.: The Calculi of λ-v-CS Conversion: A Syntactic Theory of Control and State in Imperative Higher-Order Programming Languages. PhD thesis, Computer Science Department, Indiana University, Bloomington, Indiana (August 1987)
14. Felleisen, M., Findler, R.B., Flatt, M.: Semantics Engineering with PLT Redex, The MIT Press, Flatt (2009)
15. García-Pérez, Á: Operational Aspects of Full Reduction in Lambda Calculi. PhD thesis, Departamento de Lenguajes, Sistemas Informáticos e Ingeniería de Software, Universidad Politecnica de Madrid, Madrid, Spain (2014)
16. García-Pérez, Á., Nogueira, P.: A syntactic and functional correspondence between reduction semantics and reduction-free full normalisers. In: Albert, E., Mu, S.C. (eds.) Proceedings of the ACM SIGPLAN 2013 Workshop on Partial Evaluation and Semantics-Based Program Manipulation (PEPM 2013). pp. 107–116. ACM Press, Rome, Italy (2013)
17. Geupel, O.: Overlap closures and termination of term rewriting systems. Technical Report MIP-8922, Fakultät für Mathematik und Informatik, Universität Passau, Passau, Germany (July 1989)
18. Guttag, J.V., Kapur, D., Musser, D.R.: On proving uniform termination and restricted termination of rewriting systems. SIAM Journal on Computing **12**(1), 189–214 (1983)
19. Hermann, M., Privera, I.: On nontermination of Knuth-Bendix algorithm. In: Kott, L. (ed.) Automata. Languages, and Programming, 13th International Colloquium, ICALP86, number 226 in Lecture Notes in Computer Science, pp. 146–156. Springer-Verlag, Rennes, France (1986)
20. Jiresch, E.: A Term Rewriting Laboratory with Systematic and Random Generation and Heuristic Test Facilities. PhD thesis, Vienna University of Technology, Vienna, Austria (June 2008)
21. Lankford, D.S.: Canonical inference. Technical Report ATP-32, Department of Mathematics, Southwestern University, Georgetown, Texas (December 1975)
22. Meseguer, J.: Twenty years of rewriting logic. Journal of Logic and Algebraic Programming **81**, 721–781 (2012)
23. Millikin, K.: A Structured Approach to the Transformation, Normalization and Execution of Computer Programs. PhD thesis, BRICS PhD School, Aarhus University, Aarhus, Denmark (May 2007)
24. Munk, J.: A study of syntactic and semantic artifacts and its application to lambda definability, strong normalization, and weak normalization in the presence of state. Master's thesis, Department of Computer Science, Aarhus University, Aarhus, Denmark (May 2007). BRICS research report RS-08-3
25. Pierce, B.C.: Types and Programming Languages. The MIT Press (2002)
26. Plotkin, G.D.: Call-by-name, call-by-value and the λ-calculus. Theoretical Computer Science **1**, 125–159 (1975)
27. Plotkin, G.D.: A structural approach to operational semantics. Technical Report FN-19, Department of Computer Science, Aarhus University, Aarhus, Denmark, September 1981. Reprinted in the Journal of Logic and Algebraic Programming 60–61, 17–139. (2004), with a foreword [28]
28. Plotkin, G.D.: The origins of structural operational semantics. Journal of Logic and Algebraic Programming **60–61**, 3–15 (2004)

29. Sergey, I., Clarke, D.: A correspondence between type checking via reduction and type checking via evaluation. Information Processing Letters **112**(13–20), 13–20 (2011)
30. Slagle, J.R.: Automated theorem-proving for theories with simplifiers, commutativity, and associativity. Journal of the ACM **21**(4), 622–642 (1974)
31. Terese. Term Rewriting Systems. Cambridge Tracts in Theoretical Computer Science, vol. 55. Cambridge University Press (2003)
32. Wright, A.K., Felleisen, M.: A syntactic approach to type soundness. Information and Computation **115**, 38–94 (1994)
33. Zerny, I.: The Interpretation and Inter-derivation of Small-step and Big-step Specifications. PhD thesis, Department of Computer Science, Aarhus University, Aarhus, Denmark (June 2013)

Towards Erlang Verification by Term Rewriting

Germán Vidal[✉]

MiST, DSIC, Universitat Politècnica de València,
Camino de Vera s/n, 46022 Valencia, Spain
gvidal@dsic.upv.es

Abstract. This paper presents a transformational approach to the verification of Erlang programs. We define a stepwise transformation from (first-order) Erlang programs to (non-deterministic) term rewrite systems that compute an overapproximation of the original Erlang program. In this way, existing techniques for term rewriting become available. Furthermore, one can use narrowing as a symbolic execution extension of rewriting in order to design a verification technique. We illustrate our approach with some examples, including a deadlock analysis of a simple Erlang program.

1 Introduction

The concurrent functional language Erlang [3] has a number of distinguishing features, like dynamic typing, concurrency via asynchronous message passing or hot code loading, that make it especially appropriate for distributed, fault-tolerant, soft real-time applications. The success of Erlang is witnessed by the increasing number of its industrial applications. For instance, Erlang has been used to implement Facebook's chat back-end, the mobile application Whatsapp or Twitterfall—a service to view trends and patterns from Twitter—, to name a few. The success of the language, however, will also require the development of powerful testing and verification techniques.

In this work, we present a transformational approach to the verification of Erlang programs. We define a stepwise transformation from (first-order) Erlang programs to (non-deterministic) term rewrite systems that compute an overapproximation of the original Erlang programs. In contrast to direct approaches, one can reuse the large body of techniques and tools for term rewriting in order to design a verification tool for Erlang. The transformation, however, is far from trivial. Previous work along these lines include, for instance, [19], where a translation from Erlang to *rewriting logic* [16]—a unified semantic framework for concurrency—is introduced. In this case, though, the aim was to provide an executable specification of the language semantics (as a basis for the development of verification tools). Therefore, in this approach, Erlang programs are

This work has been partially supported by the Spanish *Ministerio de Economía y Competitividad (Secretaría de Estado de Investigación, Desarrollo e Innovación)* under grant TIN2013-44742-C4-1-R and by the *Generalitat Valenciana* under grant PROMETEO/2011/052.

© Springer International Publishing Switzerland 2014
G. Gupta and R. Peña (Eds.): LOPSTR 2013, LNCS 8901, pp. 109–126, 2014.
DOI: 10.1007/978-3-319-14125-1_7

seen as data objects manipulated by a sort of interpreter implemented in rewriting logic. In contrast, our aim is to produce *plain* rewrite systems that keep the structure of the original Erlang program as much as possible, so that they can be accurately analyzed using existing techniques. To be precise, we produce a number of rewrite rules—a constant factor of the size of the original program—that mimic the reductions of the original Erlang programs, and only a few fixed number of *state* reductions rules that deal with global concurrency actions (process spawning, message sending and receiving, etc.), which are common to every transformed system. In particular, if an Erlang program contains no concurrency actions, we produce a purely functional rewrite system so that the state reductions rules are not necessary.

The usefulness of our approach is illustrated by using it to verify safety properties with a *symbolic execution* extension of rewriting. Luckily, such an extension already exists and has been extensively studied. It is called *narrowing* [23], and represents a conservative extension of rewriting to deal with non-determinism and logic variables—representing missing information. In fact, the rewrite systems produced by our transformation are steadily executable in a so-called functional logic language like Curry [12], which opens up many possibilities for verifying safety properties. Furthermore, there already exist well studied subsumption and abstraction operators for guaranteeing the termination of narrowing while still producing a sound overapproximation (see, e.g., the narrowing-driven partial evaluation approach of [2]). Therefore, one could define a narrowing-based model checker by using similar operators. A first step towards this direction can be found in [18], where a technique for building finite narrowing trees is introduced (though reducing the number of states to avoid a combinatorial explosion is still a challenge).

2   Erlang Syntax and Semantics

In this section, we present the basic syntax and semantics of a significant subset of Erlang. In particular, we consider a simplified version of the language where some features are excluded (mainly higher-order calls, predefined functions, modules and exceptions) and some other features are slightly simplified. This is similar to the language considered by Huch [13] or Noll [19], and still includes the main features of Erlang: pattern matching, process creation, message sending and receiving, etc.

The basic objects of the language are variables (denoted by $X, Y, \ldots$), atoms (denoted by a, b, $\ldots$), process identifiers –pids– (denoted by p, p$'$, $\ldots$), constructors (which are fixed in Erlang to lists, tuples and atoms), and defined functions (denoted by $f/n, g/m, \ldots$). The syntax for programs and expressions obeys the rules shown in Figure 1.

Programs are sequences of function definitions. Each function f/n is defined by a rule $f(X_1, \ldots, X_n) \to s$. where $X_1, \ldots, X_n$ are distinct variables and the body of the function, s, can be an expression, a sequence of expressions, a case distinction, message sending (e.g., main ! {hello, world} sends a message

$$pgm ::= f(X_1, \ldots, X_n) \rightarrow s. \mid pgm \; pgm$$
$$\mathsf{ErlangExp} \ni s ::= e \mid s_1, s_2 \mid \mathsf{case} \; e \; \mathsf{of} \; clauses \; \mathsf{end} \mid e_1 \, ! \, e_2$$
$$\mid \mathsf{receive} \; clauses \; \mathsf{end} \mid pat = e \mid pat = \mathsf{self}$$
$$\mid pat = \mathsf{spawn}(f(e_1, \ldots, e_n))$$
$$\mathsf{Exp} \ni e ::= f(e_1, \ldots, e_n) \mid [e_1 | e_n] \mid [\,] \mid \{e_1, \ldots, e_n\} \mid \mathsf{a} \mid \mathsf{p} \mid X$$
$$clauses ::= pat_1 \rightarrow s_1; \ldots; pat_n \rightarrow s_n$$
$$\mathsf{Pat} \ni pat ::= [pat_1 | pat_2] \mid [\,] \mid \{pat_1, \ldots, pat_n\} \mid \mathsf{a} \mid \mathsf{p} \mid X$$
$$\mathsf{Value} \ni v ::= [v_1 | v_2] \mid [\,] \mid \{v_1, \ldots, v_n\} \mid \mathsf{a} \mid \mathsf{p}$$

Fig. 1. Erlang syntax rules

$\{\mathsf{hello}, \mathsf{world}\}$ to the process with pid main) and receiving (e.g., receive $\{A, B\} \rightarrow A$ end reads a message from the process queue that matches the pattern $\{A, B\}$ and returns A), pattern matching where the right-hand side can be an expression, the primitive self (that returns the pid of the current process) or a process creation (e.g., spawn($foo(1, 2)$) creates a new process[1] initialized to $foo(1, 2)$). Expressions can contain function calls, lists, tuples, atoms, pids and variables. Patterns are made of lists, tuples, atoms, pids and variables. Values are similar to patterns but cannot contain variables. Note that we only allow occurrences of self and spawn in the right-hand side of pattern matching. This is not a serious restriction since occurrences in other positions can be *flattened* by introducing fresh variables and pattern matching.

The domain of pids, Pid, and that of atoms, Atom, must be disjoint. For simplicity, we consider that pids are natural numbers starting from 1.

Example 1. Consider the following program which simply creates a new process and sends a message. The new process receives the message and does the same. Finally, the third process receives the message and returns ok.

$$proc1 \rightarrow Pid1 = \mathsf{spawn}(proc2), \qquad proc3 \rightarrow \mathsf{receive}$$
$$Pid1 \, ! \, \mathsf{a}. \qquad\qquad\qquad X \rightarrow \mathsf{ok}$$
$$proc2 \rightarrow Pid2 = \mathsf{spawn}(proc3), \qquad\qquad \mathsf{end}.$$
$$\mathsf{receive}$$
$$X \rightarrow Pid2 \, ! \, X$$
$$\mathsf{end}.$$

In the past, there have been several attempts to formalize the semantics of Erlang (e.g., [6,7,13,17,19,20,24]). In the following, we present an operational semantics for Erlang programs that mainly follows the approach of [13].

Erlang states are denoted by the parallel composition of their processes, where each process $\langle p, e, q \rangle$ consists of a process identifier, an expression and a message queue: $\mathsf{Proc} ::= \mathsf{Pid} \times \mathsf{ErlangExp} \times \mathsf{Value}^*$. An *initial* state has the form $\langle p, f(v_1, \ldots, v_n), [\,] \rangle$ where f is a defined function, $v_1, \ldots, v_n$ are values, p is some initial pid and $[\,]$ denotes an empty message queue; we will use lists to denote message queues, where $[\,]$ denotes an empty list and $(x : xs)$ denotes a list with

---

[1] Note that we consider spawn($foo(1, 2)$) rather than the original Erlang notation spawn($foo, [1, 2]$) which is sensible since we do not allow higher order functions.

$$(\text{seq}) \quad \frac{}{\langle \mathrm{p}, C[v,s], q\rangle \ \& \ \Pi \longrightarrow \langle \mathrm{p}, C[s], q\rangle \ \& \ \Pi}$$

$$(\text{self}) \quad \frac{}{\langle \mathrm{p}, C[\mathsf{self}], q\rangle \ \& \ \Pi \longrightarrow \langle \mathrm{p}, C[\mathrm{p}], q\rangle \ \& \ \Pi}$$

$$(\text{fun}) \quad \frac{f(X_1,\ldots,X_n) \to s. \in prog}{\langle \mathrm{p}, C[f(v_1,\ldots,v_n)], q\rangle \ \& \ \Pi \longrightarrow \langle \mathrm{p}, C[\widehat{s}\{X_1 \mapsto v_1,\ldots,X_n \mapsto v_n\}], q\rangle \ \& \ \Pi}$$

$$(\text{match}) \quad \frac{\exists \sigma. \ pat\sigma = v}{\langle \mathrm{p}, C[pat = v], q\rangle \ \& \ \Pi \longrightarrow \langle \mathrm{p}, (C[v])\sigma, q\rangle \ \& \ \Pi}$$

$$(\text{case}) \quad \frac{\exists i. \ pat_i\sigma = v \text{ for some } \sigma \wedge \not\exists \sigma'. \ pat_j\sigma' = v \text{ for any } j < i}{\langle \mathrm{p}, C[\mathsf{case}\ v\ \mathsf{of}\ pat_1 \to s_1;\ldots;pat_n \to s_n\ \mathsf{end}], q\rangle \ \& \ \Pi \longrightarrow \langle \mathrm{p}, (C[s_i])\sigma, q\rangle \ \& \ \Pi}$$

$$(\text{spawn}) \quad \frac{\mathrm{p'} \text{ is a fresh pid}}{\langle \mathrm{p}, C[\mathsf{spawn}(f(v_1,\ldots,v_n))], q\rangle \ \& \ \Pi \longrightarrow \langle \mathrm{p}, C[\mathrm{p'}], q\rangle \ \& \ \langle \mathrm{p'}, f(v_1,\ldots,v_n), [\,]\rangle \ \& \ \Pi}$$

$$(\text{send}) \quad \frac{v_1 = \mathrm{p'} \in Pid}{\langle \mathrm{p}, C[v_1 \,!\, v_2], q\rangle \ \& \ \langle \mathrm{p'}, s, q'\rangle \ \& \ \Pi \longrightarrow \langle \mathrm{p}, C[v_2], q\rangle \ \& \ \langle \mathrm{p'}, s, q'\!+\!+[v_2]\rangle \ \& \ \Pi}$$

$$(\text{receive}) \quad \frac{v_k \text{ is the first message such that} \atop (\exists i. \ pat_i\sigma = v \text{ for some } \sigma \wedge \not\exists \sigma'. \ pat_j\sigma' = v \text{ for any } j < i)}{\langle \mathrm{p}, C[\mathsf{receive}\ pat_1 \to s_1;\ldots;pat_n \to s_n\ \mathsf{end}], [v_1,\ldots,v_k,\ldots,v_m]\rangle \ \& \ \Pi \atop \longrightarrow \langle \mathrm{p}, (C[s_i])\sigma, [v_1,\ldots,v_{k-1},v_{k+1},\ldots,v_m]\rangle \ \& \ \Pi}$$

Fig. 2. Basic Erlang Semantics

head x and tail xs. A *final* state has the form $\langle \mathrm{p}_1, v_1, q_1\rangle \ \& \ \cdots \ \& \ \langle \mathrm{p}_n, v_n, q_n\rangle$ where $v_1,\ldots,v_n$ are values and "&" denotes the parallel composition operator. Computations start with an initial state and proceed until a final state is reached or the computation is blocked (otherwise, it proceeds forever).

The operational semantics is formalized by a state transition relation $\longrightarrow:$ State $\times$ State. Erlang follows a leftmost innermost operational semantics. Every expression can be decomposed into a context $C[\,]$ with a (single) hole and a subexpression s where the next reduction can take place:[2]

$$C ::= [] \ | \ C, s \ | \ \mathsf{case}\ C\ \mathsf{of}\ clauses\ \mathsf{end} \ | \ C \,!\, e \ | \ v \,!\, C \ | \ pat = C$$
$$| \ \mathsf{spawn}(f(v_1,\ldots,v_i,C,e_{i+2},\ldots,e_n)) \ | \ f(v_1,\ldots,v_i,C,e_{i+2},\ldots,e_n)$$
$$| \ [v_1,\ldots,v_i,C|e] \ | \ \{v_1,\ldots,v_i,C,e_{i+2},\ldots,e_n\}$$

The definition of the operational semantics is shown in Figure 2. Let us briefly explain the rules of the semantics:

- States are denoted by sequences of processes of the form $\Gamma = \langle \mathrm{p}, e, q\rangle \ \& \ \Pi$ where Π denotes a (possibly empty) parallel composition of processes. The order of processes is not relevant here (i.e., $\langle \mathrm{p}, s, q\rangle$ might appear in any position within the pool of processes Γ).
- Rule self reduces the predefined atom self to the process identifier of the current process.
- Rule fun performs a function unfolding, where $\widehat{s}$ denotes an expression s in which the free variables of patterns (if any) have been replaced by fresh variables to avoid name conflicts.

---

[2] This is similar to the reduction contexts of [9] and allows us to deterministically identify the next expression to be reduced.

$\langle 1, \underline{proc1}, [\,] \rangle$

$\longrightarrow_{\text{fun}}$ $\langle 1, Pid1 = \mathsf{spawn}(\underline{proc2}), Pid1 \,!\, \mathsf{a}, [\,] \rangle$

$\longrightarrow_{\text{spawn}}$ $\langle 1, \underline{Pid1 = 2}, Pid1 \,!\, \mathsf{a}, [\,] \rangle$ & $\langle 2, proc2, [\,] \rangle$

$\longrightarrow_{\text{match}}$ $\langle 1, 2, 2 \,!\, \mathsf{a}, [\,] \rangle$ & $\langle 2, proc2, [\,] \rangle$

$\longrightarrow_{\text{seq}}$ $\langle 1, 2 \,!\, \mathsf{a}, [\,] \rangle$ & $\langle 2, \underline{proc2}, [\,] \rangle$

$\longrightarrow_{\text{fun}}$ $\langle 1, 2 \,!\, \mathsf{a}, [\,] \rangle$ & $\langle 2, Pid2 = \mathsf{spawn}(\underline{proc3}), \mathsf{receive}\ X \to Pid2 \,!\, X\ \mathsf{end}, [\,] \rangle$

$\longrightarrow_{\text{spawn}}$ $\langle 1, 2 \,!\, \mathsf{a}, [\,] \rangle$ & $\langle 2, \underline{Pid2 = 3}, \mathsf{receive}\ X \to Pid2 \,!\, X\ \mathsf{end}, [\,] \rangle$
    & $\langle 3, proc3, [\,] \rangle$

$\longrightarrow_{\text{match}}$ $\langle 1, 2 \,!\, \mathsf{a}, [\,] \rangle$ & $\langle 2, 3, \mathsf{receive}\ X \to 3 \,!\, X\ \mathsf{end}, [\,] \rangle$
    & $\langle 3, proc3, [\,] \rangle$

$\longrightarrow_{\text{seq}}$ $\langle 1, \underline{2 \,!\, \mathsf{a}}, [\,] \rangle$ & $\langle 2, \mathsf{receive}\ X \to 3 \,!\, X\ \mathsf{end}, [\,] \rangle$ & $\langle 3, proc3, [\,] \rangle$

$\longrightarrow_{\text{send}}$ $\langle 1, \mathsf{a}, [\,] \rangle$ & $\langle 2, \underline{\mathsf{receive}\ X \to 3 \,!\, X\ \mathsf{end}}, [\mathsf{a}] \rangle$ & $\langle 3, proc3, [\,] \rangle$

$\longrightarrow_{\text{receive}}$ $\langle 1, \mathsf{a}, [\,] \rangle$ & $\langle 2, 3 \,!\, \mathsf{a}, [\,] \rangle$ & $\langle 3, \underline{proc3}, [\,] \rangle$

$\longrightarrow_{\text{fun}}$ $\langle 1, \mathsf{a}, [\,] \rangle$ & $\langle 2, \underline{3 \,!\, \mathsf{a}}, [\,] \rangle$ & $\langle 3, \mathsf{receive}\ X \to \mathsf{ok}\ \mathsf{end}, [\,] \rangle$

$\longrightarrow_{\text{send}}$ $\langle 1, \mathsf{a}, [\,] \rangle$ & $\langle 2, \mathsf{a}, [\,] \rangle$ & $\langle 3, \underline{\mathsf{receive}\ X \to \mathsf{ok}\ \mathsf{end}}, [\mathsf{a}] \rangle$

$\longrightarrow_{\text{receive}}$ $\langle 1, \mathsf{a}, [\,] \rangle$ & $\langle 2, \mathsf{a}, [\,] \rangle$ & $\langle 3, \mathsf{ok}, [\,] \rangle$

Fig. 3. Computation for the program of Example 1

- Rules match and case deal with pattern matching. In both cases, we assume σ to be the minimal matching substitution and restricted to the variables of the pattern. For case expressions, we should select the *first* matching branch. Observe that we do not have rules for pattern matching failures, which are considered program errors and left out of this work.
- Rule spawn creates a new process with a fresh pid.
- Finally, rules send and receive deal with message passing and receiving. Note that receive should select the *first* message in the process queue that matches some pattern.

The semantics is clearly deterministic in the sense that, given a single process, there is only one applicable rule. However, we can define different strategies for selecting processes when there are more than one reducible process. In this paper, a fair selection strategy is assumed (e.g., a round-robin scheduling).

Example 2. Consider again the program of Example 1. A computation with this program is shown in Figure 3, where the reduced subexpression is underlined for clarity; moreover, we label the transitions with the applied rule. Therefore, the computation terminates and reaches a final state.

3   From Erlang Processes to Term Rewriting

In this section, we present a stepwise transformation from Erlang programs to term rewrite systems.

3.1   Term Rewriting

Here, we recall some basic notions and notations of term rewriting (see, e.g., [5] for more details). A *signature* $\mathcal{F}$ is a set of function symbols. Given a set of variables $\mathcal{V}$ with $\mathcal{F} \cap \mathcal{V} = \emptyset$, we denote the domain of *terms* by $\mathcal{T}(\mathcal{F}, \mathcal{V})$. Positions are used to address the nodes of a term viewed as a tree. A *position* p in a term t is represented by a finite sequence of natural numbers, where ϵ denotes the root position. We let $t|_p$ denote the *subterm* of t at position p and $t[s]_p$ the result of *replacing the subterm* $t|_p$ by the term s. $\mathcal{V}ar(t)$ denotes the set of variables appearing in t. A *substitution* $\sigma : \mathcal{V} \mapsto \mathcal{T}(\mathcal{F}, \mathcal{V})$ is a mapping from variables to terms such that $\mathcal{D}om(\sigma) = \{x \in \mathcal{V} \mid x \neq \sigma(x)\}$ is its domain. Substitutions are extended to morphisms from $\mathcal{T}(\mathcal{F}, \mathcal{V})$ to $\mathcal{T}(\mathcal{F}, \mathcal{V})$ in the natural way. We denote the application of a substitution σ to a term t by $t\sigma$ rather than $\sigma(t)$. The identity substitution is denoted by *id*.

A set of rewrite rules $l \rightarrow r$ such that l is a nonvariable term and r is a term whose variables appear in l is called a *term rewriting system* (TRS for short); terms l and r are called the left-hand side and the right-hand side of the rule, respectively. We restrict ourselves to finite signatures and TRSs. Given a TRS $\mathcal{R}$ over a signature $\mathcal{F}$, the *defined* symbols $\mathcal{D}_{\mathcal{R}}$ are the root symbols of the left-hand sides of the rules and the *constructors* are $\mathcal{C}_{\mathcal{R}} = \mathcal{F} \setminus \mathcal{D}_{\mathcal{R}}$. *Constructor terms* of $\mathcal{R}$ are terms over $\mathcal{C}_{\mathcal{R}}$ and $\mathcal{V}$. We sometimes omit $\mathcal{R}$ from $\mathcal{D}_{\mathcal{R}}$ and $\mathcal{C}_{\mathcal{R}}$ if it is clear from the context.

For a TRS $\mathcal{R}$, we define the associated rewrite relation $\rightarrow_{\mathcal{R}}$ as follows: given terms $s, t \in \mathcal{T}(\mathcal{F}, \mathcal{V})$, we have $s \rightarrow_{\mathcal{R}} t$ iff there exists a position p in s, a rewrite rule $l \rightarrow r \in \mathcal{R}$ and a substitution σ with $s|_p = l\sigma$ and $t = s[r\sigma]_p$; the rewrite step is often denoted by $s \rightarrow_{p,l \rightarrow r} t$ to make explicit the position and rule used in this step. The instantiated left-hand side $l\sigma$ is called a *redex*.

A *derivation* is a (possibly empty) sequence of rewrite steps. Given a binary relation $\rightarrow$, we denote by $\rightarrow^*$ its reflexive and transitive closure. Thus $t \rightarrow_{\mathcal{R}}^* s$ means that t can be reduced to s in $\mathcal{R}$ in zero or more steps.

3.2   The Transformation

Our transformation is driven by the following principles:

Keep the original structure. We try to keep the structure of the original Erlang programs as much as possible. In particular, an Erlang program without concurrent features would be mostly untouched. This is useful to keep the analyses performed on the transformed rewrite system as accurate as possible.

Overapproximate. Several Erlang constructs cannot be translated to a rewrite system while preserving their original semantics (unless a number of rather complex functions are introduced, which would be a drawback for the analysis of the resulting system). This is the case, for instance, of a case expression. While Erlang only considers the first matching clause, our translation will produce an auxiliary function that considers *all* matching clauses. Therefore, in general, we will produce rewrite systems that represent overapproximations of the original Erlang programs.

As a consequence, we cannot ensure that a bug detected in the transformed rewrite system is an actual bug of the original Erlang program, i.e., false positives may occur. For instance, the analysis of the rewrite system may point out that a deadlock may occur, while this is not the case in the original Erlang program. On the other hand, if the analysis of the rewrite system allows one to conclude that no deadlock can occur, then this is surely the case in the original Erlang program.

Concurrent actions by continuation functions. Loosely speaking, our transformation replaces every concurrent operator with a new constructor: SPAWN, SEND, RECEIVE and SELF. Then, we define a set of rewrite rules that deal with states and take care of concurrent actions. The challenge here is to always have these constructors in a topmost position of a process so that a rule can be applied without requiring complex context rules (in contrast to, e.g., [19]).

For this purpose, we introduce some auxiliary functions that can be seen as *continuations* of the original functions (see below).

We formalize our transformation $[\![\ ]\!]$ as follows. Given an Erlang program P, we have:

$$[\![P]\!] = \{f(x_1, \ldots, x_n) \to [\![s.]\!]^V \mid f(x_1, \ldots, x_n) \to s. \in P\}$$

where $V = \{x_1, \ldots, x_n\} \cap \mathcal{F}Var(s)$ is used for introducing auxiliary functions with appropriate parameters. In the following, $\mathcal{F}Var(s)$ denotes the free variables of s. Now, we define the transformation function $[\![\ ]\!]$ on every program construct.

Case Expressions. Let us first consider the transformation of case expressions. This can easily be transformed by introducing an auxiliary function as follows:

$$[\![\text{case } e \text{ of } p_1 \to s_1, \ldots, p_n \to s_n \text{ end.}]\!]^V = f(e, \overline{V})$$

where f is a fresh function symbol and $\overline{V}$ denotes a list with the variables of set V. Here, the auxiliary function f is defined as follows:

$$f(p_1, \overline{V}) \to [\![s_1.]\!]^{V_1} \quad \cdots \quad f(p_n, \overline{V}) \to [\![s_n.]\!]^{V_n}$$

where $V_i = Var(f(p_i, \overline{V})) \cap \mathcal{F}Var(s_i)$, $i = 1, \ldots, n$. When the case expression is not the last statement in the right-hand side, we proceed analogously as follows:

$$[\![\text{case } e \text{ of } p_1 \to s_1, \ldots, p_n \to s_n \text{ end}, s.]\!]^V = f(e, \overline{V})$$

where the auxiliary function f is now defined by

$$f(p_1, \overline{V}) \to [\![s_1, s.]\!]^{V_1} \quad \cdots \quad f(p_n, \overline{V}) \to [\![s_n, s.]\!]^{V_n}$$

where $V_i = Var(f(p_i, \overline{V})) \cap \mathcal{F}Var(s_i, s)$, $i = 1, \ldots, n$.

Observe that this transformation implies that, in general, the transformed function will compute an *overapproximation* of the original Erlang program when there are overlapping patterns (since rewriting considers *all* matching patterns).

Message Passing. In this case, we transform an expression $p \, ! \, e$ using a new constructor $\mathsf{SEND}(i, p, e, vars)$, where i is a unique identifier and $vars$ is a list of variables. We distinguish the following cases:

$$\llbracket e_1 \, ! \, e_2. \rrbracket^V = \mathsf{SEND}(i, e_1, e_2, [\,]) \qquad \text{with } send(i, v, \_) \to v$$

where i is a fresh constant symbol (e.g., a number), v is a fresh variable, and "$\_$" denotes an *anonymous* variable (i.e., a variable whose name is not relevant because it does not occur in the right-hand side).

In contrast to ordinary functions and the auxiliary functions introduced when transforming a case expression, SEND is a constructor symbol that will require the (global) system rules to be dealt with. Roughly speaking, the system rules will rewrite $\mathsf{SEND}(i, e_1, e_2, [\,])$ to $send(i, e_2, [\,])$—the continuation of SEND—and will also store e_2 in the mailbox of the process with pid e_1.

When the message passing is not the last construct of the sequence, we have

$$\llbracket e_1 \, ! \, e_2, s. \rrbracket^V = \mathsf{SEND}(i, e_1, e_2, \overline{V}) \qquad \text{with } send(i, \_, \overline{V}) \to \llbracket s. \rrbracket^{V'}$$

where i is a fresh constant symbol and $V' = V \cap \mathcal{F}\mathit{Var}(s)$. In this case, the system rules will proceed analogously but the value of e_2 is lost (as it will happen in the original Erlang program).

Message Reception. Here, we introduce a new constructor $\mathsf{AREC}(i, list, vars)$, where i is a unique identifier, $list$ is the list of messages already processed (initially empty), and $vars$ is a list of variables. We transform Erlang expressions as follows:

$$\llbracket \mathsf{receive} \; p_1 \to s_1, \ldots, p_n \to s_n \; \mathsf{end}. \rrbracket^V = \mathsf{AREC}(i, [\,], \overline{V})$$

where i is a fresh a constant symbol (e.g., a number). The following auxiliary functions are added to the program:

$$\begin{array}{ll}
brec(i, p_1) \to \mathsf{True} & rec(i, p_1, \overline{V}) \to \llbracket s_1. \rrbracket^{V_1} \\
\cdots & \cdots \\
brec(i, p_n) \to \mathsf{True} & rec(i, p_n, \overline{V}) \to \llbracket s_n. \rrbracket^{V_n}
\end{array}$$

where $V_j = \mathit{Var}(rec(i, p_j, \overline{V})) \cap \mathcal{F}\mathit{Var}(s_j)$, $j = 1, \ldots, n$. When the receive construct is not the last expression of a sequence, we proceed analogously as follows:

$$\llbracket \mathsf{receive} \; p_1 \to s_1, \ldots, p_n \to s_n \; \mathsf{end}, s. \rrbracket^V = \mathsf{AREC}(i, [\,], \overline{V})$$

with

$$\begin{array}{ll}
brec(i, p_1) \to \mathsf{True} & rec(i, p_1, \overline{V}) \to \llbracket s_1, s. \rrbracket^{V_1} \\
\cdots & \cdots \\
brec(i, p_n) \to \mathsf{True} & rec(i, p_n, \overline{V}) \to \llbracket s_n, s. \rrbracket^{V_n}
\end{array}$$

where $V_j = \mathit{Var}(rec(i, p_j, \overline{V})) \cap \mathcal{F}\mathit{Var}(s_j, s)$, $j = 1, \ldots, n$.

Loosely speaking, the system reduction rules will rewrite $\mathsf{AREC}(i, [\,], \overline{V})$ to $rec(i, m, \overline{V})$—the continuation of AREC—when $brec(i, m)$ is true, where m is the first message in the process mailbox; otherwise, the message m is moved to the second parameter of AREC and the traversal of the mailbox continues. When the mailbox is empty (i.e., no message matched the patterns of the receive clause), we restore the mailbox and move the process to the end of the list.

Similar to the case statements, the transformed TRS will compute an overapproximation of the original Erlang program when there are overlapping patterns.

Pattern Matching. First, we consider a pattern matching in which the right-hand side is an expression not including calls to spawn nor self. In this case, it is transformed analogously to a case statement with a single case:

$$[\![p = e.]\!]^V = f(e, \overline{V}) \qquad \text{with } f(p, \overline{V}) \to p.$$

where f is a fresh function symbol. When the pattern matching is not the last element of a sequence, we proceed as follows:

$$[\![p = e, s.]\!]^V = f(e, \overline{V}) \qquad \text{with } f(p, \overline{V}) \to [\![s.]\!]^{V'}$$

where f is a fresh function symbol and $V' = \mathcal{V}ar(f(p, \overline{V})) \cap \mathcal{F}\mathcal{V}ar(s)$.

Process Creation. Process are created using the predefined function spawn. Here, we introduce a new constructor $\mathsf{SPAWN}(i, exp, vars)$, where i is a unique identifier, exp is the function call that starts the new process, and $vars$ is a list of variables. First, we distinguish the following case:

$$[\![p = \mathsf{spawn}(e).]\!]^V = \mathsf{SPAWN}(i, e, [\,]) \qquad \text{with } spawn(i, p, \_) \to p$$

where i is a fresh constant. Basically, the auxiliary function $spawn$—the continuation of SPAWN—will be called from the system reduction rules with a second argument that contains the pid of the new process. When the pattern matching is not the last element in a sequence, we proceed as follows:

$$[\![p = \mathsf{spawn}(e), s.]\!]^V = \mathsf{SPAWN}(i, e, \overline{V}) \qquad \text{with } spawn(i, p, \overline{V}) \to [\![s.]\!]^{V'}$$

where i is a fresh constant and $V' = \mathcal{V}ar(spawn(i, p, \overline{V})) \cap \mathcal{F}\mathcal{V}ar(s)$.

The Primitive self. We replace the occurrences of self with a new constructor $\mathsf{SELF}(i, vars)$, where i is a unique identifier and $vars$ is a list of variables. We distinguish the following cases:

$$[\![p = \mathsf{self}.]\!]^V = \mathsf{SELF}(i, [\,]) \qquad \text{with } self(i, p, \_) \to p$$

where i is a fresh constant symbol. Here, the system reduction rules will check the pid of the process and will call the auxiliary function $self$—the continuation of SELF—with this pid as a second parameter. When the pattern matching is not the last element in a sequence, we proceed as follows:

$$[\![p = \mathsf{self}, s.]\!]^V = \mathsf{SELF}(i, \overline{V}) \qquad \text{with } self(i, p, \overline{V}) \to [\![s]\!]^{V'}$$

where i is a fresh constant symbol and $V' = \mathcal{V}ar(self(i, p, \overline{V})) \cap \mathcal{F}\mathcal{V}ar(s)$.

$$(\langle 0, k, [\,]\rangle : (\langle i, \mathsf{SPAWN}(n, e, vs), m\rangle : s)) \;\to\; (\langle 0, k+1, [\,]\rangle : s)$$
$$++[\langle i, spawn(n, k, vs), m\rangle, \langle k, e, [\,]\rangle]$$

$$(s_0 : (\langle i, \mathsf{SEND}(n, j, e, vs), m\rangle : s)) \qquad \to (s_0 : send\_msg(j, e, s++[\langle i, send(n, e, vs), m\rangle]))$$
$$(s_0 : (\langle i, \mathsf{SEND}(n, j, e, vs), m\rangle : s)) \qquad \to (s_0 : (s++[\langle i, \mathsf{SEND}(n, j, e, vs), m\rangle]))$$

$$(s_0 : (\langle i, \mathsf{AREC}(n, ms_2, vs), m : ms\rangle : s)) \;\to\; (s_0 : s++[\langle i, rec(n, m, vs), (ms_2++ms)\rangle])$$
$$\text{if } brec(n, m)$$
$$(s_0 : (\langle i, \mathsf{AREC}(n, ms_2, vs), m : ms\rangle : s)) \;\to\; (s_0 : (\langle i, \mathsf{AREC}(n, ms_2++[m], vs), ms\rangle : s))$$
$$\text{if } not(brec(n, m))$$
$$(s_0 : (\langle i, \mathsf{AREC}(n, ms_2, vs), [\,]\rangle : s)) \quad \to (s_0 : s)++[\langle i, \mathsf{AREC}(n, ms_2, vs), [\,]\rangle]$$

$$(\langle 0, k, [\,]\rangle : (\langle i, \mathsf{SELF}(n, vs), m\rangle : s)) \qquad \to (\langle 0, k, [\,]\rangle : s)++[\langle i, self(n, i, vs), m\rangle]$$

$$(s_0 : (\langle i, p, m\rangle : s)) \qquad\qquad\qquad \to (s_0 : (s++[\langle i, p, m\rangle]))$$

Fig. 4. State reduction rules

Sequences. Most of the sequences are transformed away using the previous transformations. However, some of them may still remain in the transformed program. In this case, they are transformed as follows:

$$[\![s_1, s_2.]\!]^V = [\![\text{case } s_1 \text{ of } \_ \to s_2 \text{ end.}]\!]^V$$

so that all remaining sequences are removed from the transformed program.

Expressions. For the remaining expressions, we have $[\![e.]\!] = e$. Note that we assumed that no occurrence of the concurrency primitives: !, receive, self, etc., can occur in expressions. Note that this is not a real restriction since these statements could be flattened by introducing fresh variables and pattern matching.

3.3   State Reduction Rules

Processes are denoted by tuples $\langle p, e, q\rangle$, which consists of a process identifier p, an expression e, and a message queue q, as introduced in Section 2. We consider natural numbers as pids, starting from 1. Also, we have an *artificial* (first) process of the form $\langle 0, n, [\,]\rangle$ that is only used for storing the first free pid n, so that we do not need to compute it every time spawn is called.

Basically, a *system* is represented by a list of processes, where the first process is always the one that stores the current free pid number. We consider the usual notation for lists: [] and $(\_ : \_)$, where $++$ denotes list concatenation. We consider a breadth-first exploration of the search space regarding concurrent actions (so that the considered process is always moved to the end of the current list). Let us briefly describe the rules:

SPAWN. A process with a constructor call $\mathsf{SPAWN}(n, e, vars)$ is reduced by creating a new process initialized with the expression e, and replacing the constructor call with a call to the auxiliary function $spawn(n, k, vars)$, where k is the pid number of the new process (which is then updated to $k + 1$). Note also

$$
\begin{array}{llll}
proc1 & \rightarrow \mathsf{SPAWN}(1, proc2, [\,]) & proc2 & \rightarrow \mathsf{SPAWN}(3, proc3, [\,]) \\
spawn(1, pid1, [\,]) & \rightarrow \mathsf{SEND}(2, pid1, \mathsf{A}, [\,]) & spawn(3, pid2, [\,]) & \rightarrow \mathsf{AREC}(4, [\,], [pid2]) \\
send(2, e, \_) & \rightarrow e & brec(4, x) & \rightarrow \mathsf{True} \\
& & rec(4, x, [pid2]) & \rightarrow \mathsf{SEND}(5, pid2, x, [\,]) \\
proc3 & \rightarrow \mathsf{AREC}(6, [\,], [\,]) & send(5, e, \_) & \rightarrow e \\
brec(6, x) & \rightarrow \mathsf{True} & & \\
rec(6, x, [\,]) & \rightarrow \mathsf{Ok} & &
\end{array}
$$

Fig. 5. TRS associated to the Erlang program of Example 1

that both the reduced process and the newly created one are moved to the end of the list.

SEND. Here, and in order to explore all possible schedulings, we consider two non-deterministic alternatives. The first rule sends the message (using the auxiliary function *send_msg*), while the second rule just moves the process to the end of the queue thus delaying the message delivery. In this way, we can explore all possible process schedulings. The definition of the auxiliary function *send_msg* is straightforward (and can be found in the next section).

AREC. For receiving a message, we consider three possibilities. First, we check whether the first message in the mailbox matches any of the receive clauses. If so, we process the message using a call to the auxiliary function *rec*. Otherwise, we move the first message to the second parameter of AREC and continue inspecting the mailbox. When the mailbox is empty (either because no message has been received or because we have already inspected all of them), the mailbox is restored and the process is moved to the end of the list.

Finally, we also include a rule that just moves a *finished* process to the end of the list. One could also remove it from the pool of processes, but we prefer to keep it for analysis and debugging purposes. Note that this rule does not overlap with the previous rules since the process must have a *value* not including the special constructors SPAWN, SEND, etc. For simplicity, in this work we assume that there are no non-terminating functions that are purely functional (i.e., without occurrences of SPAWN, SEND, AREC or SELF). Otherwise, one would also need to ensure that no expression in a process is rewritten infinitely.

Example 3. Let us consider the Erlang program of Example 1. This program is transformed into the TRS shown in Fig. 3, where functions and variables start with a lowercase letter, and constructors start with an uppercase letter.

The computation shown in Example 2 for the Erlang program proceeds now as shown in Fig. 6.[3] Here, we reach exactly the same final state of Fig. 3. Note, however, that due to non-determinism, other computations are also possible.

Proving that the transformed program computes an overapproximation (i.e., that every computation of the original program can be mimicked in the transformed one) is not difficult.

---

[3] We underline either the expression or the selected process involved in a reduction step.

$[\langle 0,2,[\,]\rangle, \langle 1,\underline{proc1},[\,]\rangle]$
$\rightarrow [\langle 0,2,[\,]\rangle, \langle 1,\overline{\mathsf{SPAWN}}(1,proc2,[\,]),[\,]\rangle]$
$\rightarrow [\langle 0,3,[\,]\rangle, \langle 1,\underline{spawn}(1,2,[\,]),[\,]\rangle, \langle 2,proc2,[\,]\rangle]$
$\rightarrow [\langle 0,3,[\,]\rangle, \langle 1,\overline{\mathsf{SEND}}(2,2,\mathsf{A},[\,]),[\,]\rangle, \langle 2,proc2,[\,]\rangle]$
$\rightarrow [\langle 0,3,[\,]\rangle, \langle 2,proc2,[\mathsf{A}]\rangle, \langle 1,\underline{send}(2,\mathsf{A},[\,]),[\,]\rangle]$
$\rightarrow [\langle 0,3,[\,]\rangle, \langle 2,\overline{\mathsf{SPAWN}}(3,proc3,[\,]),[\mathsf{A}]\rangle, \langle 1,send(2,\mathsf{A},[\,]),[\,]\rangle]$
$\rightarrow [\langle 0,4,[\,]\rangle, \langle 1,\underline{send}(2,\mathsf{A},[\,]),[\,]\rangle, \langle 2,spawn(3,3,[\,]),[\mathsf{A}]\rangle, \langle 3,proc3,[\,]\rangle]$
$\rightarrow [\langle 0,4,[\,]\rangle, \langle 1,\overline{\mathsf{A}},[\,]\rangle, \langle 2,spawn(3,3,[\,]),[\mathsf{A}]\rangle, \langle 3,proc3,[\,]\rangle]$
$\rightarrow [\langle 0,4,[\,]\rangle, \langle 2,\underline{spawn}(3,3,[\,]),[\mathsf{A}]\rangle, \langle 3,proc3,[\,]\rangle, \langle 1,\mathsf{A},[\,]\rangle]$
$\rightarrow [\langle 0,4,[\,]\rangle, \langle 2,\overline{\mathsf{AREC}}(4,[\,],[3]),[\mathsf{A}]\rangle, \langle 3,proc3,[\,]\rangle, \langle 1,\mathsf{A},[\,]\rangle]$
$\rightarrow [\langle 0,4,[\,]\rangle, \langle 3,\underline{proc3},[\,]\rangle, \langle 1,\mathsf{A},[\,]\rangle, \langle 2,rec(4,\mathsf{A},[3]),[\,]\rangle]$
$\rightarrow [\langle 0,4,[\,]\rangle, \langle 3,\overline{\mathsf{AREC}}(6,[\,],[\,]),[\,]\rangle, \langle 1,\mathsf{A},[\,]\rangle, \langle 2,rec(4,\mathsf{A},[3]),[\,]\rangle]$
$\rightarrow [\langle 0,4,[\,]\rangle, \langle 1,\underline{\mathsf{A}},[\,]\rangle, \langle 2,rec(4,\mathsf{A},[3]),[\,]\rangle, \langle 3,\mathsf{AREC}(6,[\,],[\,]),[\,]\rangle]$
$\rightarrow [\langle 0,4,[\,]\rangle, \langle 2,\underline{rec}(4,\mathsf{A},[3]),[\,]\rangle, \langle 3,\mathsf{AREC}(6,[\,],[\,]),[\,]\rangle, \langle 1,\mathsf{A},[\,]\rangle]$
$\rightarrow [\langle 0,4,[\,]\rangle, \langle 2,\overline{\mathsf{SEND}}(5,3,\mathsf{A},[\,]),[\,]\rangle, \langle 3,\mathsf{AREC}(6,[\,],[\,]),[\,]\rangle, \langle 1,\mathsf{A},[\,]\rangle]$
$\rightarrow [\langle 0,4,[\,]\rangle, \langle 3,\underline{\mathsf{AREC}}(6,[\,],[\,]),[\mathsf{A}]\rangle, \langle 1,\mathsf{A},[\,]\rangle, \langle 2,send(5,\mathsf{A},[\,]),[\,]\rangle]$
$\rightarrow [\langle 0,4,[\,]\rangle, \langle 1,\underline{\mathsf{A}},[\,]\rangle, \langle 2,send(5,\mathsf{A},[\,]),[\,]\rangle, \langle 3,rec(6,\mathsf{A},[\,]),[\,]\rangle]$
$\rightarrow [\langle 0,4,[\,]\rangle, \langle 2,\underline{send}(5,\mathsf{A},[\,]),[\,]\rangle, \langle 3,rec(6,\mathsf{A},[\,]),[\,]\rangle, \langle 1,\mathsf{A},[\,]\rangle]$
$\rightarrow [\langle 0,4,[\,]\rangle, \langle 2,\underline{\mathsf{A}},[\,]\rangle, \langle 3,rec(6,\mathsf{A},[\,]),[\,]\rangle, \langle 1,\mathsf{A},[\,]\rangle]$
$\rightarrow [\langle 0,4,[\,]\rangle, \langle 3,\underline{rec}(6,\mathsf{A},[\,]),[\,]\rangle, \langle 1,\mathsf{A},[\,]\rangle, \langle 2,\mathsf{A},[\,]\rangle]$
$\rightarrow [\langle 0,4,[\,]\rangle, \langle 3,\overline{\mathsf{Ok}},[\,]\rangle, \langle 1,\mathsf{A},[\,]\rangle, \langle 2,\mathsf{A},[\,]\rangle]$

Fig. 6. Example of reduction

4   The Transformation in Practice

In this section, we show the usefulness of our transformation in the context of program verification. An implementation of the transformation has been undertaken and can be used through a web interface that can be found here:

http://users.dsic.upv.es/~gvidal/erlang2trs/

For verifying safety properties, we consider the execution of the rewriting system using *narrowing* [14,23], a conservative extension of term rewriting for dealing with non-determinism and logic variables. Narrowing can be seen as a symbolic execution version of rewriting where pattern matching is replaced with unification (as in logic programming). Basically, given a TRS $\mathcal{R}$ and two terms $s,t \in \mathcal{T}(\mathcal{F},\mathcal{V})$, we have that $s \rightsquigarrow_{\mathcal{R}} t$ is a *narrowing step* iff there exist a nonvariable position p of s, a variant $l \rightarrow r$ of a rule in $\mathcal{R}$, and a substitution σ which is a most general unifier of $s|_p$ and l, with $t = (s[r]_p)\sigma$. E.g., narrowing has been used as the basis of a partial evaluation framework for rewrite systems [2].

In particular, in order to produce executable programs, we consider the language Curry [12] (a conservative extension of Haskell to deal with logic variables and non-determinism).

Example 4. Consider the following Erlang program:

$main \rightarrow Pid2 = \mathsf{spawn}(proc2),$ $proc1(Pid) \rightarrow \mathsf{receive}$
$\qquad\qquad Pid1 = \mathsf{spawn}(proc1(Pid2)),$ $X \rightarrow Pid\,!\,X$
$\qquad\qquad Pid1\,!\,\mathsf{hello},$ $\mathsf{end}.$
$\qquad\qquad Pid2\,!\,\mathsf{world}.$ $proc2 \rightarrow \mathsf{receive}$
$\qquad\qquad\qquad\qquad\qquad\qquad\qquad\qquad\qquad\qquad X \rightarrow \mathsf{ok}$
$\qquad\qquad\qquad\qquad\qquad\qquad\qquad\qquad\qquad\qquad \mathsf{end}.$

Our transformation tool `erlang2trs` returns the following program (we use a curried notation for functions as in Curry):

```
main = (SPAWN 1 proc2 [])
spawn 1 pid2 []  = (SPAWN 2 (proc1 pid2) (pid2:[]))
spawn 2 pid1 (pid2:[]) = (SEND 3 pid1 Hello (pid1:(pid2:[])))
send 3 e (pid1:(pid2:[])) = (SEND 4 pid2 World [])
send 4 e fresh  = e

proc1 pid = (AREC 5 [] (pid:[]))
brec 5 x = True
rec 5 x (pid:[])  = (SEND 6 pid x [])
send 6 e fresh  = e

proc2 = (AREC 7 [] [])
brec 7 x  = True
rec 7 x []  = Ok
```

together with data declarations, the system reduction rules and a few auxiliary functions:

```
data State = State Int Exp [Exp]
data Exp = I Int | SPAWN Int Exp [Exp] | SEND Int Exp Exp [Exp]
           | AREC Int [Exp] [Exp] | SELF Int [Exp]
           | World | Hello | Ok

reduce (s0 : (State i (AREC n ms2 args) (m:ms)) : s) visited
  = if (brec n m)
    then reduce (s0:(s++[State i (rec n m args) (ms2++ms)])) visited
      else reduce (s0:(State i (AREC n (ms2++[m]) args) ms):s) visited
reduce (s0 : (State i (AREC n ms2 args) []) : s) visited
  = reduce ((s0 : s) ++ [State i (AREC n ms2 args) []]) visited

reduce (State o (I k) 12 : (State i (SPAWN n e args) m : s)) visited
  = reduce ((State o (I (k+1)) 12 : s)
    ++ [State i (spawn n (I k) args) m, State k e []]) visited

reduce (s0 : (State i (SEND n (I j) e args) m : s)) visited
  = reduce (s0:(send_msg j e (s++[State i (send n e args) m]))) visited
reduce (s0 : (State i (SEND n (I j) e args) m : s)) visited
  = reduce (s0 : (s ++ [State i (SEND n (I j) e args) m])) visited
```

```
send_msg _ _ [] = []
send_msg j e (State i b m : s)
          | i==j      = State i b (m++[e]) : s
          | otherwise = State i b m : (send_msg j e s)

brec 5 fresh = case fresh of
                  x -> True
                  _ -> False

brec 7 fresh = case fresh of
                  x -> True
                  _ -> False
```

The complete code of the transformed program can be found at http://users.
dsic.upv.es/~gvidal/erlang2trs/. Consider now that we are interested in verifying
whether the message "World" can arrive to *proc3* before the message "Hello".
We can easily check this property in Curry using the following test function:

```
init = reduce [State 0 (I 2) [], State 1 main []] []

test = wrongState init
wrongState (s:ss) = case s of
                      State _ Ok [Hello] -> True
                      _ -> wrongState ss
```

where the state reduction rules are implemented by function **reduce** and states
are represented using the constructor **State**. Here, function **init** denotes the
initial state and function **test** checks if there exists a reachable final state (i.e.,
where the main expression is reduced to **Ok**) with the message **Hello** in the
mailbox.

Example 5. Consider now the Erlang program shown in Fig. 7. This program
represents a simplified version of the well-known dining philosophers problem.
Basically, there are two processes (*left* and *right*) that compete for a couple of

$$
\begin{aligned}
main \rightarrow\ &Res = \mathsf{spawn}(res(\{\mathsf{f},\mathsf{f}\})), &res(St) \rightarrow\ &\mathsf{receive}\ \{\mathsf{q},S\} \rightarrow S\ !\ St \\
&L = \mathsf{spawn}(left(Res)), & &\mathsf{end},\ \mathsf{receive}\{\mathsf{u},NSt\} \rightarrow res(NSt) \\
&R = \mathsf{spawn}(right(Res)). & &\mathsf{end}.
\end{aligned}
$$

$$
\begin{aligned}
left(Res) \rightarrow\ &S = \mathsf{self}, &right(Res) \rightarrow\ &S = \mathsf{self}, \\
&Res\ !\ \{\mathsf{q},S\}, & &Res\ !\ \{\mathsf{q},S\}, \\
&\mathsf{receive}\ \{\mathsf{f},B\} \rightarrow & &\mathsf{receive}\ \{A,\mathsf{f}\} \rightarrow \\
&\quad Res\ !\ \{\mathsf{u},\{\mathsf{b},B\}\}, & &\quad Res\ !\ \{\mathsf{u},\{A,\mathsf{b}\}\}, \\
&\quad Res\ !\ \{\mathsf{q},S\}, & &\quad Res\ !\ \{\mathsf{q},S\}, \\
&\quad left2(Res) & &\quad right2(Res) \\
&\mathsf{end}. & &\mathsf{end}.
\end{aligned}
$$

$$
\begin{aligned}
left2(Res) \rightarrow\ &\mathsf{receive}\{\mathsf{b},\mathsf{f}\} \rightarrow &right2(Res) \rightarrow\ &\mathsf{receive}\{\mathsf{f},\mathsf{b}\} \rightarrow \\
&\quad Res\ !\ \{\mathsf{u},\{\mathsf{b},\mathsf{b}\}\} & &\quad Res\ !\ \{\mathsf{u},\{\mathsf{b},\mathsf{b}\}\} \\
&\mathsf{end}. & &\mathsf{end}.
\end{aligned}
$$

Fig. 7. A simple Erlang program to perform deadlock analysis

shared resources, which are managed by process *res*—that accepts both queries {q, *pid*} and updates {u, *new_state*}—which is initialized with the state {f, f} (i.e., both resources are free). The *left* process takes the resources from left to right and the *right* process from right to left. We have not considered freeing the resources since we are only interested in illustrating the definition of a deadlock analysis in this example.

Our transformation tool `erlang2trs` returned a TRS of more than 100 lines of code (available at http://users.dsic.upv.es/~gvidal/erlang2trs/). Consider now that we are interested in verifying whether a deadlock is possible. We can easily verify this property in Curry using the following test function:

```
init = reduce [State 0 (I 2) [], State 1 main []] []

test = wrongState init
where
  wrongState s = and (map ws s) == True
  ws s = case s of
            State _ (I _) _ -> True
            State _ (AREC _ _ _) _ -> True
            _ -> False
```

where the function `test` checks if there exists a reachable state where all processes are either finished with a process identifier or are waiting for a message simultaneously (a deadlock):

```
deadlock> test
[(State 0 (I 5) []),(State 1 (I 4) []),
 (State 4 (AREC 19 [] [(I 2)]) []),
 (State 3 (AREC 12 [(T2 B B)] [(I 2)]) []),
 (State 2 (AREC 6 [(T2 C (I 4))] []) [])]
Result: True
More solutions? [Y(es)/n(o)/a(ll)] n
deadlock>
```

Therefore, our analysis concludes that a deadlock may occur in the original Erlang program.

Of course, for more contrived examples with an infinite number of states, narrowing has an infinite search space. Fortunately, there already exist techniques for ensuring the termination of narrowing while still producing overapproximations of the original program in the context of partial evaluation (see, e.g., [2]). Therefore, we can adapt such an approach to perform symbolic execution of infinite-state systems. A first step towards this direction can be found in [18].

Actually, our tool `erlang2trs` already produces a TRS that includes a simple memoization to avoid reducing the same state once and again.

5   Related Work

Giesl and Arts [11] present a verification of Erlang processes by using dependency pairs. They propose a similar idea—transforming Erlang programs to (conditional) rewrite systems—but no transformation is formalized; rather, the process is done manually. Moreover, no verification of safety properties is considered. In fact, the authors mainly focus on proposing general improvements to the termination prover for TRSs and CTRSs. Another related approach—though for different source and target languages—is that of Albert *et al.* [1], where a transformation from a concurrent object-oriented programming language based on message passing to a rule-based logic-like programming language is introduced.

Noll [19] introduces an implementation of Erlang in *rewriting logic* [16], a unified semantic framework for concurrency. Although we share some ideas with this paper, the aim is different. Noll's aim was to provide an executable specification of the language semantics that is tailored to the *Specification Language Compiler Generator* [15] in order to automatically translate the description into a verification front-end that implements the transition rules. Therefore, in this approach, Erlang programs are seen as data objects manipulated by a sort of interpreter implemented in rewriting logic. In contrast, we aim at producing plain rewrite systems that can be analyzed using existing technologies. Other approaches are based on abstract interpretation (e.g., [13]) or the use of equations to define abstraction mappings (e.g., [20]). We can also find some approaches where Erlang is translated to π-calculus [22] or μCRL [4].

More specific tools for Erlang verification include EVT [21], a theorem prover that requires user intervention, and the model checker McErlang [10], which implements a big-step operational semantics for dealing with concurrency as a run-time Erlang system. The main strengths of McErlang are that it is a robust tool that covers most of the Erlang language, explores all possible schedulings for concurrent actions, includes debugging facilities, provides mechanisms for reducing the state space, etc. On the other hand, McErlang is not intended to analyze sequential programs and, moreover, it does not allow the use of symbolic input data (in contrast to other similar tools for other programming languages, e.g., Java Pathfinder [26], a model checker for Java). This extension is far from trivial, and a symbolic execution semantics should be carefully designed. Actually, we are only aware of the approach of [8] to symbolic execution in Erlang, though no formalization is introduced in the paper (it is only explained informally). Hence we think that our approach is a promising step towards defining a symbolic execution mechanism for Erlang.

6   Discussion

We have introduced a novel approach to Erlang verification based on translating the original program to a term rewriting system. By keeping the original program structure as much as possible, we can effectively analyze the rewrite system and infer useful information regarding the original Erlang program using standard techniques and tools for rewrite systems.[4] We have illustrated the practicality of the approach with some examples.

---

[4] Nevertheless, although our syntax-directed transformation is tailored to the functional language Erlang, one could also extend it to other programming languages by using a *semantics-driven* transformation, similarly to that of [25].

As a future work, we would like to deal with scalability issues, e.g., defining an appropriate partial order reduction. We would also like to extend our approach to deal with the remaining features of Erlang (mainly higher-order functions, guards, modules, etc). Finally, we will also consider the generation of Prolog programs instead. In this case, we would have more mature environments available as well as a flurry of analysis techniques that could be applied to the transformed programs.

Acknowledgments. We thank the anonymous reviewers and the participants of LOP-STR 2013 for their useful comments to improve this paper.

References

1. Albert, E., Arenas, P., Gómez-Zamalloa, M.: Symbolic Execution of Concurrent Objects in CLP. In: Russo, C., Zhou, N.-F. (eds.) PADL 2012. LNCS, vol. 7149, pp. 123–137. Springer, Heidelberg (2012)
2. Albert, E., Vidal, G.: The narrowing-driven approach to functional logic program specialization. New Generation Computing **20**(1), 3–26 (2002)
3. Joe, A., Robert, V., Williams, M.: Concurrent programming in ERLANG. Prentice Hall (1993)
4. Arts, T., Earle, C.B., Derrick, J.: Development of a verified Erlang program for resource locking. STTT **5**(2–3), 205–220 (2004)
5. Baader, F., Nipkow, T.: Term Rewriting and All That. Cambridge University Press (1998)
6. Caballero, R., Martin-Martin, E., Riesco, A., Tamarit, S.: A Declarative Debugger for Sequential Erlang Programs. In: Veanes, M., Viganò, L. (eds.) TAP 2013. LNCS, vol. 7942, pp. 96–114. Springer, Heidelberg (2013)
7. Claessen, K., Svensson, H.: A semantics for distributed Erlang. In: Sagonas, K.F., Armstrong, J. (eds.). In: Proc. of the 2005 ACM SIGPLAN Workshop on Erlang, pp. 78–87. ACM (2005)
8. Earle, C.B.: Symbolic program execution using the Erlang verification tool. In: Alpuente, M. (eds.) Proc. of the 9th International Workshop on Functional and Logic Programming (WFLP 2000), pp. 42–55 (2000)
9. Felleisen, M., Friedman, D.P., Kohlbecker, E.E., Duba, B.F.: A syntactic theory of sequential control. Theor. Comput. Sci. **52**, 205–237 (1987)
10. Fredlund, L.-A., Svensson, H.: McErlang: a model checker for a distributed functional programming language. In: Hinze, R., Ramsey, N. (eds). In: Proc. of ICFP 2007, pp. 125–136. ACM (2007)
11. Giesl, J., Arts, T.: Verification of Erlang Processes by Dependency Pairs. Appl. Algebra Eng. Commun. Comput. **12**(1/2), 39–72 (2001)
12. Hanus, M. (ed.): Curry: An integrated functional logic language (vers. 0.8.3) (2012), http://www.curry-language.org
13. Huch, F.: Verification of Erlang Programs using Abstract Interpretation and Model Checking. In: Rémi, D., Lee, P. (eds.) Proc. of ICFP 1999, pp. 261–272. ACM (1999)
14. J.-M., H.: Canonical forms and unification. In: Bibel, W., Kowalski, R. (eds.) 5th Conference on Automated Deduction Les Arcs. LNCS, pp. 318–334. Springer, Heidelberg (1980)

15. Leucker, M., Noll, T.: Rewriting Logic as a Framework for Generic Verification Tools. Electr. Notes Theor. Comput. Sci. **36**, 121–137 (2000)
16. Meseguer, J.: Conditioned Rewriting Logic as a United Model of Concurrency. Theor. Comput. Sci. **96**(1), 73–155 (1992)
17. Neuhäußer, M.R., Noll, T.: Abstraction and Model Checking of Core Erlang Programs in Maude. Electr. Notes Theor. Comput. Sci. **176**(4), 147–163 (2007)
18. Nishida, N., Vidal, G.: A finite representation of the narrowing space. In: Proc. of the 23th International Symposium on Logic-Based Program Synthesis and Transformation (LOPSTR 2013). Technical Report TR-11-13, Universidad Complutense de Madrid, pp. 113–128 (To appear in Springer LNCS, 2013). http://users.dsic.upv.es/~gvidal/
19. Noll, T.: A Rewriting Logic Implementation of Erlang. Electr. Notes Theor. Comput. Sci. **44**(2), 206–224 (2001)
20. Noll, T.: Equational Abstractions for Model Checking Erlang Programs. Electr. Notes Theor. Comput. Sci. **118**, 145–162 (2005)
21. Noll, T.G., Fredlund, L., Gurov, D.: The Erlang Verification Tool. In: Margaria, T., Yi, W. (eds.) TACAS 2001. LNCS, vol. 2031, pp. 582–586. Springer, Heidelberg (2001)
22. Roy, C.K.: Thomas Noll, Banani Roy, and James R. Cordy. Towards automatic verification of Erlang programs by pi-calculus translation. In: Feeley,M., Trinder, P.W. (eds.) Proc. of the 2006 ACM SIGPLAN Workshop on Erlang, pp. 38–50. ACM (2006)
23. Slagle, J.R.: Automated theorem-proving for theories with simplifiers, commutativity and associativity. Journal of the ACM **21**(4), 622–642 (1974)
24. Svensson, H., Fredlund, L.-A.: A more accurate semantics for distributed Erlang. In: Thompson, S.J., Fredlund. L.-A., (eds.) Proceedings of the 2007 ACM SIGPLAN Workshop on Erlang, pp. 43–54. ACM (2007)
25. Vidal, G.: Closed symbolic execution for verifying program termination. In: Proc. of the 12th IEEE International Working Conference on Source Code Analysis and Manipulation (SCAM 2012), pp. 34–43. IEEE (2012)
26. Visser, W., Havelund, K., Brat, G.P., Park, S., Lerda, F.: Model checking programs. Autom. Softw. Eng. **10**(2), 203–232 (2003)

Extending Co-logic Programs
for Branching-Time Model Checking

Hirohisa Seki[✉]

Department of Computer Science, Nagoya Institute of Technology,
Showa-ku, Nagoya 466-8555, Japan
seki@nitech.ac.jp

Abstract. Co-logic programming is a programming language allowing
each predicate to be annotated as either inductive or coinductive. Assuming the *stratification restriction*, a condition on predicate dependency in
co-logic programs (co-LPs), a top-down procedural semantics (*co-SLD
derivation*) as well as an alternating fixpoint semantics has been given. In
this paper, we present some extensions of co-LPs, especially focusing on
the relationship with the existing alternating tree automata approaches
to branching-time model checking. We first consider the *local* stratification restriction to allow a more general class of co-LPs, so that we can
encode the CTL satisfaction relation as a co-LP, which is a direct encoding of the standard alternating automata by Kupferman et al. Next, we
consider non-stratified co-LPs based on the Horn μ-calculus. We give a
proof procedure, *co-SLD derivation with the parity acceptance condition*,
for non-stratified co-LPs, and show that it is sound and complete for a
class of non-stratified co-LPs. Its application to a goal-directed top-down
proof procedure for normal logic programs is also discussed.

1 Introduction

Co-logic programming, proposed by Gupta et al. [16] and Simon et al. [35,36], is an
extension of logic programming, where each predicate in definite programs is annotated as either *inductive* or *coinductive*. Assuming the *stratification restriction*,
a condition on predicate dependency in co-logic programs (co-LPs), a top-down
procedural semantics, *co-SLD derivation*, as well as the declarative semantics by
an alternating fixpoint semantics has been given: the least fixpoints for inductive
predicates and the greatest fixpoints for coinductive predicates.

Predicates in a co-LP are defined over infinite structures such as infinite trees
or infinite lists as well as finite ones, and co-LPs allow us to represent and reason
about properties of programs over such infinite structures. Co-logic programming
therefore has interesting applications to reactive systems and verifying properties
such as safety and liveness in model checking and so on.

This work was partially supported by JSPS Grant-in-Aid for Scientific Research (C)
24500171.

G. Gupta and R. Peña (Eds.): LOPSTR 2013, LNCS 8901, pp. 127–144, 2014.
DOI: 10.1007/978-3-319-14125-1_8

Recently, there has been reported some work [17] on understanding and extending the procedural semantics of co-LPs, in terms of the existing automata theory such as tree automata with the Büchi or Rabin acceptance conditions.

In this paper, we also consider some extensions of co-LPs, especially focusing on the relationship with the existing automata-theoretic approaches to branching-time model checking. We first consider the *local* stratification restriction to allow a more general class of co-LPs, so that we can encode the CTL satisfaction relation as a co-LP, which is a direct and natural encoding of the standard alternating automata by Kupferman et al. [21], including, among others, *weak alternating automata* (WAAs) [29].

Next, we consider non-stratified co-LPs based on the Horn μ-calculus by Charatonik et al. [2] We show that the notion of priorities in the Horn μ-calculus captures nesting of inductive/coinductive computations in co-LPs. We then give a proof procedure, *co-SLD derivation with the parity acceptance condition* (*co-SLD^{+p}* for short), for non-stratified co-LPs, and show its soundness and completeness for a class of non-stratified co-LPs. Its application to a goal-directed top-down proof procedure for a class of normal logic programs is also discussed.

The organization of this paper is as follows. In Section 2, we summarise some preliminary definitions on co-LPs and CTL [3]. In Section 3, we describe an encoding schema for weak alternating automata into co-LPs, and introduce the notion of the *local* stratification restriction. As an example, we consider an encoding of WAAs for CTL model checking into co-LP. In Section 4, we consider non-stratified co-LPs based on the Horn μ-calculus, and describe a proof procedure, co-SLD^{+p}, with its application to a top-down proof procedure for normal logic programs. Finally, we discuss about the related work and give a summary of this work in Section 5. [1]

Throughout this paper, we assume that the reader is familiar with the basic concepts of logic programming, which are found in [24].

2   Preliminaries

In this section, we first recall some basic definitions and notations concerning co-logic programs (co-LPs). The details and more examples are found in [16,35,36]. Then, we also explain some preliminaries on CTL.

Since co-logic programming can deal with infinite terms such as infinite lists or trees like $f(f(\dots))$ as well as finite ones, we consider the *complete* (or *infinitary*) Herbrand base [18,24], denoted by HB_P^*, where P is a program.

A *co-logic program* (co-LP) is a constraint definite program, where predicate symbols are annotated as either inductive or coinductive. There is one restriction on co-LP, referred to as the *stratification restriction*: Inductive and coinductive predicates are not allowed to be mutually recursive. An example which violates

---

[1]   Due to space constraints, we omit most proofs and some details, which will appear in the full paper.

the stratification restriction is $\{p \leftarrow q;\ q \leftarrow p\}$, where p is inductive, while q is coinductive.

When a co-LP P satisfies the stratification restriction, it is possible to decompose the set $\mathcal{P}$ of all predicates in P into a collection (called a *stratification*) of mutually disjoint sets $\mathcal{P}_0, \ldots, \mathcal{P}_r$ $(0 \leq r)$, called *strata*, so that, for every clause $p(\tilde{x}_0) \leftarrow c \,[\!]\, p_1(\tilde{x}_1), \ldots, p_n(\tilde{x}_n)$ in P, we have that $\sigma(p) \geq \sigma(p_i)$ if p and p_i have the same inductive/coinductive annotations, and $\sigma(p) > \sigma(p_i)$ otherwise, where $\sigma(q) = i$, if the predicate symbol q belongs to $\mathcal{P}_i$.

The following is an example of co-LPs.

Example 1. [36]. Suppose that predicates *member* and *drop* are annotated as inductive, while predicate *comember* is annotated as coinductive.

$$member(H, [H|\_]) \leftarrow \qquad\qquad drop(H, [H|T], T) \leftarrow$$
$$member(H, [\_|T]) \leftarrow member(H, T) \qquad drop(H, [\_|T], T_1) \leftarrow drop(H, T, T_1)$$

$$comember(X, L) \leftarrow drop(X, L, L_1), comember(X, L_1)$$

The definition of *member* is a conventional one; its meaning is defined in terms of the least fixpoint, since it is an inductive predicate. So, the prefix ending in the desired element H must be finite. The same applies to predicate *drop*.

On the other hand, predicate *comember* is coinductive, whose meaning is defined in terms of the greatest fixpoint. Therefore, it is true if and only if the desired element X occurs an infinite number of times in the list L. Hence it is false when the element does not occur in the list or when the element only occurs a finite number of times in the list.

For example, the goal $\leftarrow X = 1, L = [0, 1|L], comember(X, L)$ is true, while the goal $\leftarrow X = 1, L = [0, 1, 0, 1], comember(X, L)$ is false. Note that $L = [0, 1|L]$ represents an infinite list L consisting of 0s and 1s. $\qquad\square$

A meta-interpreter for co-logic programming has been developed and available [19], and recent SWI-Prolog (version 6.5.1) has also offered a module for supporting coinduction.[2]

The declarative semantics of a co-logic program is a stratified interleaving of the least fixpoint semantics and the greatest fixpoint semantics.

In this paper, we consider the complete Herbrand base HB_P^* as the set of elements in the domain of a *structure* $\mathcal{D}$. Given a structure $\mathcal{D}$ and a constraint c, $\mathcal{D} \models c$ denotes that c is true under the interpretation for constraints provided by $\mathcal{D}$. Moreover, if θ is a ground *substitution* (i.e., a mapping of variables on the domain $\mathcal{D}$, namely, HB_P^* in this case) and $\mathcal{D} \models c\theta$ holds, then we say that c is *satisfiable*, and θ is called a *solution* (or ground *satisfier*) of c, where $c\theta$ denotes the application of θ to the variables in c. We refer to [5] for an algorithm for checking constraint satisfiability.

Let P be a co-logic program with a stratification $\mathcal{P}_0, \ldots, \mathcal{P}_r$ $(0 \leq r)$. Let Π_i $(0 \leq i \leq r)$ be the set of clauses whose head predicates are in $\mathcal{P}_i$. Then,

---

[2] http://www.swi-prolog.org/pldoc/doc/swi/library/coinduction.pl

$P = \Pi_0 \uplus \ldots \uplus \Pi_r$. Similar to the "immediate consequence operator" T_P in the literature, our operator $T_{\Pi,S}$ assigns to every set I of ground atoms a new set $T_{\Pi,S}(I)$ of ground atoms as

$$T_{\Pi,S}(I) = \{A \in HB_\Pi^* \mid \text{there is a ground substitution } \theta \text{ and a clause in } \Pi$$
$$H \leftarrow c \, [\![\, B_1, \cdots, B_n, \ n \geq 0, \text{ such that}$$
$$\text{(i) } A = H\theta, \text{ (ii) } \theta \text{ is a solution of } c, \text{ and}$$
$$\text{(iii) for every } 1 \leq i \leq n, \text{ either } B_i\theta \in I \text{ or } B_i\theta \in S\}.$$

In the above, the atoms in S are treated as facts. S is intended to be a set of atoms whose predicate symbols are in lower strata than those in the current stratum Π. We consider $T_{\Pi,S}$ to be the operator defined on the set of all subsets of HB_Π^*, ordered by standard inclusion. Then, $T_{\Pi,S}$ admits a least and a greatest fixpoints denoted by $lfp(T_{\Pi,S})$ and $gfp(T_{\Pi,S})$, respectively.

Finally, the model $M(P)$ of a co-logic program $P = \Pi_0 \uplus \ldots \uplus \Pi_r$ is defined inductively as follows: Let $M(\Pi_{-1}) = \emptyset$. For $k \geq 0$, $M(\Pi_k) = lfp(T_{\Pi_k, M_{k-1}})$ if $\mathcal{P}_i$ is inductive; $gfp(T_{\Pi_k, M_{k-1}})$ if $\mathcal{P}_i$ is coinductive, where M_{k-1} is the model of lower strata than Π_k, i.e., $M_{k-1} = \cup_{i=-1}^{k-1} M(\Pi_i)$.

Then, the *model* of P is $M(P) = \cup_{i=0}^r M(\Pi_i)$, the union of all models $M(\Pi_i)$.

Syntax and Semantics of CTL. We first briefly recall the syntax and the semantics of CTL (see [3, 21]). The Computation Tree Logic (CTL) is a branching-time temporal logic for expressing properties of events. It is a subset of CTL* in which each of temporal operators X ("next time"), U ("until") and $\tilde{U}$, a dual of the U operator (often denoted by R ("release")), must be immediately preceded by a path quantifier, either A ("for all paths") or E ("for some path").

Definition 1. Given a set AP of atomic propositions, a *CTL formula* is either a *state formula* φ or a *path formula* ψ defined as follows:

$$\langle state\ formulas \rangle \ \varphi ::= \textbf{true} \mid \textbf{false} \mid p \mid \neg p \mid \varphi_1 \wedge \varphi_2 \mid \varphi_1 \vee \varphi_2 \mid A\psi \mid E\psi$$
$$\langle path\ formulas \rangle \ \psi ::= X\varphi_1 \mid \varphi_1 U \varphi_2 \mid \varphi_1 \tilde{U} \varphi_2$$

where $p \in AP$. □

Some familiar temporal operators F and G will be used as the following abbreviations: $F\varphi = \textbf{true} U \varphi$ ("eventually") and $G\varphi = \textbf{false} \tilde{U} \varphi$ ("always"). We assume that CTL formulas are given in *positive normal form* [3], that is, negations are applied only to atomic propositions. This can be assumed without the loss of generality, since we have conjunction, disjunction, both U and $\tilde{U}$ operators, and both quantifiers A and E. A formula φ is a $\tilde{U}$-*formula* if it is of the form $A\varphi_1 \tilde{U} \varphi_2$ or $E\varphi_1 \tilde{U} \varphi_2$.

Given a CTL formula φ, let $cl(\varphi)$, called the *closure*, be the set of all CTL state subformulas of φ (including φ, but excluding $\textbf{true}$ and $\textbf{false}$), and the *size* $\|\varphi\|$ of φ is defined to be the number of elements in $cl(\varphi)$.

The semantics of CTL is defined with respect to a *Kripke structure* $K = \langle AP, W, R, w^0, L \rangle$, where AP is a set of atomic propositions, W is a set of states, $R \subseteq W \times W$ is a transition relation that must be total (i.e., for every $w \in W$ there exists $w \in W$ such that $\langle w, w' \rangle \in R$), w^0 is an initial state, and $L : W \to 2^{AP}$ maps each state to the set of atomic propositions true in that state. A path in K is an infinite sequence of states, $\pi = w_0, w_1, \ldots$ such that for every $i \geq 0$, $\langle w_i, w_{i+1} \rangle \in R$. We denote the suffix $w_i, w_{i+1}, \ldots$ of π by π^i.

The notation $K, w \models \varphi$ indicates that a CTL state formula φ holds at the state w of the Kripke structure K. Similarly, $K, \pi \models \psi$ indicates that a CTL path formula ψ holds on a path π of K.

Definition 2. Given a Kripke structure $K = \langle AP, W, R, w^0, L \rangle$, the relation $\models$ is inductively defined as follows:

- For all w, we have $K, w \models$ **true** and $K, w \not\models$ **false**.
- $K, w \models p$ for $p \in AP$ iff $p \in L(w)$.
- $K, w \models \neg p$ for $p \in AP$ iff $p \notin L(w)$.
- $K, w \models \varphi_1 \wedge \varphi_2$ iff $K, w \models \varphi_1$ and $K, w \models \varphi_2$.
- $K, w \models \varphi_1 \vee \varphi_2$ iff $K, w \models \varphi_1$ or $K, w \models \varphi_2$.
- $K, w \models A\psi$ iff for every path π starting from w, we have $K, \pi \models \psi$.
- $K, w \models E\psi$ iff there exists a path π from w such that $K, \pi \models \psi$.
- $K, \pi \models \varphi$ for a state formula φ, iff $K, w_0 \models \varphi$ where $\pi = w_0, w_1, \ldots$.
- $K, \pi \models X\varphi$ iff $K, \pi^1 \models \varphi$.
- $K, \pi \models \varphi_1 U \varphi_2$ iff there exists $i \geq 0$ such that $K, \pi^i \models \varphi_2$ and for all $0 \leq j < i$, we have $K, \pi^j \models \varphi_1$.
- $K, \pi \models \varphi_1 \tilde{U} \varphi_2$ iff for all $i \geq 0$ such that $K, \pi^i \not\models \varphi_2$, there exists $0 \leq j < i$ such that $K, \pi^j \models \varphi_1$.

Note that $K, \pi \models \varphi_1 \tilde{U} \varphi_2$ if and only if $K, \pi \not\models (\neg \varphi_1) U (\neg \varphi_2)$.

3 Encoding CTL Model Checking in Co-LP

In this section, we first briefly explain the automata-theoretic approach to CTL model checking by Kupferman et al. [21], and we then present our encoding of the CTL satisfaction relation into co-LPs.

3.1 The Automata-Theoretic Approach to CTL Model Checking

Kupferman et al. [21] have proposed the automata-theoretic approach to branching-time model checking, where *weak alternating automata* (WAAs) [29] and *hesitant alternating automata* (HAAs) are shown to play an essential role. In the following, we will present an encoding method of WAAs into co-LPs, although similar encoding of HAAs into co-LPs is also possible.

Let $\mathcal{A} = \langle \Sigma, Q, \delta, q_0, F \rangle$ be an alternating tree automaton, where Σ is the input alphabet, Q is a finite set of states, δ is a transition function, $q_0 \in Q$ is an initial state and F specifies the acceptance condition of $\mathcal{A}$.

In WAAs, Q and δ satisfy the following conditions: (i) there exists a partition of Q into disjoint sets, $Q_1, \ldots, Q_m$, equipped with a partial order $\leq$, and (ii) transitions from a state in Q_i lead to states in either the same Q_i or a lower one; that is, for every $q \in Q_i$ and $q' \in Q_j$ for which q' occurs in $\delta(q, s)$, for some $s \in \Sigma$, we have $Q_j \leq Q_i$.

In addition, a WAA has the Büchi acceptance condition $F \subseteq Q$, that is, a run is accepting iff all its infinite paths π satisfy the Büchi acceptance condition: $inf(\pi) \cap F \neq \emptyset$, where $inf(\pi)$ is the set of all the states that appear infinitely often on π. Moreover, we have that for each set Q_i, either $Q_i \subseteq F$ (Q_i is said to be *accepting*), or $Q_i \cap F = \emptyset$ (Q_i is said to be *rejecting*).

The following definition (Def. 3) shows a 1-letter weak alternating word automaton $\mathcal{A}_{\mathcal{K}, \varphi}$ for CTL model checking due to Kupferman et al. [21]; For a given Kripke structure $\mathcal{K} = \langle AP, W, R, w^0, L \rangle$ and a CTL formula φ, the language accepted by $\mathcal{A}_{\mathcal{K}, \varphi}$ is nonempty iff $\mathcal{K}, w^0 \models \varphi$.

Definition 3. For a given Kripke structure $\mathcal{K} = \langle AP, W, R, w^0, L \rangle$ and a CTL formula φ, $\mathcal{A}_{\mathcal{K}, \varphi} = \langle \{a_0\}, W \times cl(\varphi), \delta, \langle w^0, \varphi \rangle, F \rangle$, where δ and F are defined as follows:

1. $\delta(\langle w, p \rangle, a_0) = \mathbf{true}$ if $p \in L(w)$.
   $\delta(\langle w, p \rangle, a_0) = \mathbf{false}$ if $p \notin L(w)$.
2. $\delta(\langle w, \neg p \rangle, a_0) = \mathbf{true}$ if $p \notin L(w)$.
   $\delta(\langle w, \neg p \rangle, a_0) = \mathbf{false}$ if $p \in L(w)$.
3. $\delta(\langle w, \varphi_1 \wedge \varphi_2 \rangle, a_0) = \delta(\langle w, \varphi_1 \rangle, a_0) \wedge \delta(\langle w, \varphi_2 \rangle, a_0)$.
4. $\delta(\langle w, \varphi_1 \vee \varphi_2 \rangle, a_0) = \delta(\langle w, \varphi_1 \rangle, a_0) \vee \delta(\langle w, \varphi_2 \rangle, a_0)$.
5. $\delta(\langle w, AX\varphi \rangle, a_0) = \bigwedge_{i=1}^{k}(w_i, \varphi)$.
6. $\delta(\langle w, EX\varphi \rangle, a_0) = \bigvee_{i=1}^{k}(w_i, \varphi)$.
7. $\delta(\langle w, A\varphi_1 U\varphi_2 \rangle, a_0) = \delta(\langle w, \varphi_2 \rangle, a_0) \vee (\delta(\langle w, \varphi_1 \rangle, a_0) \wedge \bigwedge_{i=1}^{k}(w_i, A\varphi_1 U\varphi_2))$.
8. $\delta(\langle w, E\varphi_1 U\varphi_2 \rangle, a_0) = \delta(\langle w, \varphi_2 \rangle, a_0) \vee (\delta(\langle w, \varphi_1 \rangle, a_0) \wedge \bigvee_{i=1}^{k}(w_i, E\varphi_1 U\varphi_2))$.
9. $\delta(\langle w, A\varphi_1 \tilde{U}\varphi_2 \rangle, a_0) = \delta(\langle w, \varphi_2 \rangle, a_0) \wedge (\delta(\langle w, \varphi_1 \rangle, a_0) \vee \bigwedge_{i=1}^{k}(w_i, A\varphi_1 \tilde{U}\varphi_2))$.
10. $\delta(\langle w, E\varphi_1 \tilde{U}\varphi_2 \rangle, a_0) = \delta(\langle w, \varphi_2 \rangle, a_0) \wedge (\delta(\langle w, \varphi_1 \rangle, a_0) \vee \bigvee_{i=1}^{k}(w_i, E\varphi_1 \tilde{U}\varphi_2))$,

where, for each $w \in W$ in $\mathcal{K}$, w has k successors: $succ_R(w) = \{w_1, \ldots, w_k\}$. The set F of accepting states consists of all pairs in $W \times F_\varphi$, where F_φ is all the $\tilde{U}$-formulas in $cl(\varphi)$. $\qquad\square$

3.2   An Encoding Schema of WAAs into Stratified Co-LPs

We now describe an encoding schema of WAAs into co-LPs, and explain it using the above-mentioned WAA for CTL model checking.

Talbot [37] pointed out that the encoding of the WAAs into the *alternation-free Horn μ-calculus* is straightforward. Rephrased in the co-LP framework, our *encoding schema* of a WAA, $\mathcal{A}$, into the corresponding co-LP, P, is as follows: (i) the states of $\mathcal{A}$ are translated to predicates in P, (ii) the transitions are simulated by clauses in P, and (iii) the predicates for accepting (rejecting) states are annotated as coinductive (inductive), respectively.

A Co-LP with the Local Stratification Restriction. Following the above-mentioned encoding schema, we now translate $\mathcal{A}_{\mathcal{K},\varphi}$ into the corresponding co-LP $P_{\mathcal{K}}$. Since we need literals representing the states in $\mathcal{A}_{\mathcal{K},\varphi}$, we first extend the notion of the stratification restriction to the *local* stratification restriction. A local stratification is a function σ from ground literals to natural numbers such that given $A \in HB_P^*$, we define $\sigma(\neg A) = \sigma(A) + 1$. A co-logic program P satisfies the *local stratification restriction*, if there exists a local stratification σ such that, for every ground clause γ of the form $h \leftarrow l_1, \ldots, l_m$, (i) $\sigma(h) \geq \sigma(l_i)$ when either l_i has the same annotation of h or l_i is a negative literal, and (ii) $\sigma(h) > \sigma(l_i)$ when l_i has the different annotation from h. We note that for a co-LP P with the local stratification restriction, the alternating fixpoint semantics $M(P)$ is defined similarly.

Recall in Def. 3 that for each $w \in W$ in $\mathcal{K}$, w has k successors: $succ_R(w) = \{w_1, \ldots, w_k\}$. In accordance with this, a transition relation R in a Kripke structure $\mathcal{K} = \langle AP, W, R, w^0, L \rangle$ is assumed to be specified by a finite disjunction of k formula ($k \geq 1$); for every $w, w' \in W$, $\langle w, w' \rangle \in R$ iff $\models tr(w, w')$, where $tr(X, Y) \equiv t_1(X, Y) \vee \cdots \vee t_k(X, Y)$, and each t_i is a function of its first argument, namely, $\forall X, Y, Z\ (t_i(X,Y) \wedge t_i(X,Z) \rightarrow Y = Z)$, and $\forall X(X \in W \rightarrow \exists Y\ t_i(X,Y))$.

When writing terms encoding CTL formulas, we will use the function symbols a, e, x, u and $\tilde{u}$ for the CTL operator symbols A, E, X, U and $\tilde{U}$, respectively. Moreover, for saving space we write $H \leftarrow M, \bigwedge_{i=1}^{k} L_i$ for a clause $H \leftarrow M, L_1, \ldots, L_k$ and $H \leftarrow M, \bigvee_{i=1}^{k} L_i$ for the set of clauses $\{H \leftarrow M, L_1; \ldots; H \leftarrow M, L_k\}$ ($k \geq 1$), where M and L_i are (possibly empty) sequences of atoms.

Definition 4. Encoding Program for the CTL Satisfaction Relation

Given a Kripke structure $\mathcal{K} = \langle AP, W, R, w^0, L \rangle$, the *encoding program* $P_{\mathcal{K}}$ is the following co-logic program:

1. $sat(S, F) \leftarrow elem(F, S)$
2. $sat(S, not(F)) \leftarrow \neg elem(F, S)$
3. $sat(S, and(F_1, F_2)) \leftarrow sat(S, F_1), sat(S, F_2)$
4.1 $sat(S, or(F_1, F_2)) \leftarrow sat(S, F_1)$
4.2 $sat(S, or(F_1, F_2)) \leftarrow sat(S, F_2)$
5. $sat(S, ax(F)) \leftarrow \bigwedge_{i=1}^{k}(t_i(S, S_i), sat(S_i, F))$
6. $sat(S, ex(F)) \leftarrow \bigvee_{i=1}^{k}(t_i(S, S_i), sat(S_i, F))$
7.1 $sat(S, au(F_1, F_2)) \leftarrow sat(S, F_2)$
7.2 $sat(S, au(F_1, F_2)) \leftarrow sat(S, F_1), \bigwedge_{i=1}^{k}(t_i(S, S_i), sat(S_i, au(F_1, F_2)))$
8.1 $sat(S, eu(F_1, F_2)) \leftarrow sat(S, F_2)$
8.2 $sat(S, eu(F_1, F_2)) \leftarrow sat(S, F_1), \bigvee_{i=1}^{k}(t_i(S, S_i), sat(S_i, eu(F_1, F_2)))$
9.1 $sat(S, a\tilde{u}(F_1, F_2)) \leftarrow sat(S, F_2), sat(S, F_1)$
9.2 $sat(S, a\tilde{u}(F_1, F_2)) \leftarrow sat(S, F_2), \bigwedge_{i=1}^{k}(t_i(S, S_i), sat(S_i, a\tilde{u}(F_1, F_2)))$
10.1 $sat(S, e\tilde{u}(F_1, F_2)) \leftarrow sat(S, F_2), sat(S, F_1)$
10.2 $sat(S, e\tilde{u}(F_1, F_2)) \leftarrow sat(S, F_2), \bigvee_{i=1}^{k}(t_i(S, S_i), sat(S_i, e\tilde{u}(F_1, F_2)))$

together with the clauses defining the predicate t_i's and $elem$, where $elem(p, w)$ holds iff $p \in L(w)$ for every property $p \in AP$ and $w \in W$.

In P_K, every ground atom of the form $sat(\cdot, a\tilde{u}(\cdot, \cdot))$ or $sat(\cdot, e\tilde{u}(\cdot, \cdot))$ is annotated as coinductive, while the other atoms are annotated as inductive.   □

P_K satisfies the local stratification restriction w.r.t. the stratification function σ defined as follows; For every ground terms e, s and s', $\sigma(elem(e, s)) = \sigma(t_i(s, s')) = 0$, and $\sigma(sat(s, \varphi)) = \|\varphi\|$.

Each clause in P_K is defined to simulate the corresponding transition rule of δ in $\mathcal{A}_{K,\varphi}$, and the predicate annotation in P_K is given in accordance with the accepting condition of $\mathcal{A}_{K,\varphi}$. Since the construction of P_K follows the encoding schema of WAAs into co-LPs, the following proposition holds:

Proposition 1. Correctness of Encoding the CTL Satisfaction Relation
Given a Kripke structure $\mathcal{K} = \langle AP, W, R, w^0, L \rangle$, let P_K be the encoding co-logic program and φ a CTL state formula. Then, $\mathcal{K}, w^0 \models \varphi$ iff $sat(w^0, \varphi) \in M(P_K)$.

Example 2. Let us consider: (i) the set $AP = \{a, b\}$ of elementary properties, and (ii) the Kripke structure $\mathcal{K} = \langle AP, \{s_0, s_1, s_2\}, R, s_0, L \rangle$, where R is the transition relation $\{(s_0, s_0), (s_0, s_1), (s_1, s_1), (s_1, s_2), (s_2, s_1)\}$ and L is the function such that $L(s_0) = \{a\}$, $L(s_1) = \{b\}$, and $L(s_2) = \{a\}$. Consider the CTL formula $\varphi = E(aUA(\mathbf{false}\tilde{U}(E\mathbf{true}Ub)))$, which is often abbreviated as $E(aUAGEFb)$. The encoding program P_K consists of the clauses $1 - 10$ in Definition 4 defining the predicates sat, together with the following clauses:

$tr(s_0, s_0) \leftarrow \quad tr(s_0, s_1) \leftarrow \quad tr(s_1, s_1) \leftarrow \quad tr(s_1, s_2) \leftarrow tr(s_2, s_1) \leftarrow$
$elem(a, s_0) \leftarrow \quad elem(b, s_1) \leftarrow \quad elem(a, s_2) \leftarrow$

□

Given the encoding program P_K and a CTL formula φ, the model checking problem is reduced to the problem of checking whether or not $sat(w^0, \varphi_0) \in M(P)$. This is done in co-logic programming by two methods of query evaluation: the bottom-up evaluation based on the fixpoint semantics and the top-down evaluation based on the *co-SLD* resolution [35].

In summary, our encoding schema of WAAs/HAAs into the corresponding co-LPs is quite simple and general; it is simple, because the definition of clauses is given to precisely simulate the transition rules in δ, while the predicate annotations are given to exactly correspond to the acceptance conditions of WAAs/HAAs. Our encoding schema is also general; there is no need to individually consider co-LPs for other branching-time logics such as alternating μ-calculus and CTL*, because it is shown [21] that model checking for alternating μ-calculus is done by WAAs, while CTL* model checking is done by HAAs.

3.3   Handling Negation in Co-LP

The definition of the encoding program P_K contains a negative literal in the body of clause 2 (Def. 4). In the following, we thus examine the existing proposal for handling negation in co-LP.

In [27,28], Min and Gupta proposed an extension of co-SLD resolution, called *co-SLDNF* resolution. The *declarative* semantics of co-SLDNF is given in terms of the work by Fitting [11] and Fages [10], while the *procedural* semantics is given in terms of co-SLDNF resolution. Following Min [28], we denote by $HM^{\mathrm{R}}(P)$ the (Rational) Herbrand model semantics given by co-SLDNF.

Example 3 ($M(P)$ and $HM^{\mathrm{R}}(P)$). In the following, predicates p and q are annotated as inductive.

1. First, let $P_1 = \{p \leftarrow p\}$. Then, P_1 has two models in $HM^{\mathrm{R}}(P_1)$: $M_1 = \{p\}$ and $M_2 = \emptyset$. Since p is an inductive predicate, the standard co-LP semantics $M(P_1) = M_2$. Therefore, $HM^{\mathrm{R}}(P)$ does not coincide with $M(P)$, the original semantics by Simon et al. [35].
2. We then consider the following program: $P_2 = \{p \leftarrow p; q \leftarrow not\ p\}$. P_2 also has two models in $HM^{\mathrm{R}}(P_2)$: $M_1 = \{p\}$ and $M_2 = \{q\}$. P_2 satisfies the local stratification restriction, and it is a stratified program in the standard sense. Therefore, $M(P_2) = M_2$.
3. Finally, consider the program $P_3 = \{p \leftarrow not\ q; q \leftarrow not\ p\}$. P_3 has two models in $HM^{\mathrm{R}}(P_3)$: $M_1 = \{p\}$ and $M_2 = \{q\}$. In this case, $HM^{\mathrm{R}}(P_3)$ coincides with the answer set semantics $AS(P_3)$. $M(P_3)$ is not defined, since P_3 does not satisfy the local stratification restriction.   □

It is noted in [28] that the semantics $HM^{\mathrm{R}}(P_1)$ by co-SLDNF for P_1 seems to be counter-intuitive, whereas it was also argued that such a behaviour of co-SLDNF is advantageous for Answer Set Programming (ASP), as shown in program P_3 in the above.

From the viewpoint of modular programming, however, the semantics $HM^{\mathrm{R}}(P)$ or co-SLDNF would make it difficult for a co-LP programmer to define a new predicate (such as q in P_2) by using given predicates (such as p in P_1) as building blocks, since her/his intended semantics of p will be the standard semantics $M(P)$ for *definite* co-LPs. In the following, we therefore present an alternative approach to co-LPs with negation. In our approach, the semantics coincides with $M(P)$ for a co-LP P with the (local) stratification restriction, whereas it coincides with the answer set $AS(P)$ for a class of non-stratified programs.

4   Towards an Operational Semantics for Non-Stratified Co-LP

4.1   Negation Elimination for Co-LP

Our approach to handling negation in co-LP is based on *negation elimination* (*NE* for short), a familiar program transformation technique, tailored to co-logic programs [34]. Given a co-LP P, NE [34] derives a set P^* of definite clauses from P such that (i) each predicate symbol p in P has one-to-one correspondence with a new predicate symbol, not_p, in P^* with the same arity, and (ii) for any ground atom $p(\tilde{t})$ and $not\_p(\tilde{t})$, $M(P) \models \neg p(\tilde{t})$ iff $M(P^*) \models not\_p(\tilde{t})$, by the following two steps:

(step 1) for each clause in P, we replace each occurrence $not\ p$ of naf-literals (if any) by not_p, where not_p is a new predicate not appearing elsewhere;

(step 2) we then derive the definition of each not_p from the completed definition of p in P, denoted by $comp(p)$. Namely,

(i) [Definition Derivation] Suppose that $comp(p)$ is of the form: $p \leftrightarrow B_1 \lor \cdots \lor B_n$. Then, negating both sides of $comp(p)$, and replacing every negative occurrence $\neg p$ by not_p, we obtain $not\_p \leftrightarrow \neg(B_1 \lor \cdots \lor B_n)$.

Next, transforming the right-hand side in the above to a disjunctive form, using De Morgan's laws, replacing each occurrence of $\neg not\ q$ by q, and each occurrence of $\neg q$ by not_q, we obtain the completed definition of not_p, i.e., $not\_p \leftrightarrow NB_1 \lor \cdots \lor NB_{n'}$, where each NB_i is a conjunction of positive literals. Finally, we transform $comp(not\_p)$ to a set of clauses: $\{not\_p \leftarrow NB_1; \ldots ; not\_p \leftarrow NB_{n'}\}$.

(ii) [Annotation Inversion] Annotate the derived predicate not_p as "coinductive" (resp. "inductive") if the annotation of the original predicate p is inductive (resp. coinductive).

(iii) Apply the above steps (i) to (iv) to all remaining predicates not_p appearing in this transformation process.

Let P be a co-LP with the local stratification restriction σ, and P^* be the set of all clauses obtained by applying the above NE transformation. We define the stratification function σ^* for P^* as follows: $\sigma^*(p) = \sigma(p)$ for all predicates defined in P, and $\sigma^*(not\_p) = \sigma(p) + 1$ for all predicates not_p newly introduced in τ. Then, we can show that P^* satisfies the local stratification restriction w.r.t. σ^*.

Example 4. Continued from Example 3. Recall that both p and q are annotated as inductive.

1. We first reconsider $P_1 = \{p \leftarrow p\}$. Then, $P_1^* = P_1 \cup \{not\_p \leftarrow not\_p\}$, and not_p is annotated as coinductive. Therefore, the goal not_p has a successful co-SLD derivation in P_1^*, which coincides with the original semantics $M(P_1)$.

2. Next, consider again $P_2 = \{p \leftarrow p; q \leftarrow not\ p\}$. Then, $P_2^* = \{p \leftarrow p; q \leftarrow not\_p\} \cup \{not\_p \leftarrow not\_p; not\_q \leftarrow p\}$. Then, the goal not_q, for example, has no successful co-SLD derivation in P_2^*, since p is annotated as inductive. Our approach thus coincides with the original semantics $M(P_2)$.

3. Finally, consider the *non-stratified* program $P_3 = \{p \leftarrow not\ q; q \leftarrow not\ p\}$. Then, $P_3^* = \{p \leftarrow not\_q; q \leftarrow not\_p; not\_p \leftarrow q; not\_q \leftarrow p\}$. P_3^* does not satisfy the stratification restriction.   □

As the above example shows, NE works well for locally stratified co-LPs, while NE will derive non-stratified co-LPs from non-stratified programs as P_3.

In co-logic programming, one of the challenging issues is to extend the framework to handle *non-stratified* co-LPs, e.g., [17]. Moreover, for the purpose of model checking, the conventional co-LPs with the stratification restriction are not expressive enough to handle temporal logics such as the general μ-calculus. The general μ-calculus, which allows arbitrary nesting of least (μ) and greatest

(ν) fixpoint operators, cannot be translated into WAAs; it can be translated to alternating *Rabin* automata [9].

In the following, we will therefore focus on the *Horn μ-calculus* [2], which is an extension of co-LPs. We present a practical proof procedure for Horn μ-calculus à la co-SLD derivation and its application for co-LP with negation.

4.2   Horn μ-Calculus and Its Proof Procedure

Charatonik et al. [2] have proposed the *Horn μ-calculus*, an extension of logic programs by allowing nesting of least and greatest fixpoints, in terms of a priority of each predicate for specifying whether its semantics has to be computed as a least or a greatest fixpoint. They have given to the Horn μ-programs the semantics based on ground proof trees as well as the nested fixpoints semantics.

A *Horn μ-program* (P, Ω) is a set of definite clauses in which every predicate symbol p in P is associated with a non-negative number $\Omega(p)$, called the *priority* of p. The priority of an atom $p(\tilde{t})$ is defined to be the priority of the predicate p, and denoted by $\Omega(p(\tilde{t}))$.

The semantics for a Horn μ-program is given in terms of ground proof trees. A proof tree for a ground atom $p(\tilde{t})$ is a (possibly) infinite tree with the root labeled by $p(\tilde{t})$, which is defined in a usual way: Given a logic program P, r is a proof tree if for each node n in r, labeled by some ground atom h, there exists a ground instance $h \leftarrow b_1, \ldots, b_m$ ($m \geq 0$) of a clause in P, and n has m children nodes, each of which is labeled by b_i ($0 \leq i \leq m$). When $m = 0$, the node n has no children nodes.

For an infinite path π in a proof tree, we denote by $Inf(\pi)$ the set of all priorities occurring infinitely often on the path π. We say that a proof tree *accepts* the atom $p(\tilde{t})$ if its root is labeled by $p(\tilde{t})$ and for all paths π starting from the root, the maximal element of $Inf(\pi)$ is even.

For a Horn μ-program (P, Ω) we define $[\![(P, \Omega)]\!]$ the set of all ground atoms $p(\tilde{t})$ such that there exists a proof tree which accepts the atom $p(\tilde{t})$.

Note that predicates of even (odd) priority are given greatest (least) fixpoint meanings, respectively. In particular, if all predicates have priority 0 then $[\![(P, \Omega)]\!] = gfp(T_P)$. On the other hand, if all predicates have priority 1, then $[\![(P, \Omega)]\!] = lfp(T_P)$.

Example 5. (Adapted and simplified from [2].) Let $P_0 = \{p \leftarrow p;\ p \leftarrow q;\ q \leftarrow p\}$ be a set of clauses, where p is an inductive predicate, while q is a coinductive predicate. Since P_0 does not satisfy the stratification restriction, its meaning is not determined in co-logic programming.

In the Horn μ-calculus, however, the semantics of P_0 can be determined in terms of priorities assigned to the predicates. Suppose, for example, that the coinductive predicate q has a higher priority than the inductive predicate p. We thus define: $\Omega(p) = 1$ and $\Omega(q) = 2$.

The acceptance condition for $[\![(P_0, \Omega)]\!]$ implies that any infinite path in an accepting proof tree must have an infinite number of occurrences of the predicate q. Since the priority of p is odd, it must "terminate" in the use of a higher priority

$$P_0 = \{c_1, c_2, c_3\},$$
$$\Omega(p) = 1, \ \Omega(q) = 2$$

$$c_1 : p \leftarrow p$$
$$c_2 : p \leftarrow q$$
$$c_3 : q \leftarrow p$$

$$q : 2$$
$$| \ c_3$$
$$p : 1$$
$$| \ c_1^* c_2$$
$$q : 2$$
$$| \ c_3$$
$$\vdots$$

$$q : 2$$
$$| \ c_3$$
$$p : 1$$
$$| \ c_1$$
$$p : 1$$
$$| \ c_1^\omega$$
$$\vdots$$

Fig. 1. Proof Trees for a Horn μ-Program in Ex. 5: an accepting proof tree (left) and a not-accepting one (right). In the above, each node is depicted with the priority of its labelled predicate, while each edge is depicted with the clause used for deriving a child node from its parent node.

predicate, i.e., q in this case. In contrast, since the priority of q is even, it need not terminate; it can call itself infinitely often, possibly through terminating calls to p. If we assume that q expresses some "good" property, then the program describes a kind of liveness property: from any point q will eventually happen.

In Fig. 1 (left), the node labelled with p is obtained from the root q using clause c_3. Then, applying clause c_1 finitely often and then c_2, denoted by $c_1^* c_2$, we obtain the node labelled with q, which is the same as the root node. We then repeat this process infinitely often. For this unique infinite path π, we have that $\max\{Inf(\pi)\} = 2$, implying that the proof tree is accepting: $q \in [\![(P_0, \Omega)]\!]$. Similarly, we have that $p \in [\![(P_0, \Omega)]\!]$.

On the other hand, the proof tree in Fig. 1 (right) is not accepting: it is constructed from the root node, applying clause c_1 infinitely often. Then, for the resulting infinite path π', we have that $\max\{Inf(\pi')\} = 1$, implying that the tree is not accepting.   □

It is easy to see that the notion of priorities in a Horn μ-program exactly captures the notion of *strong/weak_inductive* annotations, which have recently been proposed by Gupta et al. [17] to specify priorities of inductive predicates to coinductive ones.

Operational Semantics of Non-Stratified Co-LP. The proof tree for defining the semantics of Horn μ-programs can be considered an "ideal" proof procedure, since the proof tree will be an infinite tree in general.

One simple way to realize an *effective* procedure for Horn μ-programs will be use co-SLD derivation by Simon [36]. Recall that *coinductive hypothesis rule (CHR)* in co-SLD derivation states that during execution, the current call $p(\tilde{t})$ succeeds if it unifies with one of its ancestor calls $p(\tilde{t}')$. Our proof procedure, called *co-SLD derivation with the parity condition* (*co-SLD^{+p}* for short), simply incorporate checking the parity condition into CHR; for a path π from the the current call $p(\tilde{t})$ to its ancestor $p(\tilde{t}')$, we check whether $\max\{Inf(\pi)\}$ is even or not.

The above-mentioned simple modification allows us to handle non-stratified co-LPs, as the following proposition shows. To prove the completeness, we need to restrict atoms that have rational proof trees. A tree is *rational* ([5,6]) if the cardinality of the set of all its subtrees is finite.

Proposition 2. Correctness of co-SLD^{+p}
(Soundness) If an atom A has a successful co-SLD^{+p} in a Horn μ-program (P, Ω), then any ground instance of $E(A)$ is true in (P, Ω), where E is the resulting variable bindings for the derivation.
(Completeness) If $A \in [\![(P, \Omega)]\!]$ has a rational proof tree, then A has a successful co-SLD^{+p} in program a Horn μ-program (P, Ω).   □

It is easy to show that Horn μ-calculus is an extension of co-logic programming and that co-SLD^{+p} derivation extends co-SLD derivation. In fact, let P be a co-LP with a stratification σ. Then, we call a priority function Ω *consistent* with σ, if it satisfies the following: (i) $\sigma(p) \leq \sigma(q)$ iff $\Omega(p) \leq \Omega(q)$ for any predicates p and q, and (ii) $\Omega(p)$ is even (odd) if p is a coinductive (inductive) predicate, respectively. Then, we have the following:

Theorem 1. Let P be a co-logic program with a stratification σ, and Ω be a priority function consistent with σ. Then, the semantics of Horn μ-calculus coincides with that of co-logic programming: $[\![(P, \Omega)]\!] = M(P)$. Moreover, co-SLD^{+p} derivation in (P, Ω) coincides with co-SLD derivation in P.   □

4.3   An Alternative Approach to LPs with Negation

As an application of co-SLD^{+p}, we will utilize it for interpreters for logic programs with negation. In the following, we consider *normal* programs, where a clause γ is of the form: $p \leftarrow q_1, \ldots, q_m, not\ r_1, \ldots, not\ r_n$ $(m \geq 0, n \geq 0)$. Each of p and q_i $(1 \leq i \leq m)$ is a literal, each $not\ r_j$ $(1 \leq i \leq n)$ is a naf-literal (negation as failure or *default negation*). We call p the *head* of the clause γ, denoted by $hd(\gamma)$. A goal G is a conjunction consisting of positive literals and naf-literals, and we can assume without loss of generality that G consists of a single positive literal. For ease of explanation, we restrict ourselves to only propositional programs in what follows.

Let P be a program and G be a goal. Suppose that a program P^* is obtained from P by NE transformation. Then, we consider a Horn μ-program (P^*, Ω), where we define the priority Ω as $\Omega(not\_p) = 2$ and $\Omega(p) = 1$ for each predicate p appearing in P^*. We call (P^*, Ω) the *dual* Horn μ-program of P.

Our method is based on a "generate-and-filter" approach, i.e., first generating a candidate model by constructing a proof tree with the root G using co-SLD^{+p}, and then filtering it by checking some constraints on that model, depending on the syntax and the intended semantics of P. We consider the following three cases depending on whether a given program P is strict or not.

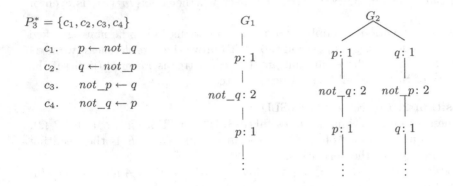

$P_3^* = \{c_1, c_2, c_3, c_4\}$

$c_1.$ $p \leftarrow not\_q$

$c_2.$ $q \leftarrow not\_p$

$c_3.$ $not\_p \leftarrow q$

$c_4.$ $not\_q \leftarrow p$

Fig. 2. The Proof Trees of G_1 in P_3^* (left) and G_2 in P_3^* (right) (Ex. 4): G_1 is defined by $G_1 \leftarrow p$, and G_2 is defined by $G_2 \leftarrow p, q$

1. Strict Answer Set Programs. First, we consider a simple case; we assume that P and G satisfy the *strictness* condition ([1,20]), which roughly says that no predicate depends on another predicate both positively and negatively. When p depends on q positively (negatively), we denote it by $p \geq_{+1} q$ ($p \geq_{-1} q$), respectively. We assume that $\geq_{+1}$ is reflexive, i.e., $p \geq_{+1} p$. Then, P is called *strict* iff we never have $p \geq_{+1} q$ and $p \geq_{-1} q$. We call P *strict with respect to G* iff for no predicate letter p, do we have both $G \geq_{+1} p$ and $G \geq_{-1} p$.

In this case, a model generated by co-SLD^{+p} is indeed a model of P, and no filtering is necessary as the following proposition shows:

Proposition 3. Correctness of Co-SLD^{+p} for Strict ASP
Let P be a strict answer set program and G a goal such that P is strict w.r.t. G. Let (P^*, Ω) be the dual Horn μ-program of P. Then,
(Soundness) If G has a successful co-SLD^{+p} in Horn μ-program (P^*, Ω), then G is true in an answer set of P.
(Completeness) If G is true in an answer set of P, then G has a successful co-SLD^{+p} in Horn-μ program (P^*, Ω). $\square$

In Fig. 2, the proof tree of G_1 in P_3^* is shown, where G_1 is defined by $G_1 \leftarrow p$. P_3 is strict, and P_3 is strict w.r.t. G_1. The proof tree accepts G_1, which corresponds to the fact that p is tree in the answer set $\{p\}$ of P_3.

2. Non-strict Answer Set Programs. Next, suppose that the strictness condition of P and G is not satisfied. In this case, after generating a candidate model by co-SLD^{+p}, we filter it by checking whether it satisfies two constraints described below.

One constraint, termed the *consistency constraint*, requires that, for each atom p, p and not_p do not appear in a proof tree. In Fig. 2, for example, the proof tree of G_2 in P_3^* is shown, where G_2 is defined by $G_2 \leftarrow p, q$. Then, P_3 is *not* strict w.r.t. G_2. Since the proof tree contains both p (q) and not_p (not_q), respectively, we filter the candidate model containing G_2, which implies that $AS(P_3) \not\models G_2$.

The other constraint will be necessary, since an non-strict answer set program will contain a well-known *odd loops over negation (OLON)* clause such as $\gamma = p \leftarrow q, not\ p$. γ imposes a constraint on candidate models that either the head of γ is true through other clauses, or the body of γ must be false. Such OLON clauses can be handled by using the method by Marple et al. [26]; this constraint will be represented as the following consistency checking clauses: $\{chk\_p \leftarrow p;\ chk\_p \leftarrow not\_q\}$. Then, for each head p_i $(n \geq 0)$, we introduce a new predicate chk_p_i defining the above-mentioned consistency checking clauses, and we add each chk_p_i to a given goal G, i.e., the goal is transformed to $G, chk\_p_1, \ldots, chk\_p_n$.

Marple et al. [26] have proposed a goal-directed procedure for executing ASP, by modifying co-SLD derivation so that a coinductive recursive call can succeed only if it is in the scope of at least one negation. We note that their condition exactly corresponds to our CHR with the parity condition in co-SLD^{+p}.

3. Co-SLD^{+p} for the WFS.
Finally, we consider how to use co-SLD^{+p} for the well-founded semantics (WFS). We introduce the following constraint, termed the *well-founded constraint*, for filtering a candidate model to obtain well-founded models.

Let $T_{P^*,G}$ be a proof tree with the root node G in P^* and $cls(T_{P^*,G})$ be the set of ground clauses used to construct $T_{P^*,G}$. Then, we say that $T_{P^*,G}$ satisfies the *well-founded constraint* if the following order $\geq$ among ground atoms exists:

$$p > q_i \text{ iff } p \leftarrow b_1, \ldots, b_m, not\_q_{m+1}, \ldots, not\_q_n \in cls(T_{P^*,G}), \text{ or}$$
$$not\_p \leftarrow q_1, \ldots, q_m, not\_b_{m+1}, \ldots, not\_b_n \in cls(T_{P^*,G}),$$

where $n \geq m \geq 0$, and $>$ is well-founded, i.e., there is no infinite sequence of ground atoms $p_1 > p_2 > \ldots$.

The idea of introducing the above ordering $\geq$ will be obvious, when we consider the *dynamic stratification* in an SLS-tree [32], where, when a negative ground literal not_p in a goal is selected, the dynamic stratum of p is strictly smaller than that of the goal of the SLS-tree.

In the proof tree of G_1 in P_3^* (Fig. 2), we have that $p > q > p > \ldots$, thus the well-founded constraint is violated. In this case, we assign the truth value **u** (undefined) to G_1, which corresponds to the fact that $v(G_3) = \mathbf{u}$, where v is the truth valuation in the well-founded semantics $WFS(P_3)$.

We omit here the proof of the correctness of the above approach; the proof is done by showing the correspondence between SLS-trees and the proof trees by co-SLD^{+p} using induction w.r.t. the dynamic stratum of a given goal.

5 Related Work and Concluding Remarks

Various logic programming-based techniques and tools have been developed for CTL and other temporal logic model checking (see, e.g., an excellent overview

[23] and the references therein). For example, techniques based on tabled resolution, CLP, constraint solving, abstract interpretation, and program transformation have been proposed for performing CTL model checking of finite and infinite state systems (see, e.g., [7,12,22,30,31,33]).

In the existing work based on the least-fixpoint semantics, CTL quantifiers with the greatest fixpoint characterization such as $A\tilde{U}$ and $E\tilde{U}$ are represented using double negation: for example, $AG\varphi = \mathbf{Afalse}\tilde{U}\varphi = \neg E\mathbf{true}U\neg\varphi = \neg EF\neg\varphi$. Such negative formulas will require an extra support such as *constructive negation*. In contrast, a CTL formula in our framework is in positive normal form: negation is applied only to atomic propositions (see Def. 1). Moreover, our encoding is a "pure" co-logic program, while an extra logical predicate such as `tfindall` is used in XMC [33].

Our approach in this paper is to study the relationship between co-LPs and the standard automata-based methods. In particular, we have proposed a general encoding schema of weak alternating automata (WAAs) into co-LPs; the resultant co-LP $P_{\mathcal{K}}$ for CTL model checking is only an instance of the schema. For example, DeVries et al. [8] have recently proposed a LTL interpreter written in co-LP. Such a co-LP for LTL model checker will be also derived via our encoding schema in Sect. 3, since Gastin and Oddoux [13] have shown a translation from LTL model checking to weak alternating Büchi automata.

Gupta et al. [17] have proposed an extension of co-LPs to handle non-stratified co-LPs by introducing *strong/weak_inductive* annotations. However, they have not discussed its relationship with the Horn μ-calculus, and its declarative semantics is not known. On the other hand, the Horn μ-calculus and its fragment, *the alternating-free Horn μ-calculus* by Talbot [37], allow nesting of least and greatest fixpoints similar to co-LPs, and they have been equipped with the procedural semantics as well as the nested fixpoints semantics. However, a *practical* top-down operational semantics like co-SLD resolution have not been provided. In this paper, we have proposed a proof procedure, co-SLD^{+p}, for non-stratified co-LPs, and have shown its correctness. Based on that, we have also shown that co-SLD^{+p} can be utilized as a top-down proof procedure for a class of normal logic programs.

One direction for future work is to extend the current co-logic programming framework to allow *generalized literals* (e.g., [15,37]), which have universal quantifications on the variables occurring in the body of clauses. For example, we can write an expression: $\forall S, S'(tr(S, S') \rightarrow sat(S', \varphi))$ in the body of a co-LP clause, which will then make more succinct the encoding program in Def. 4. Furthermore, although we have restricted ourselves to only propositional AS programs in this paper, it will be interesting to extend our approach to handle a more general class of AS programs (e.g., [28]).

Acknowledgments. The author would like to thank anonymous reviewers for their constructive and useful comments on the previous version of the paper. The idea of using co-LP techniques for a proof procedure for the WFS in Sect. 4.3 came from the discussions with Gopal Gupta at LOPSTR'13 in Madrid.

References

1. Apt, K.R., Blair, H.A., Walker, A.: Towards a theory of declarative knowledge. In: J. Minker, (ed.) Foundations of Deductive Databases and Logic Programming, pp. 89–148. Kaufmann (1988)
2. Charatonik, W., McAllester, D., Niwinski, D., Podelski, A., Walukiewicz, I.: The Horn Mu-calculus. In: Proc. LICS 1998, pp. 58–69 (1998)
3. Clarke, E.M., Grumberg, O., Peled, D.A.: Model Checking. MIT Press (1999)
4. Clarke, E.M., Jha, S., Lu, Y., Veit, H.: Tree-Like Counterexamples in Model Checking. In: Proc. LICS 2002, pp. 19–29 (2002)
5. Colmerauer, A., Prolog and Infinite Trees, Logic Programming, pp. 231–251. Academic Press (1982)
6. Courcelle, B.: Fundamental Properties of Infinite Trees. Theor. Comput. Sci. **25**(2), 95–169 (1983)
7. Delzanno, G., Podelski, A.: Model Checking in CLP. In: Cleaveland, W.R. (ed.) TACAS 1999. LNCS, vol. 1579, pp. 223–239. Springer, Heidelberg (1999)
8. DeVries, B.W., Gupta, G., Hamlen, K.W., Moore, S., Sridhar, M.: ActionScript bytecode verification with co-logic programming. In: Proc. ACM SIGPLAN PLAS 2009, pp. 9–15 (2009)
9. Emerson, E.A., Jutla, C.S.: Tree Automata, Mu-Calculus and Determinacy (Extended Abstract). FOCS 1991 **91**, 368–377 (1991)
10. Fages, F.: Consistency of Clark's Completion and Existence of Stable Models. J. Methods of Logic in Comput. Sci. **1**(1), 51–60 (1994)
11. Fitting, M.: A Kripke-Kleene Semantics for Logic Programs. J. Logic Programming **2**(4), 295–312 (1985)
12. Fioravanti, F., Pettorossi, A., Proietti, M.: Verification of Sets of Infinite State Processes Using Program Transformation. In: Pettorossi, A. (ed.) LOPSTR 2001. LNCS, vol. 2372, pp. 111–128. Springer, Heidelberg (2002)
13. Gastin, P., Oddoux, D.: Fast LTL to Büchi Automata Translation. In: Proc. the 13th Int'l. Conf. on Computer Aided Verification (CAV 2001), pp. 53–65 (2001)
14. Gelfond, M., Lifschitz, V.: The Stable Model Semantics for Logic Programming. In: Proc. Joint Int. Conf. and Symp. on Logic Programming, pp. 1070–1080 (1988)
15. Gottlob, G., Grädel, E., Veith, H.: Datalog LITE: a deductive query language with linear time model checking. ACM Trans. Comput. Logic **3**(1), 42–79 (2002)
16. Gupta, G., Bansal, A., Min, R., Simon, L., Mallya, A.: Coinductive logic programming and its applications. In: Dahl, V., Niemelä, I. (eds.) ICLP 2007. LNCS, vol. 4670, pp. 27–44. Springer, Heidelberg (2007)
17. Gupta, G., Saeedloei, N., DeVries, B., Min, R., Marple, K., Kluźniak, F.: Infinite computation, co-induction and computational logic. In: Corradini, A., Klin, B., Cîrstea, C. (eds.) CALCO 2011. LNCS, vol. 6859, pp. 40–54. Springer, Heidelberg (2011)
18. Jaffar, J., Stuckey, P.: Semantics of infinite tree logic programming. Theoretical Computer Science **46**, 141–158 (1986)
19. Kluźniak, F., Meta-interpreter supporting tabling and coinduction. http://www.utdallas.edu/~gupta/meta.html
20. Kunen, K.: Signed Data Dependencies in Logic Programs. J. Logic Programming **7**, 231–245 (1989)
21. Kupferman, O., Vardi, M.Y., Wolper, P.: An Automata-Theoretic Approach to Branching Time Model Checking. J. ACM **47**(2), 312–360 (2000)

22. Leuschel, M., Massart, T.: Infinite State Model Checking by Abstract Interpretation. In: Bossi, A. (ed.) LOPSTR 1999. LNCS, vol. 1817, pp. 62–81. Springer, Heidelberg (2000)
23. Leuschel, M.: Declarative programming for verification: lessons and outlook. In: Proc. ACM SIGPLAN PPDP 2008, pp. 1–7 (2008)
24. Lloyd, J.W.: Foundations of Logic Programming, 2nd edn. Springer (1987)
25. Marek, W., Subrahmanian, V.S.: The relationship between stable, supported, default and autoepistemic semantics for general logic programs. Theoretical Computer Science **103**, 365–386 (1992)
26. Marple, K., Bansal, A., Min, R., Gupta, G.: Goal-directed execution of answer set programs. In: Proc. PPDP 2012 ACM, pp. 35–44 (2012)
27. Min, R., Gupta, G.: Coinductive Logic Programming with Negation. In: De Schreye, D. (ed.) LOPSTR 2009. LNCS, vol. 6037, pp. 97–112. Springer, Heidelberg (2010)
28. Min, R., Predicate Answer Set Programming with Coinduction. PhD thesis, University of Texas at Dallas (2010)
29. Muller, D.E., Saoudi, A., Schupp, P.E.: Alternating automata, the weak monadic theory of the tree, and its complexity, In: Kott, L. (ed.) Automata, Languages and Programming. LNCS, vol. 226, pp. 275–283. Springer, Heidelberg (1986)
30. Nilsson, U., Lübcke, J.: Constraint Logic Programming for Local and Symbolic Model-Checking. In: Palamidessi, Catuscia, Moniz Pereira, Luís, Lloyd, John W., Dahl, Véronica, Furbach, Ulrich, Kerber, Manfred, Lau, K.-K., Sagiv, Yehoshua, Stuckey, Peter J. (eds.) CL 2000. LNCS (LNAI), vol. 1861, pp. 384–398. Springer, Heidelberg (2000)
31. Pettorossi, A., Proietti, M., Senni, V.: Deciding Full Branching Time Logic by Program Transformation. In: De Schreye, D. (ed.) LOPSTR 2009. LNCS, vol. 6037, pp. 5–21. Springer, Heidelberg (2010)
32. Przymusinski, T.C.: Every Logic Program Has a Natural Stratification and an Iterated Least Fixed Point Model. In: Proc. of the 8th ACM SIGACT-SIGMOD-SIGART Symp. on Principles of Database Systems (PODS 1989), pp. 11–21 (1989)
33. Ramakrishna, Y.S., Ramakrishnan, C.R., Ramakrishnan, I.V., Smolka, S.A., Swift, T., Warren, D.S.: Efficient Model Checking Using Tabled Resolution. In: Grumberg, O. (ed.) CAV 1997. LNCS, vol. 1254, pp. 143–145. Springer, Heidelberg (1997)
34. Seki, H.: Proving Properties of Co-logic Programs with Negation by Program Transformations. In: Albert, E. (ed.) LOPSTR 2012. LNCS, vol. 7844, pp. 213–227. Springer, Heidelberg (2013)
35. Simon, L., Mallya, A., Bansal, A., Gupta, G.: Coinductive Logic Programming. In: Etalle, S., Truszczyński, M. (eds.) ICLP 2006. LNCS, vol. 4079, pp. 330–345. Springer, Heidelberg (2006)
36. Simon, L.E., Extending Logic Programming with Coinduction, Ph.D. Dissertation, University of Texas at Dallas (2006)
37. Talbot, J.-M.: On the Alternation-Free Horn μ-Calculus. In: Parigot, M., Voronkov, A. (eds.) LPAR 2000. LNCS (LNAI), vol. 1955, pp. 418–435. Springer, Heidelberg (2000)

Towards the Implementation of a Source-to-Source Transformation Tool for CHR Operational Semantics

Ghada Fakhry$^{(\boxtimes)}$, Nada Sharaf, and Slim Abdennadher

Computer Science and Engineering Department,
The German University in Cairo, Cairo, Egypt
{ghada.fakhry,nada.hamed,slim.abdennadher}@guc.edu.eg

Abstract. Constraint Handling Rules (CHR) is a high-level committed-choice language based on multi-headed and guarded rules. Over the past decades, several extensions to CHR and variants of operational semantics were introduced. In this paper, we present a generic approach to simulate the execution of a set of different CHR operational semantics. The proposed approach uses source-to-source transformation to convert programs written under different CHR operational semantics into equivalent programs in the CHR refined operational semantics without the need to change the compiler or the runtime system.

Keywords: Source-to-Source Transformation · Constraint Handling Rules · Operational Semantics

1 Introduction

Constraint Handling Rules (CHR) [4] is a high level language that was introduced for writing constraint solvers. CHR is a committed choice language based on multi-headed and guarded rules. With CHR, users can have their own defined constraints. CHR transforms constraints into simpler ones until they are solved. Over the past decade, CHR has matured into a general purpose language. In addition, the number of CHR extensions and variants has increased [10]. These extensions have operational semantics different than the refined operational semantics (w_r) of CHR [3] regarding some properties like execution control, expressivity and declarativity.

Such extensions tackle some weaknesses and limitations of CHR and offer interesting properties to its users [10]. However, users cannot use these extensions directly through SWI-Prolog [13] since their operational semantics is different than the refined operational semantics supported by SWI-Prolog (w_r). Some extensions, nevertheless, provide transformation schemes to the refined operational semantics. However, such schemes usually require accessing the compiler and using additional low level tools [6,8,11,12] .

© Springer International Publishing Switzerland 2014
G. Gupta and R. Peña (Eds.): LOPSTR 2013, LNCS 8901, pp. 145–163, 2014.
DOI: 10.1007/978-3-319-14125-1_9

The work presented here extends the tool presented in [2] which enhanced CHR with visualization features through a source-to-source transformation approach without changing the compiler or the runtime system.

The aim of this work, on the other hand, is to introduce an approach that is able to automatically simulate the execution of a set of CHR operational semantics. Such operational semantics could have different execution models than the refined operational semantics. The proposed approach uses source-to-source transformation to convert CHR programs written under different operational semantics to equivalent programs that could be used with the refined operational semantics (w_r). This process does not require any changes to the compiler or the runtime system. The paper presents the general scheme that could be used with different operational semantics. In addition, the scheme is applied on a set of the existing CHR operational semantics. Although previous approaches provided transformation techniques, the focus here is on achieving a general approach that is usable without having to change any details regarding the runtime system. The presented work thus does not aim to provide a more efficient alternative but rather a more general one.

The paper is organized as follows. Section 2 briefly discusses the syntax and semantics of Constraint Handling Rules. Section 3 introduces the general transformation approach and the structure of the transformed file. In addition, it introduces the implementation of an explicit propagation history. Section 4 shows how the transformation approach is applied to implement a set of different CHR operational semantics. Section 5 provides an sketch proof for the equivalence between source programs and the transformed programs. Finally, we conclude with a summary and a discussion of future work in Section 6.

2   Constraint Handling Rules

This section introduces the syntax of CHR. In addition, we informally explain the abstract semantics and the refined operational semantics.

2.1   Syntax

CHR programs consist of a set of guarded rules that are applied until a fixed point is reached. In CHR two types of constraints are available. The first type is the built-in constraints provided through the host language. The second type of constraints is the CHR or user-defined constraints that are defined through the rules of a CHR program [5]. A CHR program consists of a set of a so-called "simpagation" rules in the following format:

$$H^k \setminus H^r \Leftrightarrow G \mid B.$$

The head of the CHR rule, which comes before the ($\Leftrightarrow$), consists of a conjunction of CHR constraints only. The elements of H^k are the constraints that are kept after the rule is executed. On the other hand, the constraints in H^r are removed after executing the rule. G is the optional guard that consists of built-in constraints. The body (B) could contain both CHR and built-in constraints. Two other types of rules exist. They are considered as special cases of simpagation rules. The first one occurs whenever H^k is empty. Such rules are called "simplification" rules. In this case, the head of the rule consists only of CHR constraints that should be removed on executing the rule. Such rules replace constraints by simpler ones. A simplification rule thus has the following format:

$$H^r \Leftrightarrow G \mid B.$$

The second rule type is propagation rules. In a propagation rule, H^r is empty. Consequently, all head-constraints are kept after the rule is executed adding the constraints in the body to the constraint store. This may cause further simplification afterwards. Propagation rules have the following format:

$$H^k \Rightarrow G \mid B.$$

2.2 Operational Semantics

The operational semantics of CHR programs is defined by a state transition system. The complete definition of state transitions of the abstract semantics and the refined operational semantics are introduced in [5] and [3] respectively. The execution of a CHR program starts from an initial state. Rules are applied until a final state is reached. A final state is a state where no more rules are applicable.

A rule is applied if the constraints in store matches the head-constraints of the rule, and the guard of the rule succeeds. The current implementation of CHR in SWI-Prolog does not allow binding variables in the guards of rules by default [13]. According to the type of the rule, the constraints that matched the head-constraints of the rule are either removed or kept after the rule application.

Consider the following example that computes the minimum of a multiset of numbers. The numbers are represented by the CHR constraint min/1 whose argument is the value of the number.

```
find_min @ min(N) \ min(M) <=> N=<M | true.
```

The initial query for the program is a multiset of constraints representing the numbers whose minimum is to be computed. Each time the simpagation rule (find_min) is applied, two numbers are compared and the larger one is removed. This rule is applied exhaustively until the constraint store contains one min constraint representing the minimum number.

The abstract CHR operational semantics does not specify the order in which the constraints of the initial goal are processed, or the order of the application of the rules. For the initial goal min(1), min(3), min(0), min(2), the result is always min(0). However, many execution paths could be taken. The computation can start with applying the rule on the constraints min(1) and min(3), or the constraints min(1) and min(0), or any other combination of two constraints matching the head of the rule. In addition, in case the goal constraints match the heads of more than one rule, then any of them could be applied. In other words, there is no restriction on the order of application of rules.

The refined operational semantics fixes, in part, the non-determinism of the abstract semantics. It fixes the order of processing the goal constraints and the order in which the rules are applied. In the refined operational semantics, goal constraints are processed from left to right, and rules are tried in the textual order of the program. Accordingly, for the same goal in the above example, the simpagation rule will be applied first with min(1) and min(3) constraints then it will continue with the remaining constraints.

In both semantics, there is a restriction for the application of propagation rules. A propagation rule is allowed to be applied only one time with the same combination of constraints that matched the head-constraints of the rule.

3   The Transformation Approach

This section introduces a new source-to-source transformation approach that is able to transform CHR programs written under different operational semantics into CHR programs that are equivalent when executed under the refined operational semantics.

Building on the representation used in [2], the rules of the source program are transformed into a so-called "relational normal form" introduced in [7]. This normal form uses special CHR constraints that represent the components of a rule. For example, the rule find_min in Section 2.2 is represented in the relational normal form as follows:

```
head(find_min,'min(N)',keep),
head(find_min,'min(M)',remove),
guard(find_min, 'N<=M'),
body(find_min,'true')
```

The CHR solver is first parsed. The parser extracts the information of the program and represents it in the normal form. The transformer is a CHR solver that runs on the relation normal formal of the source program and writes the new rules into the transformed program file.

The idea of the presented transformation approach relies on the execution model of the operational semantics. The work in this paper involves transforming different operational semantics that have different state transitions for rule choice and rule application. In addition, inverse execution of the rules of a program [14] is also considered. The transformation allows for the simulation of different rule-choice and rule-application state transitions of different operational semantics. This is done by separating rule matching and rule application into two steps. The basic idea is to delay the application of the body of the rule. With this approach, rather than having to apply the first matched rule, the new program is able to choose from the set of all the applicable rules. The adopted candidate set resolution approach is similar to the conflict resolution mode introduced in [5].

However, the presented work provides an automated transformation methodology that is able to combine the different properties of the execution models
of the operational semantics. As a result, the execution of different operational
semantics is possible. Two rules are generated for every rule in the source program. The first is a propagation rule that replaces the body by a new CHR
constraint representing the rule name and the constraints that matched the rule
head. The second rule applies the (possibly modified) body of the rule. The
choice of the rule to be applied depends on the candidate set resolution strategy. In the current implementation, the transformer provides different resolution
strategies simulating different rule-choice state transitions. The execution model
of the operational semantics was represented through a set of properties. Such
properties encode the execution direction, the candidate set resolution strategy
and whether multiple-rules matching is allowed. Such property-set is then used
to construct the transformed program.

Through specifying the properties of the execution model, the proposed approach is able to transform different operational semantics. The proposed transformation allows forward or inverse rule application as execution strategies. It
also offers a set of candidate set resolution strategies. In addition, at each computation step, a choice of single or multiple rules matching is possible.

Section 3.1 explains the transformation of the rules according to the properties of the execution model. Section 3.2 introduces the idea of implementing an
explicit propagation history in the transformed program.

3.1   The Transformed Program Structure

Figure 1 shows the steps of constructing the transformed program. The choice
between the different construction paths depends on properties of the execution
of the semantics. The construction of the transformed file is done in four steps
as explained below.

1. The first step adds for every CHR constraint $c(X)$ in the source program, a
   simplification rule (extend) in the transformed program P^T in the form:

   ```
   extend @ c(X) <=> c(X,_).
   ```

   This way, when executing the transformed program, an extended CHR constraint $c(X,V)$ is created for each CHR constraint $c(X)$ similar to the approach used in [8]. V is a fresh Prolog variable used as an explicit identifier for
   a constraint and is also used in the implementation of the propagation history explained in Section 3.2. In addition, all the constraints in the heads of
   the rules of the transformed file are extended with an additional argument.
   This argument represents the unique identifier of the constraint.
2. The second step adds, for each rule in the source program, a propagation
   rule that differs according to execution strategy of the operational semantics.
   One of two possible rules is added. The rule Fmatch is added in the case of
   forward execution. On the other hand, the rule Imatch is added whenever
   inverse execution is needed.

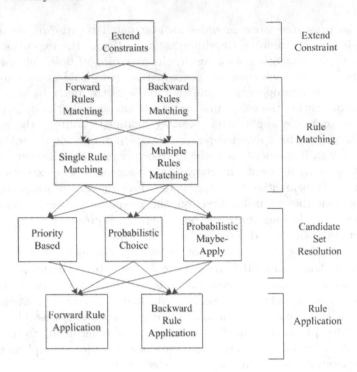

Fig. 1. Construction Steps of The Transformed File

Although both rules are propagation rules, they have different constituents. A forward match (**Fmatch**) rule has the same head-constraints and guard as the original rule of the source program rule. However, the head of an inverse match (**Imatch**) rule contains the kept head-constraints (if any) in addition to the CHR constraints of the body of the original rule. If the body of the source program rule contains built-in constraints then they are added to the guard of this new rule. The body of **Fmatch** and **Imatch** is a new CHR constraint **cand/3**. The arguments of this new constraint are the rule name, the list of identifiers of the head-constraints od the rule in addition to a number that could represent some specific property of the rule. For example, in the case of CHR with user-defined rule priorities, this number represents the priority of each rule. Since inverse execution of CHR rules is a one-to-many relationship, a **true** disjunct is added to allow backtracking for more than one result. Thus, for every CHR rule in the source program P:

$$\text{ri} @ H^k \backslash H^r \Leftrightarrow G \mid B.$$

we will have one of the following rules in the transformed program P^T:

$$\texttt{Fmatch-ri} @ H^r, H^k \Rightarrow G \mid cand(ri, Ids, p).$$
$$\texttt{Imatch-ri} @ H^k, B \Rightarrow G \mid cand(ri, Ids, p) \; ; \; true.$$

Whenever a rule is applicable, then a new constraint $\texttt{cand/3}$ is created for the applicable rule. This new constraint is added to the constraint store. The new cand/3 constraint means that the applicable rule could be fired with this specific combination of constraints. If the operational semantics allows multiple-rules matching at each computational step, an additional constraint $\texttt{id/1}$ is added to the head-constraints of the matching rules ($\texttt{Fmatch}$ or $\texttt{Imatch}$). Changing the value of the argument of $\texttt{id/1}$ allows the propagation rules to be matched with the same instances of constraints for multiple times. This behaviour is needed to reach the correct output. An example of this is the case of CHR with user-defined rule priorities, where the highest priority rule, among the applicable rules, is fired at each computational step. The rules that were not fired at one step due to the existence of higher priority rule(s), should be given a second chance of application. However, the match ($\texttt{Fmatch}$ or $\texttt{Imatch}$) rules are propagation rules that are fired for a specific combination of constraints once. Thus changing any argument of the constraints of the heads of such rules allow them to be fired again (i.e. giving the rest of the rules another chance). Since the original constraints cannot be manually modified, using the auxiliary constraint $\texttt{id/1}$ with an argument that changes with the rule application solved this problem.

In addition, a propagation rule ($\texttt{trigger}$) is added at the end of the matching rules. This new rule adds the CHR constraint $\texttt{start/0}$ to trigger the candidate set resolution step. The constraint $\texttt{trigger/0}$ is added to the end of the constraints in the original query to ensure that it is only activated at the end.

```
trigger @ trigger, id(Ni) ==> start.
```

On executing the transformed program, the result of this step is a set of all the applicable rules. Each of the candidate rules is represented by the constraint $\texttt{cand/3}$. In the refined operational semantics, head-constraints are searched from left to right. However, for simpagation rules, the head-constraints to be removed are tried before the constraints to be kept [5]. Thus to preserve the same order, removed head-constraints are added before the kept head-constraints in the $\texttt{Fmatch}$ transformed rules. $\texttt{p}$ is a property specified by the operational semantics. In the current implementation, the property is concatenated to the rule name instead of using the directive $\texttt{pragma}$ argument to give hints to the compiler.

3. In the third transformation step, rules are added to perform candidate set resolution. Only one of the candidate applicable rules is chosen to be applied according to the resolution strategy. In the current implementation of the transformer, this step is customized according to the respective CHR operational semantics. The current implementation allows probabilistic and priority based resolution strategies. More details about candidate set resolution is given in Section 4 with examples. The result of this step is the rule chosen to

be applied represented by a new CHR constraint `fire/2`. The first argument of `fire/2` is the rule name and the second argument is a list of identifiers of the head-constraints of the rule. The list of constraint identifiers is added to ensure that the rule will only be applied with this specific combination of constraints.

4. The last part in the transformed file is responsible of the actual rule application. A new rule is added for each rule in the original solver. The new rule is chosen according to the execution strategy. Thus either a forward application (`Fapply`) or an inverse application (`Iapply`) rule is added to the transformed program. In `Fapply` rules, a new constraint `fire/2` is added to the head-constraints to be removed of the source program rule (if any). This way ensures that only this rule will be applied with the specific combination of constraints in the list. The body of the rule remains unchanged. In `Iapply` rules, the `fire/2` constraint and the body of the original rule are added as head-constraints to be removed. The body of the `Iapply` rule contains the head-constraints that should be removed.

For every CHR rule in the source program P:

$$\texttt{ri } @H^k \backslash H^r \Leftrightarrow G \mid B.$$

we will have one of the following two rules in the transformed program P^T:

$$\texttt{Fapply-ri } @ H^k \backslash \texttt{ fire(ri,Ids) }, H^r \Leftrightarrow B.$$
$$\texttt{Iapply-ri } @ H^k \backslash \texttt{ fire(ri,Ids) }, B \Leftrightarrow H^r.$$

The rules are written such that the execution of one rule in the source program is done in 4 steps in the transformed program The execution of the transformed program starts by extending the original query constraints. The extended constraints (constraints with the additional identifier argument) then try to match the propagation rules in the second part of the transformed file. Among the set of the applicable rules represented by `cand/3` constraints, one rule is chosen according to the candidate set resolution strategy. Finally, the chosen rule is applied. Execution then proceeds by extending any constraints added after applying the chosen rule. The new constraints then try to match the propagation rules. Candidate set resolution is applied afterwards on the new candidate-set and so on until reaching a fixed point where no more rules are applicable. The new file depends on the textual order of the rules since it runs using the refined operational semantics implemented in SWI Prolog.

3.2 Propagation History Implementation

In some of the cases, the set of matched rules that were not applied at one computation step are given a second chance in the next computation step. This is the case when the operational semantics allows multiple-rules matching at each computation step and one rule application. In order to allow for multiple-rules matching, the constraint `id/1` is added to the head-constraints of the transformed rules.

However, this approach raised a problem with propagation rules. If a propagation rule was chosen to be fired then in the next computation step the same propagation rule will also be applicable, because the constraints that matched the head were not removed. This causes a problem of trivial non-termination in the transformed program.

This problem would be solved if every propagation rule is fired only once for each specific combination of constraint identifiers. This was done by implementing an explicit propagation history. In the proposed approach, a new constraint `history/1` is added to the original query. The argument of `history/1` is a list that contains a set of tuples `(r,I)`. Initially, the list is empty. The first argument `(r)` is the propagation rule name while the second argument `(I)` is an ordered list of the identifiers of constraints that matched the head-constraints of rule `r`. The size of the propagation history depends on the propagation rules in the program. At any computation step the length of the list of `history/1` corresponds to the number of fired propagation rules.

In the transformed program, propagation rules are modified to be applied only if the tuple containing the rule name and the list of identifiers of constraints does not exist in the propagation history, since each constraint in the executed program has a unique identifier. For example, the following rule in the original program P:

```
rule1 @ a(X) ==> b(X).
```

will have the corresponding propagation rule transformed program P^T:

```
match-rule1 @ a(X,Id1), history(L) ==> \+member((rule1,[Id1]),L)
                                      | cand(rule1,[Id1],1).
```

Moreover, in this example, the property specified to the operational semantics is the rule order. Thus for the rule `rule1`, this property is set to 1 since the program contains only one rule.

In addition, if a propagation rule is chosen to be applied, a new tuple with the name of the rule and a list with the identifiers of the matched constraints is added to the propagation history. Thus, the propagation rules are modified in the rule application part to update the propagation history. The rule in the previous example generates the following simpagation rule in the transformed program P^T:

```
apply-rule1 @ a(X,Id1) \ fire(rule1,[Id1]), history(L)
                       <=> b(X),history([(rule1,[Id1])|L]).
```

4   Source-to-Source Transformation for Different CHR Operational Semantics

This section shows, through examples, how the presented transformation approach is applied to a set of different CHR operational semantics. Transformation for Probabilistic Constraint Handling Rules is explained in Section 4.1, CHR with user defined-rule priorities is introduced in Section 4.2. Section 4.3 introduces transformation for CHRiSM. Finally, transformation for inverse CHR is introduced in Section 4.4.

4.1   Transformation for Probabilistic Constraint Handling Rules

Probabilistic Constraint Handling Rules (PCHR) [6] is an extension of CHR that allows for a probabilistic rule choice among the applicable rules. The choice of the rule is performed randomly by taking into account the relative probability associated with each rule. PCHR modifies the CHR abstract semantics (w_t) in the "Apply" transition by specifying the probability of the choices of the rules. This results in an explicit control of the chance that certain rules are applied according to their probabilities. The "Apply" transition of w_t chooses a rule from the program for execution. Constraints matching the head of the rule should exist in the store. In addition, the guard should be satisfied. PCHR rules are the same as CHR rules but with the addition of a number representing the relative probability of each rule.

PCHR is implemented using the proposed approach with forward execution strategy, multiple-rules matching. Candidate set resolution is done through a random choice after normalizing the probabilities of the probabilistic rules. The following example shows a PCHR program [6] that generates a n bit(s) random number. The number is represented as binary list of n bit(s). The list is generated bit by bit recursively and randomly. As long as N is greater than zero, the next bit will be either 0 or 1 by applying either the second or the third rules with equal probability; otherwise the non-probabilistic rule r1 will be applied and the recursion ends. The program is:

```
r1 @ rand(N,L) <=> N =:= 0 | L = [].
r2_50 @ rand(N,L) <=> N>0 | L=[0|L1], N1 is N-1, rand(N1,L1).
r3_50 @ rand(N,L) <=> N>0 | L=[1|L1], N1 is N-1, rand(N1,L1).
```

The transformation will result in the following program:

```
extend @ rand(V2,V1) <=> rand(V2,V1,_).

r1 @ rand(N,L,Id0) <=> N =:= 0 | L = [].
match-r2_50 @ rand(N,L,Id0),id(Ni)==>N>0 | cand(r2_50,[Id0],50).
match-r3_50 @ rand(N,L,Id0),id(Ni)==>N>0 | cand(r3_50,[Id0],50).
trigger @ trigger, id(Ni) ==>   start.

start     @ cand(R,IDs,N),start <=> random(0,100,Random),
                                    cand(R,IDs,0,N,N,Random).
normalize @ cand(R,IDs,N,M,UB,Random),cand(R1,IDs2,N1)
                           <=> M2 is M+N1,UB2 is UB+N1,
                               cand(R1,IDs2,M,M2,UB2,Random),
                               cand(R,IDs,N,M,UB2,Random).
drop     @ id(Ni)\ cand(R,IDs,M,M1,100,Random) <=> Random<M | true.
```

```
drop    @ id(Ni)\ cand(R,IDs,M,M1,100,Random) <=> Random>=M1| true.
choose @ cand(R,IDs,M,M1,100,Random), id(Ni)
                            <=> M=<Random,Random<M1
                            |fire(R,IDs),Ni2 is Ni+1,id(Ni2).
apply-r2_50 @ fire(r2_50,[Id0]),rand(N,L,Id0)
                  <=> N>0 | L=[0|L1] , rand(N-1,L1).
apply-r3_50 @ fire(r3_50,[Id0]),rand(N,L,Id0)
                  <=> N>0 | L=[1|L1] , rand(N-1,L1).
```

The rule **start** triggers the probability normalization by replacing the first candidate rule **cand/3** by **cand/6**. The additional arguments are a list of the constraint identifiers that were matched in the head of the rule, the lower bound and the upper bound of the rule probability interval. In addition, the last argument is a random number in the interval from 0 to the sum of all rule probabilities calculated by the built-in Prolog predicate **random/3**. The rule **normalize** keeps on replacing the rest of the candidate rules represented through **cand/3** constraints by the extended constraint **cand/6**, each with the lower and upper bound interval of the corresponding probability of the rule. The arguments of **fire/2** are the rule name and the list of constraints identifiers that were matched in the head. Otherwise, the **cand/6** constraint is replaced by true by the rules **drop** which means that this rule will not be applied.

4.2 Transformation for Constraint Handling Rules with User-Defined Rule Priorities

CHR^{r^p} extends CHR with user-defined rule priorities [8]. Rule priorities improve the expressivity of CHR as they allow for a different choice for rule application depending on the respective rule priority, resulting in a more flexible execution control. The operational semantics w_p for CHR^{r^p} only adds restrictions to the applicability of the "Apply" transition of the abstract CHR semantics w_t. The rest of state transitions are equivalent in both semantics. For CHR^{r^p} programs in which all rule priorities are equal, every execution strategy under w_t is consistent with w_p. Thus, such programs can be executed using the refined operational semantics as implemented by the current CHR implementations.

In [8], a source-to-source transformation approach that uses some of the compiler directives is presented. In [8], constraints are not activated when introduced to the store by the default transitions of the refined operational semantics, which are the "Activate" and "Reactivate" transitions [3]. Instead, they remain passive using the compiler directive **passive/1** and are scheduled for activation with the corresponding rule priority. After trying all the possible matching rules with the constraints in the query, the highest priority scheduled constraint is activated.

CHR^{r^p} is implemented using the proposed approach with forward execution, multiple-rules matching and a rule priority candidate set resolution strategy. The following example [8] shows the difference in the execution of the refined operational semantics and CHR^{r^p}. For the same initial query a, the refined operational semantics will apply the rules in the following order: 1,2,4,3. While in CHR^{r^p}, rule 3 has higher priority than rule 4. Therefore, the rules will be applied in the following order: 1,2,3. Rule 4 will not be applied anymore because constraint a is removed by rule 3. The same example is used to illustrate the proposed transformation approach.

```
r1_1 @ a ==> print('rule 1 \n'),b .
r2_2 @ a , b ==> print('rule 2 \n').
r3_3 @ a <=> print('rule 3 \n').
r4_4 @ a , b ==> print('rule 4 \n').
```

The transformation will result in the following program:

```
extend @ a <=> a(_).
extend @ b <=> b(_).

match-r1_1 @ a(Id0),id(Ni),history(L)
            ==> \+ member((r1_1,[Id0]),L)|cand(r1_1,[Id0],1).
match-r2_2 @ a(Id0),b(Id1),id(Ni),history(L)
            ==> \+ member((r2_2,[Id0,Id1]),L)|cand(r2_2,[Id0,Id1],2).
match-r3_3 @ a(Id0),id(Ni) ==> cand(r3_3,[Id0],3).
match-r4_4 @ a(Id0),b(Id1),id(Ni),history(L)
            ==> \+ member((r4_4,[Id0,Id1]),L)|cand(r4_4,[Id0,Id1],4).
trigger @ trigger,id(Ni) ==>  start.

start    @ start  <=> candList([]).
collect @ candList(L),cand(R,IDs,N) <=> candList([(N,R,IDs)|L]).
choose  @ candList(L),id(Ni) <=> sort(L,[(P,H,IDs)|T]),fire(H,IDs),
                                 N2 is Ni+1,id(N2).

apply-r1_1 @ a(Id0)\ fire(r1_1,[Id0]),history(L)
            <=> print('rule 1'), b, history([(r1_1,[Id0])|L]).
apply-r2_2 @ a(Id0),b(Id1)\ fire(r2_2,[Id0,Id1]),history(L)
            <=> print('rule 2'),history([(r2_2,[Id0,Id1])|L]).
apply-r3_3 @  fire(r3_3,[Id0]),a(Id0) <=> print('rule 3').
apply-r4_4 @ a(Id0),b(Id1) \ fire(r4_4,[Id0,Id1]), history(L)
            <=> print('rule 4'),history([(r4_4,[Id0,Id1])|L]).
```

In the transformed program, the rule with the highest priority among the set of applicable rules is chosen to be applied. The three rules start, collect, and choose are added to the transformed program to perform the candidate set resolution according to the priorities of the rules. The rule start initalizes an empty priority list such that the applicable rules represented by cand/3 are added to this list by the collect rule. After all cand/3 constraints are added to the list, the rule choose sorts the list and the rule with the highest priority is chosen for application. The chosen rule is represented by a new CHR constraint fire/2, whose first argument is the rule name and the second argument is a list of identifiers of the head-constraints of the rule. The list of constraint identifiers is added to ensure that the rule will only be applied with the specific combination of constraints. In addition, the argument of constraint is incremented to allow match rules to be tried again.

4.3   Transformation for CHRiSM

CHRiSM is a probabilistic extension of CHR that is based on CHR and PRISM [9]. The main difference between the semantics of CHRiSM and PCHR is that the rule probabilities have a localized meaning. The probability of a rule application does not depend on the other applicable rules. CHRiSM semantics adds two features to CHR [9]. First, a defined probability for the entire rule application given by the "Maybe-Apply" state transition. In the "Maybe-Apply" transition, according to the probability of the rule, it is either applied or not. However, if the rule is not applied, the propagation history is updated to prevent further rule application with the same combination of constraints that matched the rule head. The second feature is the ability to define a probability for each disjunct in the rule body in $CHR^{\vee}$ [1] given by the "Probabilistic Choice" transition. In "Probabilistic Choice", one disjunct is chosen probabilistically according to its probability relative to the other disjuncts.

In this paper, the transformation for CHRiSM implements programs with a user-defined rule probability for the entire rule application, corresponding to the "Maybe Apply" transition only. In CHRiSM operational semantics, a rule with a probability p means that whenever the rule is applicable, it will only be applied with a probability p. If the rule probability is not defined, it is set to uniform distribution 0.5. CHRiSM is implemented using the proposed approach with forward execution, single-rule matching and a probabilistic rule application choice.

The following example illustrates the transformation of CHRiSM to the refined operational semantics (w_r). Starting with initial query a, it is probable that the first rule is applied. If the first rule is applied, then the constraint b will be added to the constraints store. Consequently, there is a chance to apply the second rule with probability 0.5, removing constraint a and adding c to the constraint store. The program is:

```
r1_50 @ a ==> b .
r2_50 @ b \ a <=> c .
```

The transformation will result in the following program:

```
extend @ a <=> a(_).
extend @ b <=> b(_).
extend @ c <=> c(_).
match-r1_50 @ a(Id0)==>cand(r1_50,[Id0],50).
match-r2_50 @ a(Id0),b(Id1)==>cand(r2_50,[Id0,Id1],50).

start @ cand(R,IDs,N) <=> random(0,100,Random),cand(R,IDs,N,Random).
choose-apply  @ cand(R,IDs,M,Random) <=> Random=<M | fire(R,IDs,1).
choose-ignore @ cand(R,IDs,M,Random) <=> M<Random | fire(R,IDs,2).

apply-r1_50 @ a(Id0)\ fire(r1_50,[Id0],1) <=>b.
apply-r2_50 @ b(Id1)\ fire(r2_50,[Id0,Id1],1),a(Id0) <=> c.
```

```
ignore-r2_50 @ a(Id0),b(Id1)\ fire(r2_50,[Id0,Id1],2)<=> true.
ignore-r1_50 @ a(Id0)\ fire(r1_50,[Id0],2)<=> true.
```

In CHRiSM, each rule is given one chance for application with every combination of constraints. In order to achieve that, the constraint id(Ni) is not added to the head-constraints in the match rules similar to PCHR and CHR^{rp}. In addition, since the choice is whether to apply the rule or not, there is only one candidate rule at each computation step, therefore no need to add the rule trigger.

Whenever a rule is applicable, the cand/3 constraint of the applicable rule will fire the rule start in the candidate set resolution rules. Similar to "Maybe Apply" transition in the CHRiSM operational semantics [9], the body of the rule gets applied with a probability P. The rule start generates a random number between 0 and 1 and replaces cand/3 with cand/4. The additional argument is the randomly generated number. The rules choose-apply and choose-ignore determine whether the rule will be applied according to the randomly generated number. In both cases, the cand/4 constraint is replaced by fire/3 constraint. If the randomly generated number is less than the rule probability, the third argument in fire/3 is set to 1, otherwise it is set to 2.

For simplification and simpagation rules, if the rule is chosen not to be applied then the removed head-constraints should not be removed from the store. In order to keep the same instances of removed head-constraints in store when the probabilistic rule is not applied, each rule in the source program will have an additional ignore rule in the transformed file. The rule ignore is a simpagation rule where the head-constraints are added as kept head-constraints. Only the fire/3 constraint with the last argument set to 2 is to be removed. In addition, the body of the rule is replaced by true.

4.4 Transformation for Inverse Constraint Handling Rules

The execution of traditional CHR starts from the initial state and applies program rules until reaching a fixed point or a final state where no more rules are applicable. Inverse execution of CHR rules starts from a state and applies the inverse of program rules in order to reach the initial state. The "Apply" transition of the inverse CHR is the same as "Apply" transition of the abstract semantics of CHR but with exchanging the left and right hand side states of the transition [14]. Inverse CHR is implemented using the proposed transformation approach with inverse execution of rules, multiple-rules matching and rule priority candidate set resolution, where the rule priority is the textual rule order. Thus, the first rule in the program has the highest priority. However, different resolution strategies could be used. The following example [14] is an exchange source for elements in a list. Elements are represented by constraint a/2, the first argument is the index of the element in the list and the second argument is the value of the element.

```
eSort @ a(I,V),a(J,W) <=> I>J , V<W | a(I,W),a(J,V).
```

The transformation will result in the following program:

```
extend @ a(V2,V1) <=> a(V2,V1,_).
Imatch-eSort @ a(I,W,Id0),a(J,V,Id1),id(Ni)
                        ==> I>J,V<W | cand(eSort,[Id0,Id1],1) ; true.
trigger @ trigger, id(Ni) ==>  start.
start   @ start <=> candList([]).
collect @ candList(L),cand(R,IDs,N) <=> candList([(N,R,IDs)|L]).
choose  @ candList(L),id(Ni) <=> sort(L,[(P,H,IDs)|T]),fire(H,IDs),
                        N2 is Ni+1,id(N2) ; true.
Iapply-eSort @ fire(eSort,[Id0,Id1]),a(I,W,Id0),a(J,V,Id1)
                        <=> I>J,V<W | a(I,V) , a(J,W).
```

The current implementation of the transformer does not distinguish between user-defined and built-in constraints in reverse execution of programs. Accordingly, the transformation is limited to programs with rules whose body contain user-defined constraints only.

5   Equivalence Proof

In this section, we will show how the newly transformed file program is able to capture the needed operational semantics. In other words, we will introduce how the execution of the rules in the new solver is equivalent to the corresponding semantics. For proof of concept, we will show the equivalence of the execution of the transformed program under w_t [5] with the execution of the original program under w_p for CHR^{rp} [8].

A state in w_t [5] is a tuple in the form $\langle G, S, B, T \rangle_n$. The components of the state are defined as follows: The goal G is a multiset of all unprocessed constraints. The CHR store S is a set of numbered user-defined constraints that can be matched with rules in a given program. B is the built-in constraints store. It is the conjunction of built-in constraints that have been added to the built in constraint store. The propagation history T is a set of tuples (r, I), where r is the rule name and I is a list of identifiers of constraints that were matched in the rule head. Finally, n is a counter representing the next free integer for constraint identifying. For the sake of brevity, only G and S are shown in the proof [5].

w_r [3] provides a deterministic execution strategy for any goal. Therefore, for any program P and an initial state S, every derivation for P, $S \xmapsto[P]{w_r}{}^* S'$ corresponds to a derivation $S \xmapsto[P]{w_t}{}^* S'$. The provided sketch proof is based on the mapping between w_t and w_r.

Theorem 1. *Given a CHR program P and its corresponding transformed program $T(P)$ and two states $S_1 = \langle G, \phi \rangle$ and $S_2 = \langle G \bigcup Aux, \phi \rangle$ where G contains the initial goal constraints and Aux is a set of auxiliary constraints. Then the following holds:*

If $S_1 \xmapsto[P]{w_p}{}^* S_1'$ and $S_2 \xmapsto[T(P)]{w_t}{}^* S_1' \cup Aux'$ then $T(P)$ is equivalent to P.

Proof. (Sketch)

Table 1 shows the computational steps of executing one rule of the program P under w_p starting with S_1. On the other hand, tables 2 to 5 show the computational steps of executing the transformed program $T(P)$ under w_t starting with S_2.

Table 1. Computation steps under w_p

	$\langle G_1, \phi \rangle$	G contains initial query constraints
$\xrightarrow[\text{introduce}]{\quad}{}^*$	$\langle \phi, S \rangle$	G The store S contains all the activated goal constraints and the goal G is empty
$\xrightarrow[\text{apply}]{\quad}{}^*$	$\langle G, S \rangle$	G contains the added constraints after the rule application (if any)

Table 2. Computation of step 1 (Constraints Extending)

	$\langle G \bigcup Aux, \phi \rangle$	*Aux* contains **trigger** and **id(1)** constraints, G contains initial query constraints
$\xrightarrow[\text{introduce}]{\quad}{}^*$	$\langle \phi, S \rangle$	The store S contains all the activated goal constraints and the auxiliary constraints
$\xrightarrow[\text{apply extend}]{\quad}{}^*$	$\langle G', S \rangle$	G' contains extended constraints after applying the **extend** rules on the initial query constraints
$\xrightarrow[\text{introduce}]{\quad}{}^*$	$\langle \phi, S \rangle$	The store S contains all the activated extended constraints in addition to the auxiliary constraints

Table 3. Computation of step 2 (Rule Matching)

$\xrightarrow[\text{apply match}]{\quad}{}^*$	$\langle G'', S \rangle$	G'' contains **cand/3** constraints after applying the **match** rules
$\xrightarrow[\text{introduce}]{\quad}{}^*$	$\langle \phi, S \rangle$	S contains all the activated **cand/3** constraints
$\xrightarrow[\text{apply trigger}]{\quad}$	$\langle \{start\}, S \rangle$	The goal contains only one constraint(**start**) since it is the only applicable rule
$\xrightarrow[\text{introduce}]{\quad}$	$\langle \phi, S \rangle$	S contains all the activated **cand/3** constraints in addition to **start** constraint

A rule in w_p is fired through the "apply" transition. However, the "apply" transition is applicable only to a state with an empty goal. This transition also ensures that the highest priority rule among the set of applicable rules is the one fired [8].

Table 4. Computation of step 3 (Candidate Set Resolution)

$\xrightarrow{\text{apply start}}$	$\langle\{\texttt{candList([])}\}, S\rangle$	
$\xrightarrow{\text{introduce}}$	$\langle\phi, S\rangle$	
$\xrightarrow{\text{apply collect}}$	$\langle\{\texttt{candList(L)}\}, S\rangle$	
$\xrightarrow{\text{introduce}}$	$\langle\phi, S\rangle$	This computation step and the one above are repeated till no more $\texttt{cand/3}$ constraints are available in store
$\xrightarrow{\text{apply choose}}$	$\langle\{\texttt{fire(R,IDs)},\texttt{id(N)}\}, S\rangle$	
$\xrightarrow{\text{introduce}}$	$\langle\phi, S\rangle$	The store S at this step contains only one $\texttt{fire/2}$ constraint

Table 5. Computation of step 4 (Rule Application)

$\xrightarrow{\text{apply apply}}$	$\langle G, S\rangle$	The goal G contains the body of the fired rule, and the store S contains the rest of the activated constraints after applying the chosen rule

As shown in the last step of each of table 3 and 4, the result step contains an empty goal. Thus before firing any rule in table 5, the state contains an empty goal. Table 5 shows the rule application step. Therefore, the transformed program only fires the rule when the goal of the state is empty.

In Section 4.1, we showed how the rule matching step finds the set of all applicable rules. The set of rules of Table 3 on the other hand chooses the highest priority rule among this set using the built-in constraint $\texttt{sort/2}$.

The only difference between the goal of S_1 and $S2$ is the set of auxiliary constraints. As shown, the conditions required by w_p to fire a rule in P are the same as the conditions required by w_t to fire a rule in $T(P)$. Consequently, the same rules will be fired in both programs with the same order. Thus, both derivations add the same constraints to the final store. Thus, at the end of both derivations the only difference in the result states $S_1{}'$ and $S_1{}' \cup Aux'$ is auxiliary constraints. Accordingly, omitting the auxiliary constraints from $S_1{}' \cup Aux'$ will result in $S_1{}'$. Hence, we proved that the transformed program $T(P)$ is equivalent to the source program P.

□

6 Conclusion

This paper introduced a source-to-source transformation approach to implement a set of CHR operational semantics that have a different execution model than the refined operational semantics. The source programs written in different operational semantics are transformed into equivalent programs written under the refined operational semantics. Moreover, the execution of the transformed program does not need accessing the compiler or changing the runtime environment.

The transformation approach allows a different rule application choice when there is more than one applicable rule compared to the top-down program order of the refined operational semantics. Moreover, it allows forward and inverse execution of CHR programs. A sketch proof is provided to show the equivalence between the transformed programs and the source programs.

For future work, we intend to extend the transformation approach to implement a larger set of CHR operational semantics by incorporating additional properties to the current model. In addition, we intend to investigate the result of combining the properties of the execution model of different operational semantics such as combining inverse CHR and CHR^{r^p}.

References

1. Abdennadher, S., Schütz, H.: CHR$^\vee$: A Flexible Query Language. In: Andreasen, T., Christiansen, H., Larsen, H.L. (eds.) FQAS 1998. LNCS (LNAI), vol. 1495, pp. 1–14. Springer, Heidelberg (1998)
2. Abdennadher, S., Sharaf, N.: Visualization of CHR through Source-to-Source Transformation. In: Dovier, A., Costa, V.S. (eds.) Technical Communications of the 28th International Conference on Logic Programming (ICLP 2012). Leibniz International Proceedings in Informatics (LIPIcs), vol. 17, Dagstuhl, Germany, pp. 109–118. Schloss Dagstuhl-Leibniz-Zentrum fuer Informatik (2012)
3. Duck, G.J., Stuckey, P.J., Garcia de la Banda, M., Holzbaur, C.: The Refined Operational Semantics of Constraint Handling Rules. In: Demoen, B., Lifschitz, V. (eds.) ICLP 2004. LNCS, vol. 3132, pp. 90–104. Springer, Heidelberg (2004)
4. Frühwirth, T.: Theory and Practice of Constraint Handling Rules, Special Issueon Constraint Logic Programming . The Journal of Logic Programming 37(13), 95–138 (1998)
5. Frühwirth, T.: Constraint Handling Rules. Cambridge University Press (August 2009)
6. Frühwirth, T., Di Pierro, A., Wiklicky, H.: Probabilistic ConstraintHandling Rules. In: Comini, M., Falaschi, M. (eds.) WFLP 2002: Proc. 11th Intl. Workshop on Functional and (Constraint) Logic Programming, Selected Papers, vol. 76 (June 2002)
7. Frühwirth, T.W., Holzbaur, C.: Source-to-Source Transformation for a Class of Expressive Rules. In: Buccafurri, F. (eds.) APPIA-GULP-PRODE, pp. 386–397 (2003)
8. De Koninck, L., Schrijvers, T., Demoen, B.: User-definable rule priorities for chr. In: Leuschel, M., Podelski, A. (eds.) PPDP, pp. 25–36. ACM (2007)
9. Sneyers, J., Meert, W., Vennekens, J.: CHRiSM: CHance Rules induce Statistical Models. In: Proceedings of the Sixth International Workshop on Constraint Handling Rules, pp. 62–76 (2009)
10. Sneyers, J., Van Weert, P., Schrijvers, T., De Koninck, L.: As Time Goes By: Constraint Handling Rules - A Survey of CHR Research between 1998 and 2007 10(1), 1–47 (2010)

11. Sneyers, J., Van Weert, P., Schrijvers, T., Demoen, B.: Aggregates in Constraint Handling Rules. In: Dahl, V., Niemelä, I. (eds.) ICLP 2007. LNCS, vol. 4670, pp. 446–448. Springer, Heidelberg (2007)
12. Van Weert, P., Sneyers, J., Schrijvers, T., Demoen, B.: Extending CHR with Negation as Absence. Technical report CW 452, 125–140 (July 2006)
13. Wielemaker, J., Frühwirth, T., De Koninck, L., Triska, M., Uneson, M.: SWI Prolog Reference Manual 6.2.2. (September 2012)
14. Zaki, A., W. Frühwirth, T., Abdennadher, S.: Towards inverse execution of constraint handling rules. TPLP, 13 (4-5-Online-Supplement) (2013)

A Logical Encoding of Timed π-Calculus

Neda Saeedloei[✉]

Department of Computer Science, University of Minnesota Duluth, Duluth, USA
nsaeedlo@d.umn.edu

Abstract. We develop a logical encoding of the operational semantics of timed π-calculus: a real-time extension of Milner's π-calculus. This executable encoding is based on *Horn logical semantics* of programming languages and directly leads to an implementation for timed π-calculus. This implementation can be used for modeling and verification of real-time systems and cyber-physical.

1 Introduction

In previous work [17], we extended the π-calculus [14] with *real* time by adding clocks and assigning time-stamps to actions. The resulting formalism, timed π-calculus, provides a simple and novel way to annotate transition rules of π-calculus with timing constraints. The timed π-calculus provides a framework for describing systems whose components interact with each other under time constraints. It contains an algebraic language for describing processes in terms of the communication actions they can perform. The timed π-calculus can model mobility, concurrency and message exchange between processes as well as infinite computation (through the infinite replication operator '!'), while taking into account the time constraints imposed on the actions. Therefore, it is suitable for modeling real-time systems and cyber-physical systems (CPS) [7,11] and support reasoning about their behavior related to time.

Our extension of π-calculus with time unlike most of other approaches [2,4,5,12], represents time faithfully as a continuous quantity: in other words, *it does not discretize time*. Discretizing means that time is represented through finite time intervals. As a result, *infinitesimally small time intervals cannot be represented or reasoned about in these approaches*. In practical real-time systems, e.g., a nuclear reactor, two or more events *can* occur within an infinitesimally small interval. Discretizing time can miss the modeling of such behavior which may be wholly contained within this infinitesimally small interval. Some other approaches for extending π-calculus with time e.g., the work of Chen [3] miss out the replication operator of the original π-calculus. Therefore, they are unable to model infinite processes. In our approach the infinite behavior of processes is modeled through the infinite replication operator '!'.

We also developed an operational semantics as well as a notion of timed bisimilarity for the timed π-calculus and we investigated the properties of timed bisimilarity; in particular, expansion theorem for real-time, concurrent, mobile processes [17].

© Springer International Publishing Switzerland 2014
G. Gupta and R. Peña (Eds.): LOPSTR 2013, LNCS 8901, pp. 164–182, 2014.
DOI: 10.1007/978-3-319-14125-1_10

In this paper, we show how an executable operational semantics of timed π-calculus can be elegantly realized through coinductive constraint logic programming extended with coroutining. In our implementation of timed π-calculus concurrency is modeled by *coroutining*, and (rational) infinite computation in presence of constraints by *coinductive constraint logic programming over reals* (Co-CLP) [16]. The executable semantics faithfully captures real-time behaviors and allows us to prove behavioral and timing properties of a system modeled in timed π-calculus.

The work of Gupta et al. [21,22] showed how Horn logical semantics and partial evaluation can be used to generate provably correct code. From the Horn logical semantic description of the language $\mathcal{L}$, one immediately obtains an interpreter of language $\mathcal{L}$. In this paper, we apply the approach of [21,22] to our timed π-calculus to obtain an implementation of this language. First, we express the syntax of timed π-calculus in the *Definite Clause Grammar* (DCG) notation. This syntax specification trivially and naturally yields an *executable parser* for timed π-calculus. This parser can be used to parse timed π-calculus expressions and obtain their parse trees. Next, we express the semantic algebra and valuation functions of timed π-calculus in logic programming. The syntax and semantics specifications of timed π-calculus loaded into a coinductive constraint logic programming system directly leads to an interpreter for timed π-calculus. This interpreter can be executed and used for verifying properties of systems expressed as timed π-calculus processes. We illustrate our approach by applying it to the *rail road crossing problem* of Lynch and Heitmeyer [8].

Note that there is a past work on logic based implementation of the operational semantics of π-calculus, but not timed π-calculus [23]. However, this work is different from our work as it is unable to model infinite processes and infinite replication. In our implementation we are using coinductive logic programming, a more recently developed concept, which allows such modeling. Also our implementation is based on using Horn logic semantics.

2 Timed π-Calculus

Design decisions. Timed π-calculus [17] is an extension of the original π-calculus [14] with (local) clocks, clock operations and time-stamps. As in π-calculus, timed π-calculus processes use names (including clock names) to interact, and pass names to one another. We assume an infinite set $\mathcal{N}$ of names (channel names and names passing through channels), an infinite set Γ of clock names (disjoint from $\mathcal{N}$) and an infinite set Θ of variables representing time-stamps (disjoint from $\mathcal{N}$ and Γ). When a process outputs a name through a channel, it also sends the time-stamp of the name and the clock that is used to generate the time-stamp. Thus, messages are represented by triples of the form $\langle m, t_m, c \rangle$, where m is a name in $\mathcal{N}$, t_m is the time-stamp on m, and c is the clock that is used to generate t_m.

All the clocks are local clocks; however, their scope grows as they are sent among processes. Note that all the clocks advance at the same rate. At any instant, the *reading of a clock* is equal to the time that has elapsed since the last time the clock was reset. Following the semantics of timed automata [1], only non-Zeno behaviors are considered, that is, only a finite number of transitions can happen within a finite amount of time.

Clock operations and interpretations. Clock resets, represented by γ, and clock constraints, denoted by δ, are defined by the following syntactic rules,

$$\delta ::= (c \sim r)\delta \mid (c - t \sim r)\delta \mid (t - c \sim r)\delta \mid \epsilon$$
$$\gamma ::= (c_1 := 0) \dots (c_n := 0) \mid \epsilon$$

where c and $c_i, 1 \leq i \leq n$, are clock names, r is a constant in $\mathbb{R}_{\geq 0}$, t is a time-stamp and $\sim \in \{<, >, \leq, \geq, =\}$. ϵ represents an empty clock constraint or clock reset.

For a process P, $c(P)$ is the set of clock names in P. For every two processes P and Q, $c(P) \cap c(Q) = \emptyset$. A *clock interpretation* I for a set Γ of clocks is a mapping from Γ to $\mathbb{R}_{\geq 0}$. It assigns a real value to each clock in Γ. A clock interpretation I for Γ satisfies a clock constraint δ over Γ iff the expression obtained by applying I to δ evaluates to true. For $t \in \mathbb{R}_{\geq 0}$, $I + t$ denotes the clock interpretation which maps every clock c to the value $I(c) + t$. For $\gamma \subseteq \Gamma$, $[\gamma \mapsto t]I$ denotes the clock interpretation for Γ which assigns t to each $c \in \gamma$, and agrees with I over the rest of the clocks.

Syntax. The set of timed π-calculus processes is defined by the following syntactic rules in which, P, P', M and M' range over processes, x, y and z range over names in $\mathcal{N}$, c and d range over clock names in Γ, and t_y represents a time-stamp.

$$M ::= \delta\gamma\bar{x}\langle y, t_y, c\rangle.P \mid \delta\gamma x(\langle y, t_y, c\rangle).P \mid \delta\gamma\tau.P \mid 0 \mid M + M'$$
$$P ::= M \mid (P \mid P') \mid !P \mid (z)\,P \mid [x = y]\,P \mid [c = d]\,P$$

The expression $\delta\gamma\bar{x}\langle y, t_y, c\rangle.P$ represents a process that is capable of outputting name y on channel x. This process generates a time-stamp t_y using clock c and sends t_y and c along with y via the channel x, and evolves to P. The time-stamp t_y is the reading of clock c at the time of transition. *The assignment of a time-stamp to y and sending y is an atomic operation.* The clock constraint δ must be satisfied by the current value of clocks at the time of transition. γ specifies the clocks to be reset with this transition.

The expression $\delta\gamma x(\langle y, t_y, c\rangle).P$ stands for a process which is waiting for a message on channel x. When a message arrives, the process will behave like $P\{z/y, t_z/t_y, d/c\}$ (substitution is formally defined later in this section) where z is the name received; t_z is the time-stamp of z; and d is the clock of the sending process that is used to generate t_z. The time-stamp t_z must satisfy the clock constraint expressed by δ; γ specifies the clocks to be reset with the transition.

The expression $\delta\gamma\tau.P$ stands for a process that takes an internal action and evolves to P, and in doing so resets the clocks specified by γ, if the clock constraint δ is satisfied.

In each of three processes explained above, if the clock constraint δ is not satisfied by the value of clocks at the time of transition, then, the process becomes inactive. An inactive process, represented by 0, is a process that does nothing. The operators $+$ and $|$ are used for non-deterministic *choice* and *composition* of processes, just as in π-calculus [14]. The *replication* $!P$, represents an infinite composition $P \mid P \mid \ldots$, just as in π-calculus. The *restriction* $(z)P, z \in \mathcal{N}$, behaves as P with z local to P. Therefore, z cannot be used as a channel over which to communicate with other processes or the environment. $[x = y]\,P, x, y \in \mathcal{N} \cup \Gamma$, evolves to P if x and y are the same name; otherwise, it becomes inactive.

Example 1. The timed π-calculus expression $x(\langle m, t_m, c \rangle).(c - t_m \leq 5)\bar{y}\langle n, t_n, c \rangle$ represents a process that is waiting for a message on channel x. The process upon receiving a name m with time-stamp t_m and its accompanying clock c on channel x, sends a name n with time-stamp t_n on channel y with the delay of at most 5 units of time since the time-stamp of m. The process will use the clock c to choose a time t_n on c such that $c - t_m \leq 5$.

In a process of the form $\delta \gamma x(\langle y, t, c \rangle).P$ the occurrences of y, t and c are binding occurrences, and the scope of the occurrences is P. In $(n)P, n \in \mathcal{N}$ the occurrence of n is a binding occurrence, and the scope of the occurrence is P. An occurrence of a (non-clock) name n in a process is *free* if it does not lie within the scope of a binding occurrence of n, and *bound* if it is not free. All occurrences of a clock c in a process P are bound. The set of bound names and free names of P are denoted by $bn(P)$ and $fn(P)$, respectively. We write $n(P)$ for the set $fn(P) \cup bn(P)$.

Example 2. Let $P = x(\langle y, t, c \rangle).0$ and $Q = (d > 1)(d < 5)x\langle z, t', d\rangle.0$. Then, $fn(P) = \{x\}, bn(P) = \{y, t, c\}$, $fn(Q) = \{x, z, t'\}$, and $bn(Q) = \{d\}$. x is a channel that is shared between P and Q.

A *substitution* is a function θ from a set of names $\mathcal{N}$ to $\mathcal{N}$. If $x_i\theta = y_i$ for all i with $1 \leq i \leq n$ (and $x\theta = x$ for all other names x), we write $\{y_1/x_1, \ldots, y_n/x_n\}$ for θ. The effect of applying a substitution θ to a process P is to replace each free occurrence of each name x in P by $x\theta$, with change of bound names to avoid name capture (to preserve the distinction of bound names from the free names). Substitution for time-stamps can be defined similarly. A *clock substitution* is a function θ_c from a set of clock names Γ to Γ. If $c_i\theta_c = d_i$ for all i with $1 \leq i \leq n$ (and $c\theta_c = c$ for all other clock names c), we write $\{d_1/c_1, \ldots, d_n/c_n\}$ for θ_c. The effect of applying a substitution θ_c to a process P, $P\theta_c$, is to replace all occurrences of each clock name c in P by $c\theta_c$. Given a clock c, the function θ_f creates a fresh copy, f, of c (f does not appear in any process) and updates the interpretation with $I(f) = I(c)$. The application of θ_f to c is represented by $c\theta_f$. A *clock renaming* θ_r is a clock substitution $\{f_1/c_1, \ldots, f_n/c_n\}$ in which $f_i = c_i\theta_f, 1 \leq i \leq n$. The effect of applying a clock renaming $\theta_r = \{f_1/c_1, \ldots, f_n/c_n\}$ to process P, $P\theta_r$, is to replace all occurrences of each name c in P by $c\theta_r$.

Operational semantics. The actions of timed π-calculus are defined by the following syntactic rule:

$$\alpha ::= \bar{x}\langle y, t, c \rangle \mid \bar{x}\langle (y), t, c \rangle \mid x(\langle y, t, c \rangle) \mid \tau$$

The first two actions are the *bound output actions*. The expression $\bar{x}\langle y, t, c\rangle$ is used for sending a name y, time-stamp of y, t, and the (local) clock that is used to generate t, via channel x. The process that gives rise to this action can be of the form $\bar{x}\langle y, t, d\rangle.P$. In this action x, y and t are free and $c = d\theta_f$ is bound (c is a fresh copy of d). The expression $\bar{x}\langle (y), t, c\rangle$ is used by a process for sending its private name y (y is bound in the process) and its (local) clock c. The process that gives rise to this action can be of the form: $(y)\bar{x}\langle y, t, d\rangle.P$. In this action x and t are free, while y and $c = d\theta_f$ are bound (c is a fresh copy of d).

The third action is the *input action* $x(\langle y, t, c\rangle)$. This action is used for receiving any name z with its time-stamp t_z, and a clock d via x. In this action x is free, while y, t and c are bound names.

The last action is the *silent action* τ, which is used to express performing an internal action. Silent actions can naturally arise from processes of the form $\tau.P$, or from communications within a process (e.g., rule COM in Table 1).

We use $fn(\alpha)$ for set of free names of α, $bn(\alpha)$ for set of bound names of α, and $n(\alpha)$ for the union of $fn(\alpha)$ and $bn(\alpha)$. Note that $fn(\tau) = \emptyset$ and $bn(\tau) = \emptyset$.

A transition in timed π-calculus is of the form $P \xrightarrow{\langle \delta, \alpha, \gamma\rangle} P'$. This transition is understood as follows: if δ is satisfied by the current values of clocks, P evolves into P', and in doing so performs the action α and resets the clocks specified by γ. With abuse of notation, γ is used as a set of clocks to be reset. The triple $\langle \delta, \alpha, \gamma\rangle$ is called a timed action. The set of transition rules of timed π-calculus are represented in Table 1. These rules are labelled by timed actions. Note that there are two more rules for SUM and PAR where the process Q takes an action. These rules are symmetric to SUM and PAR rules of Table 1 and are eliminated.

A time sequence $w = w_1 w_2 \ldots$ is an infinite sequence of time values $w_i \in \mathbb{R}_{\geq 0}$, satisfying the following constraints:

- Monotonicity: w increases strictly monotonically; that is, $w_i < w_{i+1}$ for all $i \geq 1$.
- Progress: For every $w \in R$, there is some $i \geq 1$ such that $w_i \geq w$.

A system specified by the set of timed π-calculus processes starts with all the clocks initialized to 0. As time advances the value of all clocks advances, reflecting the elapsed time. At time w_i, a process P_{i-1} takes a timed action $\langle \delta_i, \alpha_i, \gamma_i\rangle$ and evolves to P_i, if the current values of clocks satisfy δ_i. The clocks specified by γ_i are reset to 0, and thus start counting time with respect to it. This behavior is captured by defining *runs* of timed π-calculus processes. A run for a process P records the state (process expression) and the values of all the clocks at the transition points. For a time sequence $w = w_1 w_2 \ldots$, a run r denoted by $(\bar{P}, \bar{I})$, of a process P, is a finite or an infinite sequence of the form

$$\langle P_0, I_0\rangle \xrightarrow[w_1]{\langle \delta_1, \alpha_1, \gamma_1\rangle} \langle P_1, I_1\rangle \xrightarrow[w_2]{\langle \delta_2, \alpha_2, \gamma_2\rangle} \langle P_2, I_2\rangle \xrightarrow[w_3]{\langle \delta_3, \alpha_3, \gamma_3\rangle} \ldots$$

where P_i is a process and $I_i \in [\Gamma \to \mathbb{R}]$, for all $i \geq 0$, satisfying the following requirements:

- Initiation: P_0 is the initial process expression, and $I_0(c) = 0$ for all $c \in \Gamma$.
- Consecution: for all $i \geq 1$, there is a transition of the form $P_{i-1} \xrightarrow{\langle \delta_i, \alpha_i, \gamma_i \rangle} P_i$ such that $(I_{i-1}+w_i-w_{i-1})$ satisfies δ_i and I_i equals $[\gamma_i \mapsto 0](I_{i-1}+w_i-w_{i-1})$. We assume $w_0 = 0$.

Along a run $r = (\bar{P}, \bar{I})$, the values of the clocks at time $w_i \leq w \leq w_{i+1}$ are given by the interpretation $(I_i + w - w_i)$. When the transition from P_i to P_{i+1} occurs,

Table 1. Timed π-calculus Transition Rules

$$\text{IN} \xrightarrow{\hspace{1.2cm}} \quad y \notin fn((z)P)$$
$$\delta\gamma x(\langle z, t, c \rangle).P \xrightarrow{\langle \delta\{t'/t,d/c\}, x(\langle y,t',d\rangle), \gamma\{d/c\}\rangle} P\{y/z, t'/t, d/c\}$$

$$\text{OUT} \xrightarrow{\hspace{1cm}} d = c\theta_f \qquad \text{TAU} \xrightarrow{\hspace{1cm}}$$
$$\delta\gamma\bar{x}\langle y, t, c \rangle.P \xrightarrow{\langle \delta, \bar{x}\langle y,t,d\rangle, \gamma\rangle} P \qquad\qquad \delta\gamma\tau.P \xrightarrow{\langle \delta, \tau, \gamma\rangle} P$$

$$\text{PAR} \frac{P \xrightarrow{\langle \delta,\alpha,\gamma\rangle} P'}{(P \mid Q) \xrightarrow{\langle \delta,\alpha,\gamma\rangle} (P' \mid Q)} \; bn(\alpha) \cap fn(Q) = \emptyset \qquad \text{SUM} \frac{P \xrightarrow{\langle \delta,\alpha,\gamma\rangle} P'}{P + Q \xrightarrow{\langle \delta,\alpha,\gamma\rangle} P'}$$

$$\text{COM} \frac{P \xrightarrow{\langle \delta, \bar{x}\langle z,t,c\rangle, \gamma\rangle} P' \quad Q \xrightarrow{\langle \delta', x(\langle z,t,c\rangle), \gamma'\rangle} Q'}{(P \mid Q) \xrightarrow{\langle \delta\delta', \tau, \gamma\gamma'\rangle} (P' \mid Q')}$$

$$\text{OPEN} \frac{P \xrightarrow{\langle \delta, \bar{x}\langle y,t,c\rangle, \gamma\rangle} P'}{(y)P \xrightarrow{\langle \delta, \bar{x}\langle(u),t,c\rangle, \gamma\rangle} P'\{u/y\}} \; y \neq x \wedge u \notin fn((y)P')$$

$$\text{CLOSE} \frac{P \xrightarrow{\langle \delta, \bar{x}\langle(z),t,c\rangle, \gamma\rangle} P' \quad Q \xrightarrow{\langle \delta', x(\langle z,t,c\rangle), \gamma'\rangle} Q'}{(P \mid Q) \xrightarrow{\langle \delta\delta', \tau, \gamma\gamma'\rangle} (z)(P' \mid Q')}$$

$$\text{RES} \frac{P \xrightarrow{\langle \delta,\alpha,\gamma\rangle} P'}{(z)P \xrightarrow{\langle \delta,\alpha,\gamma\rangle} (z)P'} \; z \notin n(\alpha), z \in \mathcal{N}$$

$$\text{MATCH} \frac{P \xrightarrow{\langle \delta,\alpha,\gamma\rangle} P'}{[x = x]P \xrightarrow{\langle \delta,\alpha,\gamma\rangle} P'} \; x \in \mathcal{N} \text{ or } x \in \Gamma \qquad \text{REP} \frac{P \xrightarrow{\langle \delta,\alpha,\gamma\rangle} P'}{!P \xrightarrow{\langle \delta,\alpha,\gamma\rangle} (P'\theta_r \mid !P)}$$

$$\text{REP-COM} \frac{P \xrightarrow{\langle \delta, \bar{x}\langle z,t,c\rangle, \gamma\rangle} P' \quad P \xrightarrow{\langle \delta', x(\langle z,t,c\rangle), \gamma'\rangle} P''}{!P \xrightarrow{\langle \delta\delta', \tau, \gamma\gamma'\rangle} ((P'\theta'_r \mid P''\theta''_r) \mid !P)}$$

$$\text{REP-CLOSE} \frac{P \xrightarrow{\langle \delta, \bar{x}\langle(z),t,c\rangle, \gamma\rangle} P' \quad P \xrightarrow{\langle \delta', x(\langle z,t,c\rangle), \gamma'\rangle} P''}{!P \xrightarrow{\langle \delta\delta', \tau, \gamma\gamma'\rangle} ((z)(P'\theta'_r \mid P''\theta''_r) \mid !P)}$$

the value $(I_i + w_{i+1} - w_i)$ is used to check the clock constraint. At time w_{i+1}, the value of a clock that gets reset is defined to be 0.

When the transition from $P_i = \delta\gamma x(\langle z, t, c\rangle).P$ to $P_{i+1} = P\{y/z, t'/t, d/c\}$ occurs ($\langle y, t', d\rangle$ is the received name), we check the satisfiability of the clock constraint $\delta\{d/c\}$, similarly we reset the clocks specified by $\gamma\{d/c\}$. Intuitively, this means that the value of the receiving clock should satisfy the constraint δ for the transition to take place. Moreover, the incoming clock might get reset upon arrival. When the transition from $P_i = \delta\gamma\bar{x}\langle y, t, c\rangle.P$ to $P_{i+1} = P$ occurs in which, the timed action $\langle\delta, \bar{x}\langle y, t, d\rangle, \gamma\rangle, d = c\theta_f$ takes place, the time-stamp t in $\bar{x}\langle y, t, d\rangle$ gets bound to $(I_i(c) + w_{i+1} - w_i)$. Note that at this point $I_{i+1}(d) = I_{i+1}(c)$.

3   Operational Semantics in Logic Programming

For a complete encoding of the operational semantics of timed π-calculus, we must account for the facts that: (i) clock constraints are posed over continuous time, (ii) infinite computations are defined in timed π-calculus (and also π-calculus) through the infinite replication operator '!', and (iii) we are dealing with mobile concurrent processes. We have developed an implementation of the operational semantics of timed π-calculus using coinductive constraint logic programming over reals extended with coroutining, in which channels are modeled as streams and all the three aspects are handled faithfully.

Coinductive constraint logic programming [16] is a paradigm that combines constraint logic programming (CLP) [9] and coinductive logic programming [6,18,19]. The operational semantics of coinductive CLP relies on the *coinductive hypothesis rule* and systematically computes elements of the *greatest fixed point* (*gfp*) of a program via backtracking. The coinductive hypothesis rule states that during execution, if the current resolvent R contains a call G' that unifies with an ancestor call G, and the set of accumulated constraints are satisfied, then the call G' succeeds; the new resolvent is $R'\theta$ where θ is the most general unifier of G and G', and R' is obtained by deleting G' from R. Regular constraint logic programming execution extended with the coinductive hypothesis rule is termed *co-constraint logic programming* (or co-CLP)[16]. In co-CLP, predicates can be declared as being either coinductive or inductive. Using co-CLP enables us to handle (i) real time and timing constraints; (ii) infinite computations, realized by the replication operator of timed π-calculus.

Consider the following program, where `stream/2` is a coinductive predicate and `number/1` is an inductive predicate.

```
:-coinductive(stream).
:-inductive(number).
stream([H | T], X) :- number(H), stream(T, Y), {Y - X >= 3}.
number(0).
number(s(N)) :- number(N).
```

The following is an execution trace for the query ?- `stream([0,s(0),s(s(0))` `| R], W).`, in a co-CLP system, which shows the recursion stack and also the set of generated constraints (that must be satisfied) after each recursive call (substitution is not shown).

1. $\mathtt{stream}([0, \mathtt{s}(0), \mathtt{s}(\mathtt{s}(0)) \mid \mathtt{R}], \mathtt{W}), \mathtt{C}_1 = \emptyset$
2. $\mathtt{stream}([\mathtt{s}(0), \mathtt{s}(\mathtt{s}(0)) \mid \mathtt{R}], \mathtt{U}), \mathtt{C}_2 = \{\mathtt{U} - \mathtt{W} >= 3\}$
3. $\mathtt{stream}([\mathtt{s}(\mathtt{s}(0)) \mid \mathtt{R}], \mathtt{V}), \qquad \mathtt{C}_3 = \{\mathtt{U} - \mathtt{W} >= 3, \mathtt{V} - \mathtt{U} >= 3\}$
4. $\mathtt{stream}(\mathtt{R}, \mathtt{Z}), \qquad\qquad \mathtt{C}_4 = \{\mathtt{U} - \mathtt{W} >= 3, \mathtt{V} - \mathtt{U} >= 3, \mathtt{Z} - \mathtt{V} >= 3\}$

The last goal call unifies with ancestor (1) and immediately succeeds, since $\mathtt{C}_4$ is consistent. Hence the original query succeeds with $\mathtt{R} = [0, \mathtt{s}(0), \mathtt{s}(\mathtt{s}(0)) \mid \mathtt{R}]$ with the answer constraint $\{\mathtt{U} - \mathtt{W} \geq 3, \mathtt{V} - \mathtt{U} \geq 3, \mathtt{Z} - \mathtt{V} \geq 3\}$.

The coroutining feature of logic programming deals with having logic program goals scheduled for execution as soon as some conditions are fulfilled. In LP the most commonly used condition is the instantiation (binding) of a variable. Scheduling a goal to be executed immediately after a variable is bound can be used to model the actions taken by processes as soon as a message is received in the channel specified by that variable. Coroutining can be practically realized through the delay/freeze construct supported in most Prolog systems.

Our implementation of the operational semantics of timed π-calculus consists of developing the logical denotational semantics of timed π-calculus which yields an interpreter for timed π-calculus. The denotational semantics of a language L has three components: (i) syntax specification which maps sentences of L to parse trees; (ii) semantic algebra which represents the mathematical objects whose elements are used for expressing the meaning of a program written in the language L along with associated operations to manipulate the elements; (iii) valuation functions which map parse trees to elements of the semantic algebras.

We present the names in timed π-calculus as Prolog variables. *The direct encoding of the operational semantics of timed π-calculus in logic programming is due to the similarity between the treatment of variables in the resolution procedures for logic programs and the treatment of names in the operational semantics of timed π-calculus.* Note that the same (non-clock) name may appear bound and also free in a timed π-calculus expression. For instance, in $\bar{y}\langle y, t, c\rangle.(y)\bar{z}\langle y, t, c\rangle.0$, the name y appears both free and bound. *In our encoding, we ensure (automatically) that bound names are all distinct from each other and from the free names: all top-down resolution techniques (such as SLD) rename the variables in a clause that is selected at each resolution step in order to avoid the capture of free variables. This procedure is called standardization apart.*

First, we express the grammar of timed π-calculus as a DCG. There is a one-to-one correspondence between rules in the grammar of timed π-calculus and rules in the DCG. The DCG, shown in Table 2, is a logic program and when executed in the Prolog system, leads to an automatic parser for timed π-calculus. This parser parses a timed π-calculus expression representing a process, and produces a parse tree for it. For instance the query for parsing the expression $((d < 2)\bar{x}\langle w, t_1, d\rangle.0|(c := 0)x(\langle z, t_2, c\rangle).0)$:

```
process( T, [(,(,d,<,2,),x,<,w,',',t1,',',d,>,.,0,|,
             (,c,:=,0,),x,(,<,z,',',t2,',',c,>,),.,0,)], [] ).
```

Table 2. DCG for Timed π-calculus Grammar

```
process(par(P1,P2)) --> [(],process(P1),[|],process(P2),[)].

process(rep(P)) --> [!],process(P).

process(((N),P)) --> [(],name(N),[)],[(],process(P),[)].

process(match(N1,N2,P)) --> [[],name(N1),[=],name(N2),[]],process(P).

process(P) --> mprocess(P).

mprocess([Prefix|Processes]) --> prefix(Prefix),[.],process(Processes).

mprocess([process(0)]) --> [0].

mprocess(sum(M1,M2)) --> [(],mprocess(M1),[+],mprocess(M2),[)].

prefix(in(CO,CN,N,T,C)) --> clockOp(CO),name(CN),[(],[<],name(N),[,],
                              time(T),[,],clock(C),[>],[)].

prefix(out(CO,CN,N,T,C)) --> clockOp(CO),name(CN),[<],name(N),[,],
                              time(T),[,],clock(C),[>].

prefix(tau(CO)) --> clockOp(CO),[tau].

clockOp(clockOp(C,R)) --> constraints(C),resets(R).

constraints([Constr|List]) --> constraint(Constr),constraints(List).
constraints([]) --> [].

constraint((C,Op,R)) --> [(],clock(C),op(Op),const(R),[)].
constraint((C,T,Op,R)) --> [(],clock(C),[-],time(T),op(Op),const(R),[)].
constraint((T,C,Op,R)) --> [(],time(T),[-],clock(C),op(Op),const(R),[)].

resets([Reset|List]) --> reset(Reset),resets(List).
resets([]) --> [].

reset(C) --> [(],clock(C),[:=],[0],[)].

time(T) --> [T],{atom(T); var(T)}.

const(R) --> [R],{number(R)}.

clock(C) --> [C],{atom(C); var(C)}.

name(N) --> [N],{atom(N); var(N)}.

op(O) --> [O],{O = <; O = =<; O = >; O = >=; O = =}.
```

will produce the parse tree shown below:

```
T = par([out(clockOp([(d,<,2)]), x, w, t1, d), process(0)],
        [in(clockOp([c]), x, z, t2, c), process(0)])
```

Next, we express the semantic algebra and the valuation functions as logic programs. The semantic algebra consists of two store domains: one for storing clocks, realized as an association list of the form [(Clock, Value),...], and one for storing timed events, realized as a list (initially empty). The valuation functions, realized by the coinductive predicate sem/8, shown in Table 3, take the parse tree patterns and the current stores and assign meaning to the parse tree patterns, while updating the stores. This denotational semantics can be viewed also as the logical encoding of the operational semantics of timed π-calculus. The encoded operational semantics generates the symbolic transition systems from process expressions using the sem/8 predicate. The syntax of timed π-calculus realized as a DCG along with the semantic algebras and valuation functions realized as a logic program lead to an interpreter for timed π-calculus. This interpreter when loaded into a coinductive constraint logic programming system can be run directly. Note how concurrency is realized through the coroutining facility of logic programming– in particular *freeze* construct of Prolog– while writing the rules for COM, CLOSE, REP-COM and REP-CLOSE. Note also that the *freeze* construct on variable A2 in COM corresponds to the case when the process P2 takes an output action and P1 takes an input action. In other words, P1 which is the receiving process has to wait until A2 gets bound, where A2 is an output action taken by P2. The symmetric rule in which P1 and P2 take output and input actions, respectively, is omitted. The symmetric rules for PAR and SUM where process P2 takes action are also omitted.

In our formulation of the operational semantics of timed π-calculus, clock expressions and time constraints are modeled using *constraint logic programming over reals* [9], infinite (rational) computations, realized through the replication operator of timed π-calculus (and also π-calculus) are handled using *coinductive logic programming* [6,19], and finally concurrency is simulated by *coroutining* within logic programming computations. Therefore, timed π-calculus processes are modeled as coroutined coinductive constraint logic programs.

If the above syntax and semantics rules along with the logic program p are loaded into a co-CLP system and the query: ?-main(S2,E2,E,Q,A) is posed, then S2=[(w,T),(c,T),(d,0)], T>0,T<2, E2=[out(x,y,W,d),in(x,Z,T,D)], Q=((d),par([process(0)],[process(0)])), E=[Z=y,T=W,D=d], A=tau(T).

```
p: main(S2,E2,E,Q,A) :-
                process(P,[(,(,d,<,2,),x,<,y,',',W,',',d,>,.,0,|,
                        (,c,:=,0,),x,(,<,Z,',',T,',',D,>,),.,0,)],[]),
                store(S1),
                eventStore(E1),
                sem(P, S1, E1, S2, E2, E, Q, A).

   store([ (wallClock, 0), (c, 0), (d, 0) ]).
   eventStore([]).
```

Table 3. Denotational Semantics of Timed π-calculus in Logic Programming

```
% OUTPUT
sem([ out(CO, ChN, N, T, C) | P ], C1, E1, C2, E2, Eq, P, A):-
    access(wallClock, C1, Wall),
    { W > Wall },
    clockOp(CO, C1, Ctemp, W),
    update(wallClock, W, Ctemp, C2),
    Eq = [], A = out(ChN, N, W, C),
    add_events(out(ChN, N, W, C), E1, E2).
% INPUT
sem([ in(CO, ChN, N, W, C) | P ], C1, E1, C2, E2, Eq, P, in(ChN,N,W,C)):-
    clockOp(CO, C1, C2, W),
    add_events(in(ChN, N, W, C), E1, E2),
    Eq = [].
% TAU
sem([tau(CO)|P], C1, E1, C2, E2, [], P, tau(W)):-
    access(wallClock, C1, Wall),
    { W > Wall },
    clockOp(CO, C1, C2, W),
    add_events(tau(W), E1, E2).
% MATCH
sem(match(N1, N2, P), C1, E1, C2, E2, Eq, R, A):-
    ( N1 == N2 -> sem(P, C1, E1, C2, E2, Eq, R, A)
                ;sem(P, C1, E1, C2, E2, Eq1, R, A),
                 Eq = [ N1 = N2 | Eq1 ] ).
% SUM
sem(sum(P1,_P2), C1, E1, C2, E2, Eq, R, A) :-
    sem(P1, C1, E1, C2, E2, Eq, R, A).
% PAR
sem(par(P1, P2), C1, E1, C2, E2, Eq, par(Q1, P2), A):-
    sem(P1, C1, E1, C2, E2, Eq, Q1, A).
% COM
sem(par(P1, P2), C1, E1, C2, E2, Eq, ((C), par(Q1, Q2)), tau(W)):-
    freeze(A2, sem(P1, Ctemp, Etemp, C2, E2, Eq1, Q1, A1)),
    sem(P2, C1, E1, Ctemp, Etemp, Eq2, Q2, A2),!,
    match(A1, A2, W, C, Eq4),
    append(Eq1, Eq2, Eq3), append(Eq3, Eq4, Eq).
% OPEN
sem(((N), P), C1, E1, C2, E2, Eq, Q, out(ChN, M, W, C)):-
    sem(P, C1, E1, C2, E2, Eq, Q, out(ChN, M, W, C)),
    N \== ChN, N == M,
    not_in(N, Eq).
% CLOSE
sem(par(P1, P2), C1, E1, C2, E2, Eq, ((C, N), par(Q1, Q2)), tau(W)):-
    freeze(A2, sem(P1, Ctemp, Etemp, C2, E2, Eq1, Q1, A1)),
    sem(P2, C1, E1, Ctemp, Etemp, Eq2, Q2, A2),
    match(A1, A2, N, W, C, Eq4),
    append(Eq1, Eq2, Eq3), append(Eq3, Eq4, Eq).
```

Table 4. Denotational Semantics of Timed π-calculus in Logic Programming

```
% RES
sem(((N),P), C1, E1, C2, E2, Eq, ((N), Q), A) :-
    (P, C1, 1, C2, E2, Eq, Q, A),
    not_in_act(N, A), not_in(N, Eq).
% REP
sem(rep(P), C1, E1, C2, E2, Eq, par(Q,rep(P)), A) :-
    sem(P, C1, E1, C2, E2, Eq, Q, A).
% REP-COM
sem(rep(P), C1, E1, C2, E2, Eq, par(((C), par(P1, P2)),rep(P)), tau(W)):-
    freeze(A2, sem(P, Ctemp, Etemp, C2, E2, Eq1, P1, A1)),
    sem(P, C1, E1, Ctemp, Etemp, Eq2, P2, A2),!,
    match(A1, A2, W, C, Eq4),
    append(Eq1, Eq2, Eq3),
    append(Eq3, Eq4, Eq).
% REP-CLOSE
sem(rep(P), C1, E1, C2, E2, Eq, par(((N,C),par(P1,P2)),rep(P)), tau(W)):-
    freeze(A2, sem(P, Ctemp, Etemp, C2, E2, Eq1, P1, A1)),
    sem(P, C1, E1, Ctemp, Etemp, Eq2, P2, A2),!,
    match(A1, A2, N, W, C, Eq4),
    append(Eq1, Eq2, Eq3), append(Eq3, Eq4, Eq).

not_in_act(X, in(ChN, N, W, C)) :- X \== ChN, X \== N.
not_in_act(X, out(ChN, N, W, C)) :- X \== ChN, X \== N.
not_in_act(X, tau(W)).

not_in(X, [ Y = Z | T]) :- X \== Y, X \== Z, not_in(X, T).
not_in(_X, []).

match(in(H1, N, T, D), out(H2,M , W, C), W, C, E) :-
((nonvar(H1), nonvar(H2)) -> H1 == H2, E1 =[ ]
                         ; H1 = H2, E1 = [H1 = H2]),
((nonvar(N), nonvar(M)) ->   N == M, E2 = E1
                         ; N = M, E2 = [ N = M | E1]),
((nonvar(T), nonvar(W)) ->   T == W, E3 = E2
                         ; T = W, E3 = [T = W | E2]),
((nonvar(D), nonvar(C)) ->   D == C, E = E3
                         ; D = C, E = [D = C | E3]).

match(out(H1, N, T, D), in(H2, M, W, C), M, W, C, E) :-
((nonvar(H1), nonvar(H2)) -> H1 == H2, E1 = []
                         ; H1 = H2, E1 = [H1 = H2]),
((nonvar(N), nonvar(M)) ->   N == M, E2 = E1
                         ; N = M, E2 = [N = M | E1]),
((nonvar(T), nonvar(W)) ->   T == W, E3 = E2
                         ; T = W, E3 = [T = W | E2]),
((nonvar(D), nonvar(C)) ->   D == C, E = E3
                         ; D = C, E = [ D = C | E3]).
```

Note that not_in_act/2, not_in/2, match/5 and match/6 are predicates that handle the side conditions of Table 1 and defined in Table 4. Operations for creating, accessing and updating the stores and clocks are implemented through simple predicates: access/3, update/4, add_events/3 and clockOp/3, which are not presented here.

4   Example: The Rail Road Crossing Problem

The generalized rail road crossing (GRC) problem [8] describes a rail road crossing system with several tracks and an unspecified number of trains traveling through the tracks. The gate at the rail road crossing should be operated in a way that guarantees the *safety* and *utility* properties. The *safety* property stipulates that the gate must be down while there is a train in the crossing. The *utility* property states that the gate must be up (or going up) when there is no train in the crossing. The system is composed of three components: *train*, *controller* and *gate*. The components of the system which are specified via three timed automata in Fig. 1, communicate by sending and receiving signals. We specify the components of the system in timed π-calculus.

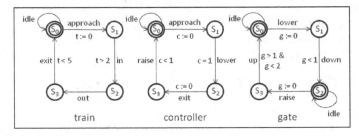

Fig. 1. Timed automata for train, controller, and gate in the rail road crossing problem

The controller at the rail road crossing might receive various signals from trains in different tracks. In order to avoid signals from different trains being mixed, each train communicates through a private channel with the controller. A new channel is established for each approaching train to the crossing area through which the communication between the train and the controller takes place. For simplicity of presentation we consider only one track in this example.

In our modeling of the rail road crossing problem in timed π-calculus each component of the system is considered as a timed π-calculus process. This model is presented in Table 5.

Note that the design of the rail road crossing problem shown in Fig. 1 (originally from [1]) does not account for the delay between the sending of *approach* (*exit*) signal by a train and receiving it by the controller. Similarly the delay between sending *lower* (*raise*) by the controller and receiving it by the gate is not taken into account. Arguably, in a correct design, the delay before *approach*

Table 5. The timed π-calculus expressions for components of the rail road crossing problem

$$
\begin{aligned}
&train \equiv & &controller \equiv \\
&\quad !(ch)\overline{ch1}\langle ch, t_c, t\rangle. & &\quad !ch1(\langle y, t_y, d\rangle).y(\langle x, t_x, c\rangle). \\
&\quad (t := 0)\overline{ch}\langle approach, t_a, t\rangle. & &\quad ([x = approach](c = 1)(e := 0)\overline{ch2}\langle lower, t_l, e\rangle + \\
&\quad (t > 2)\tau.\tau. & &\quad [x = exit](c - t_x < 1)(e := 0)\overline{ch2}\langle raise, t_r, e\rangle) \\
&\quad (t < 5)\overline{ch}\langle exit, t_e, t\rangle \\
&gate \equiv \\
&\quad !ch2(\langle x, t_x, g\rangle). \\
&\quad ([x = lower](g < 1)\tau + [x = raise](g > 1)(g < 2)\tau)
\end{aligned}
$$

$$main \equiv train \mid controller \mid gate$$

is received by the controller should be taken into account. The *lower* signal must be sent within one unit of time since the time-stamp of the *original approach* but not the time at which the controller receives the signal (note that the controller resets its clock to remember the time it receives *approach*). In contrast, in our specification of the rail road crossing problem in timed π-calculus, we are considering the delays; therefore, all the time-related reasoning in the system is performed against *train*'s clock and the time-stamp of *approach* signal (sent by *train* to *controller*).

Note that in the π-calculus expression for *train*, t is the local clock of *train* and the two consecutive τ actions correspond to *train*'s internal actions *in* and *out*. In the expression for *controller*, c is a place holder for the receiving clock t from *train*; while, e is the controller's clock that is reset before it is sent to *gate*. In the expression for *gate*, g is a place holder for the receiving clock e from *controller* and the two τ actions correspond to *gate*'s internal actions; the first τ represents *down*; while the second τ represents *up*.

Timed π-calculus allows the rail road crossing problem to be modeled faithfully. Additionally, significantly more complex systems can be modeled. The timed π-calculus specification can be used for verification of the system as well as generating the implementation [15].

The coroutined coinductive constraint logic program corresponding to the components of the system are presented next.

```
:- coinductive(train).                    t_trans(s0, approach, s1).
train(X, Y, St, W, T) :-                  t_trans(s1, in,       s2).
    ( H = approach, { T2 = W }            t_trans(s2, out,      s3).
    ; H = in, { W - T > 2, T2 = T }       t_trans(s3, exit,     s0).
    ; H = out, { T2 = T}
    ; H = exit, { W - T < 5, T2 = T}),
    { W2 > W },
    t_trans(St, H, St2),
    freeze(X, train(Xs, Ys, St2, W2, T2)),
    ( ( H = approach; H = exit )->
```

```
      Y = [ (H, W) | Ys ]
      ; Y = Ys ),
    X = [ (H, W) | Xs ].

:- coinductive(controller).
controller([ (H, W) | Xs ], Y, Sc) :-
    freeze(Xs, controller(Xs, Ys, Sc3)),
    ( H = approach, M = lower, { W2 > W, W2 - W = 1 }
    ; H = exit, M = raise, { W2 > W, W2 - W < 1 } ),
    c_trans(Sc,  H, Sc2),
    c_trans(Sc2, M, Sc3),
    Y = [ (M, W2) | Ys ].

:- coinductive(gate).
gate([ (H, W) | Xs ], Sg) :-
    freeze(Xs, gate(Xs, Sg3)),
    ( H = lower, M = down, { W2 > W, W2 - W < 1 }
    ; H = raise, M = up, { W2 > W, W2 - W > 1, W2 - W < 2} ),
    g_trans(Sg,  H, Sg2),
    g_trans(Sg2, M, Sg3).

c_trans(s0, approach, s1).    g_trans(s0, lower, s1).
c_trans(s1, lower,    s2).    g_trans(s1, down,  s2).
c_trans(s2, exit,     s3).    g_trans(s2, raise, s3).
c_trans(s3, raise,    s0).    g_trans(s3, up,    s0).
```

The first argument of train/5 is the list of timed events (a list of events with their time-stamps) generated by *train*, the second argument is the list of events sent to *controller* and the third argument is *train*'s current state. W is the current wall clock time and T is *train*'s clock. Similarly, the first argument of controller/3 is the list of timed events received from *train*; while the second argument is the list of events generated by *controller* and sent to *gate*. Sc is the current state of *controller*. Finally, the first argument of gate/2 is the list of timed events received from *controller*; while the second argument is the current state of *gate*. g_trans/3 specifies the internal transitions of *gate*; while, the internal transitions of *train* and *controller* are specified by t_trans/3 and c_trans/3, respectively. The entire system will wait for *train* to generate the initial signals and send them to *controller*; as soon as *controller* receives these signals (the first argument of controller/3 gets bound), it will send appropriate signals to *gate*. This composition of three processes is realized by the expression:

```
freeze(A, (freeze(C, gate(C,s0)),controller(B,C,s0))),train(A,B,s0,0,0).
```

This expression can be understood as follows: The entire system waits for *train* to generate the initial *approach* (captured by the first *freeze* on variable A) and send it to *controller* (via variable B), then the system will wait for appropriate signals to be sent to *gate* by *controller* (captured by the second *freeze* on variable C).

Once the system is modeled as a coinductive coroutined CLP(R) program, the model can be used to verify interesting properties of the system by posing queries. Given a property Q to be verified, we specify its negation as a logic program, notQ. If the property Q holds, the query notQ will fail w.r.t. the logic program that models the system. If the query notQ succeeds, the answer provides a counterexample to why the property Q does not hold.

To prove the *safety* property, we define unsafe/1 in which the safety property is negated: we look for any possibility that a train is in the crossing area before the gate goes down, with the gate being up initially. main(R) represents the composition of three processes in which R is the time trace of the system after the execution is done. The call to unsafe/1 fails, which proves the safety of the system. Similarly we check the *utility* property using unutilized/1 defined below. unutilized/1 looks for the possibility of a situation in which the gate is down without any train being in the crossing area. Likewise if a call to unutilized/1 fails we know that the *utility* property is satisfied. We have also verified the *liveness* property by using not_live/1 predicate, defined below. The *liveness* property states that once the gate goes down, it will not stay down forever. To verify this, we negate the liveness property and look for the possibility that *up* does not appear infinitely often in the accepting timed trace. Coinductive co_not_member/2 succeeds if *up* does not appear in R infinitely often.

```
unsafe(R) :-
          main(R),
          append(C, [ (in, _) | D ], R),
          append(A, [ (up, _) | B ], C),
          not_member((down, _), B).

unutilized(R) :-
          main(R),
          append(A, [ (down, _) | B ], R),
          find_first_up(B, C),
          not_member((in, _), C).

not_live(R) :-
          main(R),
          co_not_member((up, _), R).

find_first_up([ (X, _) | T ], [ (X, _) | R ] ) :-
          X \== up, find_first_up(T, R).

find_first_up([ (up, _) | T ], [] ).
```

5   Conclusions and Related Work

In our previous work [17], we presented an extension of π-calculus with *real-time* and we developed an operational semantics for it. Our timed π-calculus handles, mobility, concurrency and infinite computation. In contrast to other extensions, in our work the notion of time and clocks is adopted directly from the well-understood formalism of timed automata [1]. Therefore, *time is faithfully treated as a continuous quantity.*

In this paper, we developed an implementation of timed π-calculus [17]. This implementation is based on *Horn logical semantics* of programming languages and directly leads to an implementation of timed π-calculus. First, we expressed the syntax of timed π-calculus in the *Definite Clause Grammar* (DCG) notation, which trivially leads to a parser for the language. Next, we expressed the semantic algebra and valuation functions of timed π-calculus in logic programming. The syntax and semantics specifications of timed π-calculus loaded into a coinductive constraint logic programming system directly yields an interpreter for timed π-calculus. This interpreter is executable and can be used for verifying properties of real-time systems and CPS, expressed as timed π-calculus processes. We illustrated our approach by applying it to the *rail road crossing problem* and verifying properties of the system.

There have been some efforts on implementing the operational semantics of π-calculus and its various extensions with time; the most notable one and closest one to our approach is the work of Yang et al. [23]. However, this work is different from our work as: (i) it is unable to model infinite processes and infinite replication. In our implementation we are using coinductive logic programming, a more recently developed concept, which allows such modeling, (ii) it models π-calculus but not timed π-calculus, (iii) it does not use Horn logic semantics.

Tiu et al. [20] specify the operational semantics and bisimulation relations for the *finite* π-calculus (but not timed π-calculus) within a logic called $FO\lambda^{\Delta\nabla}$. However, the focus of this work is to show the use of a certain logic ($FO\lambda^{\Delta\nabla}$) to specify and reason about computation in general. A major goal of this work is to illustrate how the ∇-quantifier and a second proof-level binding (introduced earlier by the same authors) [13] can be used to specify and reason about computation. The authors claim that they have chosen π-calculus because it is a small calculus in which bindings play an important role in computation. Our contribution is quite different as we develop the executable operational semantics of full (not just finite) *timed* π-calculus. Handling name-bindings in our work has been automatically done by resolution procedures for logic programming. Infinite computations as well as constraints are handled using coinduction and CLP, respectively.

Our timed π-calculus is an expressive, natural model for describing real-time, mobile, concurrent processes and our logic-based implementation of the operational semantics of this calculus provides a framework for modeling and verification of real-time systems and CPS.

As for future work, partial evaluation can be used to optimize the logical encoding of the operational semantics of timed π-calculus. It is well known that compiled code for a program $\mathcal{P}$ written in language $\mathcal{L}$ can be obtained by partially evaluating the interpreter for $\mathcal{L}$ w.r.t. the program $\mathcal{P}$ [10]. Given a partial evaluator for pure Prolog, the interpreter can be partially evaluated w.r.t. program $\mathcal{P}$ to obtain provably correct compiled code for $\mathcal{P}$ [21,22]. After obtaining an interpreter for timed π-calculus from its denotational specifications, partial evaluation can be used to obtain a more efficient encoding. We plan to use this technique with $\mathcal{L}$ being timed π-calculus and program $\mathcal{P}$ being timed π-calculus expressions, to obtain a coroutined co-CLP program Q which is equivalent to direct (and more efficient) encoding of $\mathcal{P}$.

References

1. Alur, R., Dill, D.L.: A theory of timed automata. Theor. Comput. Sci. **126**(2), 183–235 (1994)
2. Berger, M.: Towards abstractions for distributed systems. Tech. rep., Imperial College London (2004)
3. Chen, J.: Timed extensions of π calculus. Theor. Comput. Sci. **11**(1), 23–58 (2006)
4. Ciobanu, G., Prisacariu, C.: Timers for distributed systems. Electr. Notes Theor. Comput. Sci. **164**(3), 81–99 (2006)
5. Degano, P., Loddo, J.-V., Priami, C.: Mobile processes with local clocks. In: Dam, M. (ed.) LOMAPS-WS 1996. LNCS, vol. 1192, pp. 296–319. Springer, Heidelberg (1997)
6. Gupta, G., Bansal, A., Min, R., Simon, L., Mallya, A.: Coinductive Logic Programming and Its Applications. In: Dahl, V., Niemelä, I. (eds.) ICLP 2007. LNCS, vol. 4670, pp. 27–44. Springer, Heidelberg (2007)
7. Gupta, R.: Programming models and methods for spatiotemporal actions and reasoning in cyber-physical systems. In: NSF Workshop on CPS (2006)
8. Heitmeyer, C., Lynch, N.: The generalized railroad crossing: A case study in formal verification of real-time systems. In: IEEE Real-time Systems Symposium, pp. 120–131. IEEE Computer Society Press (1994)
9. Jaffar, J., Maher, M.J.: Constraint logic programming: A survey. J. Log. Program. **19**(20), 503–581 (1994)
10. Jones, N.D.: An introduction to partial evaluation. ACM Comput. Surv. **28**(3), 480–503 (1996)
11. Lee, E.A.: Cyber physical systems: Design challenges. In: IEEE Symposium on Object Oriented Real-Time Distributed Computing, ISORC 2008, pp. 363–369. IEEE Computer Society (2008)
12. Lee, J.Y., Zic, J.: On modeling real-time mobile processes. Aust. Comput. Sci. Commun. **24**(1), 139–147 (2002)
13. Miller, D., Tiu, A.: A proof theory for generic judgments. ACM Trans. Comput. Log. **6**(4), 749–783 (2005)
14. Milner, R., Parrow, J., Walker, D.: A calculus of mobile processes, parts i and ii. Inf. Comput. **100**(1), 1–77 (1992)
15. Saeedloei, N.: Modeling and Verification of Real-Time and Cyber-Physical Systems. Ph.D. thesis, University of Texas at Dallas, Richardson, Texas (2011)
16. Saeedloei, N., Gupta, G.: Coinductive constraint logic programming. In: Schrijvers, T., Thiemann, P. (eds.) FLOPS 2012. LNCS, vol. 7294, pp. 243–259. Springer, Heidelberg (2012)
17. Saeedloei, N., Gupta, G.: Timed π-calculus. In: Abadi, M., Lluch Lafuente, A. (eds.) TGC 2013. LNCS, vol. 8358, pp. 119–135. Springer, Heidelberg (2014)
18. Simon, L.: Coinductive Logic Programming. Ph.D. thesis, University of Texas atDallas, Richardson, Texas (2006)
19. Simon, L., Bansal, A., Mallya, A., Gupta, G.: Co-Logic programming: Extending logic programming with coinduction. In: Arge, L., Cachin, C., Jurdziński, T., Tarlecki, A. (eds.) ICALP 2007. LNCS, vol. 4596, pp. 472–483. Springer, Heidelberg (2007)

20. Tiu, A., Miller, D.: Proof search specifications of bisimulation and modal logics for the pi-calculus. ACM Trans. Comput. Log. 11(2) (2010)
21. Wang, Q., Gupta, G.: Provably correct code generation: A case study. Electr. Notes Theor. Comput. Sci. **118**, 87–109 (2005)
22. Wang, Q., Gupta, G., Leuschel, M.: Towards provably correct code generation via horn logical continuation semantics. In: Hermenegildo, M.V., Cabeza, D. (eds.) PADL 2004. LNCS, vol. 3350, pp. 98–112. Springer, Heidelberg (2005)
23. Yang, P., Ramakrishnan, C.R., Smolka, S.A.: A logical encoding of the pi-calculus: model checking mobile processes using tabled resolution. International Journal on Software Tools for Technology Transfer (STTT) **6**(1), 38–66 (2004)

A New Hybrid Debugging Architecture
for Eclipse

Juan González, David Insa, and Josep Silva[✉]

Universitat Politècnica de València, Valencia, Spain
{jgonza,dinsa,jsilva}@dsic.upv.es

Abstract. During many years, Print Debugging has been the most used method for debugging. Nowadays, however, industrial languages come with a trace debugger that allows programmers to trace computations step by step using breakpoints and state viewers. Almost all modern programming environments include a trace debugger that allows us to inspect the state of a computation in any given point. Nevertheless, this debugging method has been criticized for being completely manual and time-consuming. Other debugging techniques have appeared to solve some of the problems of Trace Debugging, but they suffer from other problems such as scalability. In this work we present a new hybrid debugging technique. It is based on a combination of Trace Debugging, Algorithmic Debugging and Omniscient Debugging to produce a synergy that exploits the best properties and strong points of each technique. We describe the architecture of our hybrid debugger and our implementation that has been integrated into Eclipse as a plugin.

1 Introduction

Debugging is one of the most time-consuming tasks in software engineering. However, the automatization of debugging is still far from being a reality. In fact, during many years, Print Debugging (also known as Echo Debugging) has been the most common method for debugging. Print Debugging allows us to easily know whether the computation traverses one specific point. Many bugs can be corrected with this information, and the programmer (maybe optimistically) prefers to use this method before loading a real debugger. Nevertheless, some bugs are almost impossible to detect with Print Debugging, specially in presence of random values, input, and concurrency.

Fortunately, all modern programming environments, e.g., Borland JBuilder [5], NetBeans [2], Eclipse [3], SICStus Prolog SPIDER IDE [6] or SWI-Prolog [1] include a trace debugger, which allows programmers to trace computations step by step. However, Trace Debugging is a completely manual task, and the programmer is in charge of inspecting the computations of the program at a low abstraction level. For this reason, other debugging techniques have been proposed to solve some of these problems, but they also suffer from other problems. For instance, Algorithmic Debugging [26, 27] (also known as Declarative Debugging) is semi-automatic, i.e., the search for the bug is directed by the debugger instead

© Springer International Publishing Switzerland 2014
G. Gupta and R. Peña (Eds.): LOPSTR 2013, LNCS 8901, pp. 183–201, 2014.
DOI: 10.1007/978-3-319-14125-1_11

of the programmer; and its abstraction level is so high that programs can be debugged without even seeing the code, but it suffers from scalability problems.

In this work we introduce a hybrid debugging technique that combines three different techniques, namely, Trace Debugging (TD), Omniscient Debugging (OD) and Algorithmic Debugging (AD). The combination is done exploiting the strong points of each technique, and counteracting or removing the weak points with their composition. Our method is presented for the programming language Java—our implementation is an Eclipse plugin for Java—but the technique and the architecture of our debugger could be applicable to any other programming language. In summary, the main contributions of this work are the following:

- The design of a new hybrid debugging technique that combines TD, OD and AD.
- The integration of the technique on top of the JPDA architecture—which was conceived for tracing, but not for algorithmic or omniscient debugging—.
- The implementation of the technique as a Eclipse plugin.
- The empirical evaluation of the new architecture that demonstrates the practical scalability of the technique.

The rest of the paper is structured as follows: In Section 2 we describe TD, OD, and AD, analyzing their strong and weak points. Then, in Section 3 we present our new hybrid debugging technique and explain its architecture. In Section 4 we describe our implementation, which has been integrated into Eclipse. The related work is presented in Section 5. Finally, Section 6 concludes and outlines the future work.

2   Debugging Techniques

This section describes the three debugging techniques that we use in our hybrid method: TD, OD and AD. For each technique, we also analyze its strong and weak points and its applicability to Java.

2.1   Trace Debugging

The most used method for debugging is TD. It allows the programmer to traverse the trace of a computation step by step. The programmer places a *breakpoint* in a line of the source code and the debugger stops the computation when this line is reached. Then, the programmer proceeds line by line and, at each step, the programmer can inspect the state of the computation (i.e., variables' values, exceptions, etc.). During the traversal of the trace, when a call to a method is reached, the debugger can either enter the method (*step into*) or skip it (*step over*). Modern breakpoints are conditional, i.e., the breakpoint includes conditions over the values of some variables, or over the action performed where they are defined. For instance, it is possible to define a breakpoint that only stops the computation when an exception happens, or when a class is loaded. TD has one important advantage over other debugging techniques: scalability. The debugger only needs to take control over the interpreter to execute the program normally. Hence, its scalability is the same as the one of the interpreter. On the other hand, TD has four main drawbacks:

1. The whole debugging process is done at a very low abstraction level. The programmer just follows the steps of the interpreter, and she needs to understand how variables' values change to identify an error.
2. The debugger can generate an overwhelming amount of information.
3. The debugging process is completely manual. The programmer uses her intuition to place the breakpoints. If the breakpoint is after the bug, she has to place it again before, and restart the program. If the breakpoint is placed long before the bug, then she has to manually inspect a big part of the computation.
4. The inspection of the computation is made forwards, while the natural way of discovering the bug is backwards from the bug symptom.

2.2 Omniscient Debugging

Omniscient debugging [19] solves the fourth drawback of TD with the cost of sacrificing scalability. Basically both techniques rely on the use of breakpoints and they both do exactly the same from a functional point of view. The difference is that OD allows the programmer to trace the computation forwards and backwards (chronologically). This is very useful, because it allows the programmer to perform steps backwards from the bug symptom. To do this, the debugger needs a mechanism to reconstruct every state of the computation. One of the most scalable schemas to do this is depicted in Figure 1. In this figure, we have an horizontal line representing an execution as a sequence of events. Some of these events are method invocations (represented with a white circle), and method exits (represented with a black circle). Each event is identified with a timestamp. From the execution, the omniscient debugger stores a variable history record that contains the values of all variables together with the exact timestamp where they updated each value. The omniscient debugger also stores information about the scope of variables that we omit here for clarity. With this information the debugger can reconstruct any state of the computation. For instance, in state 42, value M.N.y did not exist, and the last value of variables O.x and O.v[3] was 23 and 3 respectively.

Being able to reconstruct the complete trace also allows the programmer to start the execution at any point. Nevertheless, storing all values taken by all variables in an execution is usually impossible for realistic industrial (large) programs, and even for medium sized programs. Thus scalability is very limited in this technique.

2.3 Algorithmic Debugging

Algorithmic Debugging [26,27] is a semi-automatic debugging technique that is based on the answers of the programmer to a series of questions generated

186   J. González et al.

automatically by the algorithmic debugger. The questions are always whether a given result of a method invocation with given input values is actually correct. The answers provide the debugger with information about the correctness of some (sub)computations of a given program; and the debugger uses them to guide the search for the bug until a buggy portion of code is isolated.

Example 1. Consider the Java program in Figure 2 that simulates Tic-Tac-Toe games —we suggest the reader not to see the code now, and try to debug this program without seeing the code. This is possible with AD as it is shown in the following debugging session—. This program is buggy, and thus it does not produce the expected marks in the board. Class *Replay* reads from a file a new game and it reproduces the game using a *TicTacToe* object. The null character is represented in Java with '\u0000'.

An AD session for this program is shown below where boards are represented with a picture for clarity (e.g., {{X,",", "}{O,",", "}{",",", "}} is represented

with ⊞). For the time being ignore column Node:

```
Starting Debugging Session...
Node  Initial context        Method call        Final context        Answer

(2) [turn='X',board= ⊞ ]  game.mark('X',0,0)  [turn='O',board= ⊞ ] ?  YES

(7) [turn='O',board= ⊞ ]  game.mark('O',0,1)  [turn='X',board= ⊞ ] ?  NO

(8) [turn='X',board= ⊞ ]  game.win(0,1)=false [turn='X',board= ⊞ ] ?  YES

Bug found in method: TicTacToe.mark(char, int, int)
Discovered with the call: game.mark('O',0,1)
```

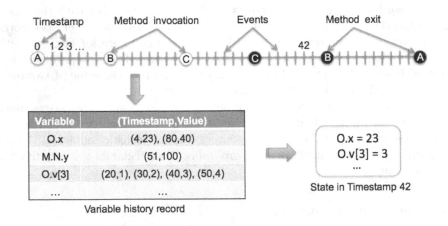

Fig. 1. Timestamps-based scheme to store traces in Omniscient Debugging

```
public class Replay {
    public static void main(String[] args) throws IOException {
        TicTacToe game = new TicTacToe();
        FileReader file = new FileReader("./game.rec");
        play(game, file);
    }
    private static void play(TicTacToe game, FileReader file) throws IOException {
        BufferedReader br = new BufferedReader(file);
        String linea = br.readLine();
        while ((linea = br.readLine()) != null) {
            char player = linea.charAt(0);
            int row = Integer.parseInt(linea.charAt(2) + "");
            int col = Integer.parseInt(linea.charAt(4) + "");
            game.mark(player, row, col);
        }
    }
}

public class TicTacToe {
    private static boolean equals(char c1, char c2, char c3) {
        return c1 == c2 && c2 == c3;
    }

    private char turn = 'X';
    private char[][] board = new char[3][3];

    public void mark(char player, int row, int col) {
        if (turn == '\u0000' || turn != player
            || row < 0 || row > 2 || col < 0 || row > 2
            || board[row][col] != '\u0000')
            return;

        board[col][row] = player; // Bug!! Correct: board[row][col] = player;
        turn = turn == 'X' ? '0' : 'X';
        if (win(row, col))
            turn = '\u0000';
    }
    private boolean win(int row, int col) {
        if (board[row][col] == '\u0000')
            return false;
        if (equals(board[row][0], board[row][1], board[row][2]))
            return true;
        if (equals(board[0][col], board[1][col], board[2][col]))
            return true;
        if (col == row && equals(board[0][0], board[1][1], board[2][2]))
            return true;
        if (col + row == 2 && equals(board[0][2], board[1][1], board[2][0]))
            return true;
        return false;
    }
}
```

Fig. 2. Example program

Note that the debugger generates questions, and the programmer only has to answer the questions with YES or NO. It is not even necessary to see the code. Each question is about the execution of a particular method invocation, and the programmer answers YES if the execution is correct (i.e., the output and the final context are correct with respect to the input and the initial context) and NO otherwise.

At the end, the debugger points out the specific call to a method in the code that revealed a bug in that method. In this case, method *TicTacToe.mark* is wrong. This method first checks whether the movement is correct (e.g, it is the player's turn, the mark is inside the board, etc.). If the movement is correct, then it places the mark in the corresponding position of `board`, it updates the next player to make a movement, and it finally checks whether this mark wins the game. Unfortunately, the programmer interchanged the row and the column producing a bug. This error can be easily corrected by replacing `board[col][row] = player` by `board[row][col] = player`.

Typically, algorithmic debuggers have a front-end that produces a data structure representing a program execution—the so-called *execution tree* (ET) [23]—and a back-end that uses the ET to ask questions and process the programmer's answers to locate the bug. Each node of the ET contains an equation that consists of a method execution with completely evaluated arguments and results. The node also contains additional information about the context of the method before and after its execution (attributes values or global variables in the scope of the method).

Essentially, AD is a two-phase process: During the first phase, the ET is built, while in the second phase, the ET is explored. The ET is constructed as follows: The root node is (usually) the *main* function of the program; for each node n with associated method m, and for each method invocation done from the definition of m, a new node is recursively added to the ET as the child of n.

Example 2. Consider again the Java program in Figure 2. Figure 3 depicts the portion of the ET associated with the execution of the method `play(game, file)` using *game.rec* as the input file. Each node contains:

- A string representing the method call (including input and output) depicted at the top of each node.
- The variables (and their values) in the scope at the beginning and at the end of the method execution. When the value of a variable is modified during the execution of the method, the node contains both values on the left and on the right of the node respectively. When the variable is not modified, it is shown only once in the middle of the node.

Once the ET is built, in the second phase, the debugger uses a strategy to traverse the ET asking an oracle to answer each question. For instance, each question in the debugging session of Example 1 corresponds to a node (see column `Node`) of the ET in Figure 3. These nodes have been selected by the strategy Divide & Query [26]. After every answer, some nodes of the ET are marked as correct or wrong. When all the children (if any) of a wrong node are correct, the node becomes buggy and the debugger locates the bug in the part of the program associated with this node [24].

Theorem 1 (Correctness of AD [24]). *Given an ET with a buggy node n, the method associated with n contains a bug.*

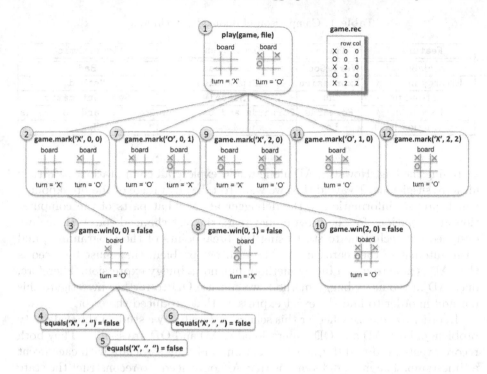

Fig. 3. ET associated with the call `play(game, file)` of the program in Figure 2

Theorem 2 (Completeness of AD [26]). *Given an ET with a bug symptom (i.e., the root is a method with a wrong final context), provided that all the questions generated by the debugger are answered, then, a bug will eventually be found.*

The most important advantage of AD is its high level of abstraction and its semi-automatic nature. The main drawbacks of this technique are:

1. Low scalability. Each ET node needs to record a part of the computation state (i.e., the context before and after the method execution). Storing the ET of the whole execution can be unpractical.
2. The strategy that traverses the ET can ask unnecessary questions until it reaches the part of the computation that contains the bug.
3. Low granularity of the error found. This technique reports a method as buggy, instead of an expression.

2.4   Comparison of the Techniques and Empirical Analysis

Table 1 summarizes the strong and weak points of the techniques.

In our hybrid technique, we want to take advantage of the high abstraction level of AD. We also want to exploit the semi-automatic nature of this technique to speed up bug finding and to avoid errors introduced by the programmer when

Table 1. Comparison of debugging techniques

Feature	Trace	Omniscient	Algorithmic
Scalability	Very Good	Very bad	Bad
Error granularity	Expression	Expression	Method
Automatized process	Manual	Manual	Semi-automatic
Execution	Forwards	Forwards and backwards	Forwards and backwards
Abstraction level	Low	Low	High

searching the bug. However, AD alone would explore all computations as if they all were suspicious. To avoid this, we want to take advantage of the breakpoints, which provide information to the debugger about what parts of the computation are suspicious for the programmer (e.g., the last changed code). Hence, we designed our technique to start using the breakpoints of the programmer, and then automatize the search using AD. Another problem that must be faced is that AD is able to find a buggy method, but not a buggy expression. Therefore, once AD has found a buggy method, we can use OD to further investigate this method in order to find the exact expression that produced the error.

In order to analyze whether this scheme is feasible, we studied the scalability problem of both AD and OD. Operationally, AD and OD are similar. They both record events produced during an execution, and they associate with each event a timestamp. The main difference is that AD only needs to reconstruct the state of the events that correspond to method invocations and method exits (white and black circles in Figure 1). Moreover, AD does not need to store information about local variables—only about attributes and global variables—, which is an important difference regarding scalability.

We conducted some experiments to measure the amount of information stored by an algorithmic debugger to produce the ET of a collection of medium/large benchmarks (e.g., an interpreter, a parser, a debugger, etc.) accessible at:

http://www.dsic.upv.es/~jsilva/DDJ/#Experiments

Results are shown in Table 2.

Table 2. Benchmark results

Benchmark	var. num.	ET size	ET depth
argparser	8.812	2 MB	7
cglib	216.931	200 MB	18
kxml2	194.879	85 MB	9
javassist	650.314	459 MB	16
jtstcase	1.859.043	893 MB	57
HTMLcleaner	3.575.513	2909 MB	17

Column `var. num.` represents the total amount of variable changes stored. Column `ET size` represents the size of the information stored. Observe that the last benchmark needs almost 3 GB. Column `ET depth` is the maximum depth of the ET (e.g., in benchmark `jtstcase`, there was a stack of 57 activation records during its execution). If we consider that this information does not include local variables, then we can guess that the amount of information needed by an omniscient debugger can be huge. Clearly, these numbers show that neither AD nor OD are scalable enough as to be used with the whole program. They should be restricted to a part of the execution. For AD, we propose to restrict its use only to the part of the execution that corresponds to a breakpoint (i.e., the execution of the method where the breakpoint is located). For OD, we propose to restrict its use only to the part of the execution that corresponds to a single method (i.e., the method where AD identified a bug). This proposal is completely aligned with the previous ideas discussed: AD will only start in a suspicious area pointed out by a breakpoint, and OD will only be used when a buggy method has been found, and thus the programmer can trace backwards the incorrect values identified at the end of this method.

3   Hybrid Debugging

In this section we present our hybrid debugger for Java (HDJ) based on the ideas discussed in the previous section. It combines TD, AD and OD to produce a synergy that exploits the best properties and strong points of each technique.

We start by describing the steps followed in a hybrid debugging session. Consider the diagram in Figure 4 that summarizes our hybrid debugging method. We see three main blocks that correspond to TD, AD and OD. These blocks contain four items that have been numbered; and these items are connected by arrows. Black arrows represent an automatic process (performed by the debugger), whereas white arrows represent a manual process (performed by the programmer):

Trace Debugging. First, after a bug symptom is identified, the user explores the code as usual with the trace debugger and she places a breakpoint b_1 in a suspicious line (probably, inside one of the last modified parts of the code).

Algorithmic Debugging. Second, the debugger identifies the method m_1 that contains breakpoint b_1, and it generates an ET whose root method is m_1. This is completely automatic. Then, the user explores the ET using AD until

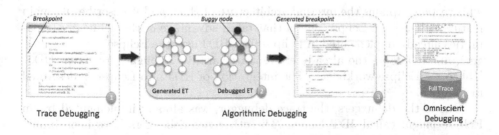

Fig. 4. Hybrid debugging with HDJ

a buggy node n is found. Note that, according to Theorem 2, if method m_1 is wrong, then it is guaranteed that AD will find a buggy node (and thus a buggy method). From n, the AD automatically generates a new breakpoint b_2. b_2 is placed in the definition of the method m_2 associated with n. And, moreover, b_2 is a *conditional* breakpoint that forces the debugger to stop at this definition, only when the bug is guaranteed to happen. The condition ensures that all values of the parameters of m_2 are exactly the same as their values in the call to m_2 associated with n.

Example 3. Consider a buggy node $\{x = 0\}\ m(42)\ \{x = 1\}$, where the definition of method m, *void m(int a)*, is located between lines 176 and 285. Then, the conditional breakpoint generated for it is $(176, \{x = 0, a = 42\})$. Alternatively, another conditional breakpoint can also be generated at the end of the method.

According to Theorem 1, because node n is buggy, then method m_2 contains a bug.

Omniscient Debugging. Third, the debugger acts as an omniscient debugger that explores method m_2 by reproducing the concrete execution where the bug showed up during AD. The user can explore the method backwards from the final incorrect result of the method. Observe that the OD phase is scalable because it only needs to record the trace of a single method. Note that all method executions performed from this method are known to be correct thanks to the AD phase.

The three phases described produce a debugging technique that takes advantage of all the best properties of each technique. However, one of the most important objectives in our debugger is to avoid a rigid methodology. We want to give the programmer the freedom to change from one technique to another at any point. For instance, if the programmer is using TD and decides to use OD in a method, she must be able to do it. Similarly, new breakpoints can be inserted at any moment, and AD can be activated when required. The architecture of our tool provides this flexibility that significantly increases the usability of the tool, and we think that it is the most realistic approach for debugging.

3.1   Architecture

This section explains the internal architecture of HDJ, and it describes its main features. HDJ is an Eclipse plugin that takes advantage of the debugging capabilities already implemented in Eclipse (i.e., HDJ uses the Eclipse's trace debugger), and it adapts the already existent Declarative Debugger for Java (DDJ) [15] to the Eclipse workbench. The integration of HDJ into Eclipse is described in Figure 5.

One of the debuggers, the trace debugger, was already implemented by an Eclipse plugin called JDT Debug. The other two debuggers have been implemented in the HDJ plugin. The tool allows the programmer to switch between three perspectives:

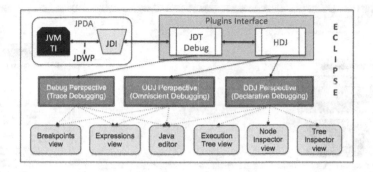

Fig. 5. Integration of HDJ into Eclipse

Debug: This perspective allows us to perform TD. It is the standard perspective of Eclipse for debugging. It is composed of several views and editors and it offers a wide functionality that includes conditional breakpoints, exception breakpoints, watch points, etc.

ODJ: This perspective allows us to perform OD. It contains the same views and editors that form the standard debug perspective. Therefore, although the programmer is using a different debugger with a totally different debugging mechanism, their GUI is exactly the same; and thus, the internal differences are transparent for her. The only difference is that ODJ allows us to explore the execution backwards. Internally, it uses a trace of the execution (as the one described in Section 2.2) that is stored in a database.

DDJ: This perspective allows us to perform AD. An usage example of this perspective interface is presented in Figure 6. In the figure we can see two of its three views and one editor. First, on the left we see the `ET view`, which contains the ET and the questions generated by the debugger. Second, on the right we see the `Node inspector`, which shows all the information associated with the selected ET node. This includes the initial context, the method invocation and the final context, where changes are highlighted with colors. Third, at the bottom we see the `Java editor`, which contains the source code and the breakpoints. This editor is shared between the three debuggers, and thus, all of them manipulate the same source code, and handle the same breakpoints of the programmer.

One of the important challenges when integrating two new debuggers into Eclipse was to allow all of them to debug the same program together (i.e., giving the programmer the freedom to change from one debugger to the other in the same debugging session). For this, all of them must have access to the same target source code (e.g., a breakpoint in the target source code should be shared by the debuggers), and use the same target Java Virtual Machine (JVM) and the same execution control over this target JVM. In the figure, this common target JVM is represented with the black box. The Java Virtual Machine Tools Interface (JVM TI) provides both a way to inspect the state and to control the

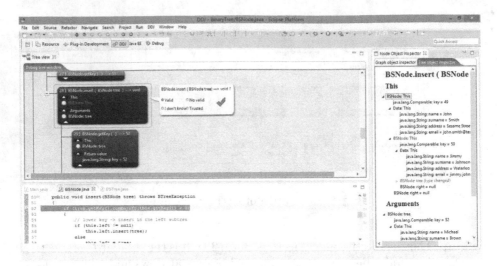

Fig. 6. Snapshot of HDJ (DDJ perspective)

execution running in the target JVM. The debuggers access it through the Java
Debug Interface (JDI) whose communication is ruled by the Java Debug Wire
Protocol (JDWP). This small architecture to control the JVM is called Java
Platform Debugger Architecture (JPDA) [21].

The integration of HDJ into Eclipse implies having three different debuggers
accessing and controlling the same JVM where the debuggee is being executed.
Therefore, our architecture uses two different JVMs that run in parallel and
communicate via JPDA. The first JVM is where the debuggee is executed. The
second JVM is where the debuggers are executed. It is important to remark
that the information of one JVM cannot be directly accessed by the other JVM.
Controlling one JVM from the other must be done through JPDA.

A first idea could be to execute the program in the target JVM and stop
it when the statement that the programmer wants to inspect is reached. However, this would imply to re-execute the program once and again every time the
programmer wants to perform a step backwards (i.e., to inspect the previous
statement). Obviously, this is a bad strategy, because every time the program
is re-executed, the state could change due to, e.g., concurrency, nondetermin-
ism, input, etc. Therefore, even if we reached the same statement, it could vary
between executions, and the information shown to the programmer would not
be confident. Hence, we need to use some memorization mechanism to store all
relevant states of a single execution.

Prior to our current implementation, our first design was conceived in such a
way that the JVM of the debugger directly controlled the JVM of the debuggee
using communication through JPDA. This implementation had to establish communication between both JVMs after every relevant event. This produced a
heavy interaction with a massive message passing that was not scalable even for

small programs. Therefore, we designed a second strategy whose key idea is to let the JVM of the debuggee to control itself. More precisely, before executing the debuggee in the JVM, we load a thread in this JVM so that, this thread directs the debugging of the program, thus, avoiding unneeded communication thorough JPDA. Figure 7 summarizes the internal architecture of the debugger to control the execution of the debuggee.

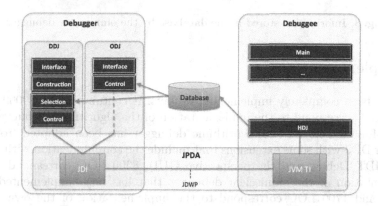

Fig. 7. Architecture of HDJ

The big boxes represent two JVMs. One for the debugger, and one for the debuggee. The debugger has two independent modules that can be executed in parallel: The algorithmic debugger DDJ, and the omniscient debugger ODJ. Each dark box represents a thread. DDJ has four threads: *interface* to control the GUI, *construction* to build the ET, *control* to control and communicate with the debuggee JVM, and *selection* to select the next question. ODJ has two threads: *interface* and *control* that perform similar tasks as in DDJ. In the debuggee, a new thread is executed in parallel with the program. This thread, called *HDJ*, is in charge of collecting all debugging information and storing it in a database. This information is later retrieved by threads *control*. Thread *HDJ* makes this approach scalable, because it allows to retrieve all the necessary information with a very reduced set of JPDA connections. In the case of OD, the information stored in the database by thread *HDJ* contains all changes of variable values occurred during the execution of the method being debugged.

Example 4. Consider again the debugging session in Example 1. In this debugging session AD determined that method mark is buggy, and that the bug shows up with the specific call game.mark('O',0,1). With this information, HDJ automatically generates a conditional breakpoint to debug this call. The information stored in the database by thread *HDJ* for this call is shown in Figure 8. Observe that only the variables that changed their value during the execution are stored.

Variable	(Timestamp, Value)
player	(0, 'O')
row	(0, 0)
col	(0, 1)
turn	(0, 'O'), (5, 'X')
board	(0, [['X',,],[,,],[,,]]), (4, [['X',,],['O',,],[,,]])

Fig. 8. Information stored in the database by the omniscient debugger

4   Implementation

HDJ has been completely implemented in Java. It contains about 29000 LOC: 19000 LOC correspond to the implementation of the algorithmic debugger (the internal functionality of the algorithmic debugger has been adapted from the debugger DDJ with some extensions that include the communication with JPDA trough JDT Debug, and the perspective GUI), 8300 LOC correspond to the implementation of the omniscient debugger that has been implemented from scratch, and 1700 LOC correspond to the implementation of the own plugin and its integration and communication with Eclipse. The debugger can make use of a database to store the information of the ET and the trace used in OD (if the database is not activated, the ET and the trace are stored in main memory). Thanks to JDBC, HDJ can interact with different databases. The current distribution includes both a MySQL and Access databases. The last release of the debugger is distributed in English, Spanish and French.

All described functionalities in this paper are completely implemented in the last stable release. This version is open and publicly available at:

http://www.dsic.upv.es/~jsilva/HDJ/

In this website, the interested reader can find installation steps, examples, demonstration videos and other useful material.

4.1   Empirical Evaluation

In order to measure the scalability of our technique, we conducted a number of experiments to achieve the time needed by the debugger to start the debugging session. The scalability of TD is ensured by the own nature of the technique that reexecutes the program up to a breakpoint, and then shows the current state. In fact, we use the Eclipse's standard trace debugger that is scalable no matter where the breakpoint is placed. In the case of AD, scalability could be compromised if the debugger is forzed to generate the ET of the whole execution. Even in this case, we are able to ensure scalability: (i) The memory problem is solved with a database. Our debugger never stores the whole ET in main memory. It uses a clustering mechanism to store and load from the database the subtrees of the ET that are dynamically needed by the GUI. (ii) The time problem is solved by allowing the debugger to start the debugging session even if the ET is not completely generated (i.e, our debugger is able to debug incomplete ETs while they are being generated) [16]. In the case of OD, we cannot ensure scalability if it is applied to the whole program. For this reason, we limit the application of OD to a single method. This is scalable as demonstrated by our empirical evaluation whose results are shown in Table 3.

Table 3. Benchmarks results for OD

Benchmark		Execution	Omniscient
Statements	Objects	Time (ms)	Time (ms)
0 - 9 (294)	0 - 1 (96)	5	2060
	2 - 5 (97)	273	3265
	6 - 18 (101)	27	4099
10 - 19 (29)	7 - 18 (10)	51	6527
	19 - 24 (10)	739	7348
	25 - 32 (9)	2062	12379
20 - 57 (10)	25 - 39 (3)	83	3999
	40 - 54 (4)	117	6347
	55 - 100 (3)	176	3757

This table summarizes the results obtained for 333 benchmarks. Each benchmark measures the time needed to generate all the information used in OD (the information stored in the database by thread *HDJ* in Figure 7). After this time, the debugger contains the state at any point in the method, and thus, the programmer can make backwards steps, jump to any point in the method and show the values of the variables at any point. These benchmarks correspond to all methods executed (333 different methods) by the loops2recursion Java library [17] applied over a collection of 25 Java projects. This library automatically transforms all loops in the Java projects to equivalent recursive methods.

All benchmarks have been grouped into three categories according to the number of statements executed in the method (0-9, 10-19, and 20-57). Inside each category, we indicate the number of benchmarks that fall on this category between parentheses. Categories have been divided in subcategories that indicate the number of objects that have changed during the execution of the method (i.e., the number of objects that must be inspected and stored in the database). We have a total of 9 subcategories. Each of them indicates the average time needed to execute the methods in that subcategory (Execution Time), and the time needed to generate the information for OD (Omniscient Time). All the information is generated between 2 and 12 seconds. The variability between the rows is dependent on the size of the objects changed. Clearly, row 6 has less objects to store than rows 7, 8 and 9, but these objects are bigger, and thus both the execution and omniscient times are higher.

5   Related Work

While a trace debugger is always present in modern development environments, algorithmic debuggers and omniscient debuggers are very unusual due to their scalability problems already discussed. There exist, however, a few attempts to implement algorithmic debuggers for Java such as the algorithmic debugger JDD [12] and its more recently reimplemented version DDJ [15]. Other debuggers exist that incorporate declarative aspects such as the Eclipse plugin JavaDD [11] or the Oracle JDeveloper's declarative debugger [10] however, they are not able to automatically produce questions and to control the search to automatically find the bug. This means that they lack the common strategies for AD implemented in standard algorithmic debuggers of declarative languages such as Haskell (Hat-Delta [8]) or Toy (DDT [7]). None of this debuggers can work with breakpoints as our debugger does.

The situation is similar in the case of omniscient debuggers. To the best of our knowledge, OmniCore CodeGuide [4] is the only development environment for Java that includes by default an omniscient debugger. Nevertheless, for the sake of scalability, this debugger uses a trace limited to the last few thousands events. Some ad-hoc implementations exist that can work stand-alone or be integrated in commercial environments [14,19,20,22,25]. Almost all these works focus on how to make OD more scalable [20,25]. For instance, by reducing the overhead of trace capture as well as the amount of information to store using partial traces that exclude certain trusted classes from the instrumentation process [19]. Other works try to enhance OD, e.g., with causality links [22] that provide the ability to jump from the point a value is observed in a given variable to the point in the past when the value was assigned to that variable. This can certainly be very valuable to resolve the chain of causes and effects that lead to a bug.

There have been several attempts to produce hybrid debuggers that combine different techniques. The debugger ODB [19] combines TD with OD. It allows the user to debug the program using TD and start recording the execution for OD when the user prefers. The debugger by Kouh et al. [18] combines AD with TD. Once the algorithmic debugger has found a buggy method, they continue the search with a trace debugger to explore this method (forwards) step-by-step. This idea is also present in our debugger, but we use OD instead of TD, and thus we also permit backwards steps. The debugger JIVE [9] combines TD, OD and dynamic slicing. It does not use AD, but allows the programmer to perform queries to the trace.

To the best of our knowledge, JHyde [13] is the only previous technique that combines TD, OD and AD. Unfortunately, we have not been able to empirically evaluate this tool (it is not publicly accessible); but considering its architecture, it is highly probable that it suffers from the same scalability problems as any other omniscient debugger. Unlike our solution, their architecture is based on program transformations that instrument the code to store the execution trace in a file as a side effect. First, this instrumentation and the execution of the trace usually takes a lot of time with an industrial program, so that the programmer has to wait for the instrumentation before starting to debug; and second, they store the trace of the whole program, while our scheme only needs the trace of a single method. The common point is that both techniques are implemented as an Eclipse plugin, and they both use the same data structure for OD and AD. This is important to reuse the trace information collected by the debugger. Another important feature implemented by both techniques is the use of a color vocabulary used in the views. This is very useful to allow the programmer to quickly see the changes in the state.

6   Conclusions and Future Work

Trace Debugging, Algorithmic Debugging and Omniscient Debugging are three of the most important debugging techniques. Some of them are more suitable for one specific kind of program, while for other programs the other techniques can be better. Furthermore, it is possible that one technique is desirable to debug one part of a program, while other technique is preferable for other part of the same program. For these reasons, in any development environment the three techniques should be available.

In this work, we introduce a new debugger called HDJ that implements and integrates the three techniques. The implementation uses a new debugging architecture that allows the three techniques to share the same target virtual machine, and the same target source code. This allows the programmer to change from one technique to the other in the same debugging session. Moreover, we present a new model for debugging that combines the three techniques. Our new debugging architecture is particularly interesting because it exploits the best properties of each technique (e.g., high precision, high abstraction level, etc.) and it minimizes the problems such as scalability. HDJ is open and freely distributed as an Eclipse plugin.

As future work, we plan to incorporate in our debugger the causality links functionality [22], which allows the programmer to click on a variable and jump to the statement that produced the value of this expression. We are also further improving the integration between AD and OD. In particular, we want to allow the programmer to select a node in the ET and automatically start an omniscient debugging session with the information of this node.

Acknowledgments. This work has been partially supported by the Spanish *Ministerio de Economía y Competitividad (Secretaría de Estado de Investigación, Desarrollo e Innovación)* under grant TIN2008-06622-C03-02 and by the *Generalitat Valenciana* under grant PROMETEO/2011/052. David Insa was partially supported by the Spanish Ministerio de Eduación under FPU grant AP2010-4415.

References

1. Swi-prolog (1987). http://www.swi-prolog.org/
2. Netbeans (1999) http://www.netbeans.org/
3. Eclipse (2003). http://www.eclipse.org/

4. Omnicore codeguide (2007). http://www.omnicore.com/en/codeguide.htm
5. Borland JBuilder (2008). http://www.embarcadero.com/products/jbuilder/
6. Sicstus prolog spider ide (2009). https://sicstus.sics.se/spider/
7. Caballero, R.: A Declarative Debugger of Incorrect Answers for Constraint Functional-Logic Programs. In: Proceedings of the 2005 ACM-SIGPLAN Workshop on Curry and Functional Logic Programming (WCFLP 2005), pp. 8–13. ACM Press, New York (2005)
8. Davie, T., Chitil, O.: Hat-delta: One Right Does Make a Wrong. In: Proceedings of the 7th Symposium on Trends in Functional Programming (TFP 2006) (April 2006)
9. Gestwicki, P., Jayaraman, B.: JIVE: Java Interactive Visualization Environment. In: Companion to the 19th Annual ACM-SIGPLAN Conference on Object-Oriented Programming Systems, Languages, and Applications (OOPSLA 2004), pp. 226–228. ACM Press, New York (2004)
10. Giammona, D.: ORACLE ADF - Putting It Together. Technical report, ADF Declarative Debugger Archives (November 2009)
11. Girgis, H., Jayaraman, B.: JavaDD: a Declarative Debugger for Java. Technical report, University at Buffalo (2006)
12. González, F., De Miguel, R., Serrano, S.: Depurador Declarativo de Programas Java. Technical report, Universidad Complutense de Madrid (2006). http://eprints.ucm.es/9114/
13. Hermanns, C., Kuchen, H.: Hybrid Debugging of Java Programs. In: Escalona, M.J., Cordeiro, J., Shishkov, B. (eds.) ICSOFT 2011. CCIS, vol. 303, pp. 91–107. Springer, Heidelberg (2013)
14. Montebello, M., Abela, C.: Design and Implementation of a Backward-In-Time. In: Chaudhri, A.B., Jeckle, M., Rahm, E., Unland, R. (eds.) NODe-WS 2002. LNCS, vol. 2593, pp. 46–58. Springer, Heidelberg (2003)
15. Insa, D., Silva, J.: An Algorithmic Debugger for Java. In: Proceedings of the 26th IEEE International Conference on Software Maintenance (ICSM 2010), pp. 1–6 (2010)
16. Insa, D., Silva, J.: Scaling Up Algorithmic Debugging with Virtual Execution Trees. In: Alpuente, M. (ed.) LOPSTR 2010. LNCS, vol. 6564, pp. 149–163. Springer, Heidelberg (2011)
17. Insa, D., Silva, J.: loops2recursion Java Library (2013). http://www.dsic.upv.es/~jsilva/loops2recursion/
18. Kouh, H.-J., Yoo, W.-H.: The Efficient Debugging System for Locating Logical Errors in Java Programs. In: Kumar, V., Gavrilova, M.L., Kenneth Tan, C.J., L'Ecuyer, P. (eds.) ICCSA 2003. LNCS, vol. 2667, pp. 684–693. Springer, Heidelberg (2003)
19. B. Lewis. Debugging Backwards in Time. Available in the Computing Research Repository 2003, (http://arxiv.org/abs/cs.SE/0310016), cs.SE/0310016
20. Lienhard, A., Gîrba, T., Wang, J.: Practical Object-Oriented Back-in-Time Debugging. In: Vitek, J. (ed.) ECOOP 2008. LNCS, vol. 5142, pp. 592–615. Springer, Heidelberg (2008)
21. S. Microsystems. Java Platform Debugger Architecture - JPDA (2010). http://java.sun.com/javase/technologies/core/toolsapis/jpda/
22. Mirghasemi, S., Barton, J., Petitpierre, C.: Debugging by lastChange. Technical report (2011). http://people.epfl.ch/salman.mirghasemi
23. Nilsson, H.: Declarative Debugging for Lazy Functional Languages. PhD thesis, Linköping, Sweden (May 1998)

24. Nilsson, H., Fritzson, P.: Algorithmic Debugging for Lazy Functional Languages. Journal of Functional Programming **4**(3), 337–370 (1994)
25. Pothier, G.: Towards Practical Omniscient Debugging. PhD thesis, University of Chile (June 2011)
26. Shapiro, E.: Algorithmic Program Debugging. MIT Press (1982)
27. Silva, J.: A Survey on Algorithmic Debugging Strategies. Advances in Engineering Software **42**(11), 976–991 (2011)

Compiling a Functional Logic Language:
The Fair Scheme

Sergio Antoy[✉] and Andy Jost

Computer Science Department, Portland State University,
Oregon, Portland, USA
antoy@cs.pdx.edu, andrew.jost@synopsys.com

Abstract. We present a compilation scheme for a functional logic programming language. The input program to our compiler is a constructor-based graph rewriting system in a non-confluent, but well-behaved class. This input is an intermediate representation of a functional logic program in a language such as Curry or $\mathcal{TOY}$. The output program from our compiler consists of three procedures that make recursive calls and execute both rewrite and pull-tab steps. This output is an intermediate representation that is easy to encode in any number of programming languages. We formally and tersely define the compilation scheme from input to output programs. This compilation scheme is the only one to date that implements a deterministic strategy for non-deterministic computations with a proof of optimality and correctness.

1   Introduction

Recent years have seen a renewed interest in the implementation of functional logic languages [16,18,23]. The causes of this trend, we conjecture, include the maturity of the paradigm [1,5,25], its growing acceptance from the programming languages community [6,13,28], and the discovery of and experimentation with new techniques [7,9,19] for handling the most appealing and most problematic feature of this paradigm—non-determinism.

Non-determinism can simplify encoding difficult problems into programs [6,10], but it comes at a price. The compiler is potentially more complicated and the execution is potentially less efficient than in deterministic languages and programs. The first issue is the focus of our work, whereas the second one is addressed indirectly. In particular, we present an easy to implement, *deterministic* strategy for *non-deterministic* computations. Our strategy is the only one to date in this class with a proof of its correctness and optimality.

Section 2 defines the *source* programs taken by our compiler as a certain class of non-confluent constructor-based graph rewriting systems. Section 3 formally defines and informally describes the design of our compiler by means of

This material is based in part upon work supported by the National Science Foundation under Grant No. 1317249.

A longer version of this paper, which include proofs of the statements, is available at http://web.cecs.pdx.edu/~antoy/homepage/publications.html

G. Gupta and R. Peña (Eds.): LOPSTR 2013, LNCS 8901, pp. 202–219, 2014.
DOI: 10.1007/978-3-319-14125-1_12

three abstract *target* procedures that can be easily implemented in any number of programming languages. Section 4 relates to each other *source* and *target* computations and states some properties of this relation. In particular, it shows that every step executed by the *target* program on an expression is needed to compute a value of that expression in the *source* program. Section 5 formalizes the strong completeness of our scheme: any value of an expression computed by the *source* program is computed by the *target* program as well. Sections 6 and 7 summarize related work and offer our conclusion.

2   Background

The class of rewrite systems that we compile is crucial for the relative simplicity, efficiency and provability of our design. Below we both describe and motivate this class. Functional logic programming languages, such as Curry [27,30] and *TOY* [21,39], offer to a programmer a variety of high-level features including expressive constructs (e.g., list comprehension), checkable redundancy (e.g., declaration of types and free variables), visibility policies (e.g., modules and nested functions), and syntactic sugaring (e.g., infix operators, anonymous functions).

A typical compiler transforms a program with these high-level features into a program that is semantically equivalent, i.e., it has the same I/O behavior, but is in a form that is easier to compile and/or execute. The details of this transformation are quite complex and include lambda lifting [34], elimination of partial applications and high-order function [41], elimination of conditions [4], transformation of non-inductively sequential functions into inductively sequential ones [4] and replacement of logic (free) variables with generator functions [11]. This transformed program, which is the input of our compilation scheme, is a graph rewriting system [22] in a class called *LOIS* (limited overlapping inductively sequential). Definitional trees characterize this class.

A *definitional tree* is a hierarchical structure consisting of *rule* nodes abstracting the rules of a program, *branch* nodes abstracting subexpressions that need to be evaluated for the application of the rules in the tree below the branch, and *exempt* nodes abstracting the incompleteness of certain definitions. Since definitional trees are a standard tool for the implementation of functional logic languages, we defer to [2,5] for the details. A defined operation is *inductively sequential* when it has a definitional tree.

Definition 1 (LOIS). *A LOIS system is a constructor-based graph rewriting system R in which every operation of the signature of R either is the binary choice operation denoted by the infix symbol "?" and defined by the rules:*

$$\begin{aligned} \text{x ? \_ = x} \\ \text{\_ ? y = y} \end{aligned} \tag{1}$$

or is inductively sequential. A LOIS system will also be called a source *program.*

All the non-determinism of a *LOIS* system is confined to the *choice* operation, which is also the only non-inductively sequential operation. While its rules can be used in a rewriting computation, the code generated by our compiler will not (explicitly) apply these rules. The reason is that the application of a rule of (1) makes an irrevocable decision in a computation. In this event, the completeness of computations can be ensured by techniques such as backtracking or copying which have undesirable aspects [7]. By avoiding the application of the *choice* rules, pull-tabbing (also bubbling [8,9]) makes no irrevocable decisions.

   LOIS systems are an ideal core language for functional logic programs for the reasons discussed below and are therefore the *source* programs of our compiler.

1. Any *LOIS* system admits a complete, sound and optimal evaluation strategy [3]. (A crucial difference between this strategy and our work is explained in Sect. 6.)
2. Any constructor-based conditional rewrite system is semantically equivalent to a *LOIS* system [4].
3. Any *narrowing* computation in a *LOIS* system is semantically equivalent to a *rewriting* computation in another similar *LOIS* system [11,36].
4. In a *LOIS* system, the order of execution of disjoint steps of an expression does not affect the value(s) of the expression [3,12].

Below we define a binary relation on nodes (and the expressions rooted by these nodes, since they are in a bijection) that with a slight abuse we call *needed*. This relation is at the core of some of our results.

Definition 2 (Needed). *Let S be a source program, e an expression of S whose root node we denote by p, and n a node of e. Node n is needed for e, and similarly needed for p, iff in any derivation of e to a constructor-rooted form the subexpression of e at n is derived to a constructor-rooted form. A node n (and the redex rooted by n, if any) of a state e of a computation in S is needed iff it is needed for some maximal operation-rooted subexpression of e.*

When a node n is needed for an expression e and n roots a redex, the subexpression at n is a needed redex of e in the classic sense of [32]. Our definition of need is well-suited for constructor-based systems and is convenient because it may "see" the need of a subexpression s before s becomes a redex and even in the case in which s will not become a redex (see the following definition of failure). The relationships between our definition and the classic one are explored in [15].

   Situations where a node n, root of an *irreducible* expression, is needed for an expression e enable aborting a possibly non-terminating computation of e which cannot produce a value. The next definition formalizes this point. An example will follow.

Definition 3 (Failure). *Let S be a source program and e an operation-rooted expression of S. Expression e is a failure iff there exists no derivation of e to a constructor-rooted form. When e is a failure, we may denote it with the symbol "$\perp$" instead of e if the nodes, labels, and other components of e are of no interest.*

In general, telling whether an expression e is a failure is undecidable, since it entails knowing whether some computation of e terminates. However, detecting failures in programming is commonplace. Indeed, in many programming languages a failure goes by the name of *exception*, a name that also denotes the mechanism for recovering from computations failing to produce a value. In functional logic programming, because of non-determinism, there are useful programming techniques based on failing computations [10] and failures are simply and silently ignored. Detecting some failures is easy, even in the presence of non-terminating computations. For example, consider the expression $e = \texttt{loop+(1/0)}$, where loop is defined below and the other symbols have their usual meaning:

$$\texttt{loop = loop} \tag{2}$$

It is immediate to see that the only redex of e is loop and consequently the computation of e does not terminate. Relying on the intuitive meaning of the symbols, since we have not defined them by rewrite rules, 1/0 is a failure, but its value would be needed to evaluate e. Hence, e itself is a failure. Thus, the computation of e can be terminated (in a failure) even though e is reducible and loop is a needed redex in the classic sense [32].

The definition of the compiler in Fig. 2 rewrites failures to the distinguished symbol "$\perp$". These rewrites are only a notational convenience to keep the presentation compact. An implementation needs not rewrite failures to the "$\perp$" symbol. Instead, the internal representation of a node may be tagged to say whether that node is the root of a failure.

3   Compilation

3.1   Preliminary Definitions

Our compilation scheme is abstract in the sense that both input and output of the compiler are programs expressed in convenient intermediate languages. The advantage of this abstraction is decoupling a concrete language from a concrete implementation so that different functional logic languages can be mapped to different run-time environments. This simplifies design and eases experimentation, which are essential for high-performance implementations.

The input of the compilation is a *LOIS* system described in the previous section. The output of the compilation consists of three procedures denoted **D** (Dispatch), **N** (Normalize) and **S** (Step). These procedures make recursive calls, and execute *rewrite* [22, Def. 23] and *pull-tab* [7, Def. 2] steps. A concrete compiler only has to represent graphs as objects of some language L and map the *target* procedures into procedures (functions, methods, subroutines, etc.) of L that execute both the recursive calls and the replacements originating from the steps.

This style of compilation for functional logic languages was pioneered in [16], where three procedures were also defined for the same purpose. We will compare these two approaches in Section 6, but in short, our strategy handles failures, avoids "don't know" non-determinism, and ensures the (strong) completeness of computations. None of these properties holds for the scheme of [16].

Pull-tabbing [7,19] is a technique for computing in graph rewriting systems that avoids making any *irrevocable* non-deterministic decisions and incurs a very modest overhead when a computation is deterministic. Informally, if e is an expression of the form $s(\ldots, x?y, \ldots)$, where s is not the *choice* symbol, then a pull-tab step of e produces $s(\ldots, x, \ldots) ? s(\ldots, y, \ldots)$. Therefore, pull-tabbing is a binary relation over the expressions of a *source* program similar to rewriting—in a graph a (sub)graph is replaced. The difference with respect to a rewrite step is that the replacement is not an instance of the right-hand side of a rewrite rule. It seems very natural for pull-tab steps, as well, to call the (sub)graph being replaced the *redex* and to denote a step as a reduction.

Definition 4 (Pull-Tab). *Let e be an expression, n a node of e, referred to as the* target, *not labeled by the* choice *symbol and $s_1 \ldots s_k$ the successors of n in e. Let i be an index in $\{1, \ldots k\}$ such that s_i, referred to as the* source, *is labeled by the* choice *symbol and let t_1 and t_2 be the successors of s_i in e. Let e_j, for $j = 1, 2$, be the graph whose root is a fresh node n_j with the same label as n and successors $s_1 \ldots s_{i-1} t_j s_{i+1} \ldots s_k$. Let $e' = e_1 ? e_2$. The* pull-tab *of e with source s_i and target n is $e[n \leftarrow e']$ and we write $e \rightarrow e[n \leftarrow e']$.*

Without some caution, however, pull-tabbing is unsound with respect to rewriting because a pull-tab step clones a *choice*, and different clones of the *same choice* could be reduced to *different* alternatives in a single expression. For example, consider the operation:

```
xor True x = not x                                           (3)
xor False x = x
```

and the expression:

```
xor x x where x = False ? True                               (4)
```

A pictorial representation of this expression is shown in the left-hand side of Fig. 1. The *choice* of this expression is pulled up along two paths creating *two pairs* of strands, one for each path, which eventually must be pair-wise combined together. Some combinations will contain mutually exclusive alternatives, i.e., subexpressions that cannot be obtained by rewriting because they combine both the left and right alternatives of the same *choice*. Fig. 1 presents an example of this situation.

The soundness of pull-tabbing computations is preserved so long as the alternatives of a *choice* are never combined in the same expression [7]. To this aim, a node n labeled by the *choice* symbol is decorated with a *choice identifier* [7, Def. 1], such as an arbitrary, unique integer created when n is "placed in service" [7, Princ. 1]. When a *choice* is pulled up, this identifier is preserved. Should a *choice* be reduced to either of its alternatives, every other *choice* with the same identifier must be reduced to the same alternative. A very similar idea in a rather different setting was proposed by Brassel et al. [17,19]. A rewriting computation that for any *choice* identifier i consistently takes either the left or

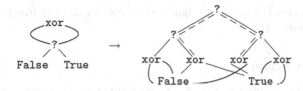

Fig. 1. Pictorial representation of two states of the computation of *(4)*: the initial state to the left, and the state after three pull-tab steps to the right. Every *choice* in every state has the same *choice identifier* which is then omitted from the representation. The dashed paths are inconsistent, since they combine the left and right alternatives of the same *choice*, and therefore should be discarded.

the right alternative of i is called a *consistent* computation. Furthermore, consistent computations with pull-tab steps are correct (i.e., sound and complete) with respect to rewriting computations [7, Th. 1]

The notion of *trace* [16], recalled below, allows us to keep track of a subgraph in a graph after the graph undergoes a sequence of replacements. The definition is non-trivial, but its application in an implementation is straightforward. We will discuss this point after defining the *target* procedures.

Definition 5 (Trace). *Let $g_0, g_1, \ldots$ be a sequence of expressions such that, for all $i > 0$, g_i is obtained from g_{i-1} by a replacement, i.e., there exist an expression r_{i-1} compatible [22, Def. 6] with g_{i-1} and a node p_{i-1} such that $g_i = g_{i-1}[p_{i-1} \leftarrow r_{i-1}]$. A node m of g_i is called a trace of a node n of g_j, for $j \leqslant i$, according to the following definition by induction on $i \geqslant 0$. Base case, $i = 0$: m is a trace of n iff $n = m$. Ind. case, $i > 0$: by assumption $g_i = g_{i-1}[p_{i-1} \leftarrow r_{i-1}]$ and by the induction hypothesis it is defined whether a node q of g_{i-1} is a trace of n. A node m of g_i is a trace of a node n of g_j iff there exists a trace q of n in g_{i-1} such that $m = q$ or m is the root of r_{i-1} and $q = p_{i-1}$.*

The trace of t captures the changes that t undergoes as it passes through *target* procedures. An implementation in which the expression being evaluated is a global, persistent datum passed to the *target* procedures by reference provides very efficient tracing. Considering traces is essential for the correctness of our approach. For example, consider the expression $f(p{:}t, p)$, where the two arguments of f are the same, i.e., t is shared. If t is evaluated to u, the resulting expression is $f(q{:}u, q)$, for some node q, i.e., the two arguments of f remain the same. Using traces preserves the sharing of arguments throughout a computation. Not only does this improve efficiency by avoiding repeated computations, it is essential to the soundness of computations. If the *same* non-deterministic expression is re-evaluated to a *different* value, the computation is unsound. Formalisms that might break this identity, e.g., because they look at expressions as trees instead of graphs, must introduce some device to preserve the identity. For example, CRWL [25] (a natural semantics) observes the call-time choice semantics [33] by reducing the function arguments to partial terms before performing the parameter-passing substitution and let-rewriting [38] (a small-step semantics) observes the call-time choice semantics by sharing function arguments through let-constructs.

3.2   The Fair Scheme

Our compiler is presented in Fig. 2. In the definition, whenever appropriate and understandable from the context, a single object may stand for a sequence containing only that object. Subsequences and/or individual objects in a sequence are separated by a semicolon. The empty sequence is denoted by *"null"*. The *target* procedures execute only two particular kinds of replacement. The graph where the replacement occurs is always the procedure argument and this argument is always the redex. Hence, we use the simpler notations introduced in Def. 6.

Definition 6 (*Target* Procedures).   *Each procedure of the target system takes a graph, or sequence of graphs in the case of D, as argument. Each procedure is defined by cases on its argument. Each case, called a rule, is selected by a liberal form of pattern matching and is defined by a possibly empty sequence of semicolon-terminated actions, where an action is either a recursive call to a target procedure, or a graph replacement [22, Def. 9] resulting from either a rewrite [22, Def. 23] or a pull-tab step [7, Def. 2]. In addition, procedure N returns a Boolean shown between curly braces in the pseudo-code. The rules are presented in Fig. 2. The rules have a priority as in common functional languages. Rules with higher priority come first in textual order. The application of a rule is allowed only if no rule of higher priority is applicable. Any reference to a node in the actions of any rule is the* trace *[16] of the node being referenced, i.e., tracing is consistently and systematically used by every rule without explicit notation. The notation* null *is a visible representation of an empty sequence of expressions, actions, steps, etc. depending on the context. The notations* REWR(p) *and* PULL(p) *are a rewrite and pull-tab steps, respectively, where p is the root of the replacement and the redex is the root of the argument of the rule where the notations occur. Graphs are written in* linear notation *[22, Def. 4], e.g., in $p{:}e$, p is the root node of the pattern expression e, with the convention that nodes are explicitly written only when they need to be referenced.*

Procedure **D** manages a queue of expressions being evaluated. If the queue is not empty, it examines the expression, e, at the front of the queue. Depending on the form of e, e may be removed from queue or it may undergo some evaluation steps and be placed back at the end of the queue. Initially, the queue contains only the top-level expression. Pull-tabbing steps pull *choices* toward the root. If the front of the queue is a *choice*-rooted expression e, e is removed from the queue and its two alternatives are placed at the end of the queue (rule **D**.1). Their order does not matter because any call to **N** terminates. Therefore, any expression in the queue is a subexpression of a state of computation of the top-level expression. Since we use pull-tab steps, some of these expressions could be inconsistent. Thus, we will refine this rule, after introducing the notion of fingerprint, to discard inconsistent expressions. If the expression at the front of

$$\mathbf{D}(g; \bar{G}) =$$

case g of

when $x\,?\,y$: $\mathbf{D}(\bar{G}; x; y)$;		D.1
when $\perp$: $\mathbf{D}(\bar{G})$;		D.2
when g is a value: $\mathbf{D}(\bar{G})$;	-- *yield g*	D.3
default: $\mathbf{N}(g)$;		
if $\mathbf{vn}(g)$ then $\mathbf{D}(\bar{G})$; else $\mathbf{D}(\bar{G}; g)$;		D.4
$\mathbf{D}(null) = null$;	-- *program ends*	D.5

$\mathbf{N}(c(\ldots, \perp, \ldots)) = null;\ \{\textbf{return } true\}$	N.1
$\mathbf{N}(c(\ldots, p\,{:}\,?(\_, \_), \ldots)) = \text{PULL}(p);\ \{\textbf{return } false\}$	N.2
$\mathbf{N}(c(x_1, \ldots, x_k)) = \mathbf{N}(x_1); \ldots; \mathbf{N}(x_k);$	
$\qquad \{\textbf{return } \mathbf{vn}(x_1) \vee \ldots \vee \mathbf{vn}(x_k)\}$	N.3
$\mathbf{N}(n) = \mathbf{S}(n);\ \{\textbf{return } false\}$	N.4

compile $\mathcal{T}$

case $\mathcal{T}$ of

when $rule(\pi, l \rightarrow r)$:	
$\quad$ output $\mathbf{S}(l) = \text{REWR}(r)$;	S.1
when $exempt(\pi)$:	
$\quad$ output $\mathbf{S}(\pi) = \text{REWR}(\perp)$;	S.2
when $branch(\pi, o, \bar{\mathcal{T}})$:	
$\quad \forall \mathcal{T}' \in \bar{\mathcal{T}}$ compile $\mathcal{T}'$	
$\quad$ output $\mathbf{S}(\pi[o \leftarrow \perp]) = \text{REWR}(\perp)$;	S.3
$\quad$ output $\mathbf{S}(\pi[o \leftarrow p\,{:}\,?(\_, \_)]) = \text{PULL}(p)$;	S.4
$\quad$ output $\mathbf{S}(\pi) = \mathbf{S}(\pi|_o)$;	S.5
$\mathbf{S}(c(\ldots)) = null$	S.6

Fig. 2. Compilation of a *source* program with signature Σ into a *target* program consisting of three procedures: **D**, **N**, and **S**. The rules of **D** and **N** depend only on Σ. The rules of **S** are obtained from the definitional trees of the operations of Σ with the help of the procedure compile. The structure of the rules and the meaning of symbols and notation are presented in Def. 6. The notation $\mathbf{vn}(x)$ stands for the value returned by $\mathbf{N}(x)$. The symbol c stands for a generic constructor of the *source* program and $\perp$ is the fail symbol. A symbol of arity k is always applied to k arguments. Line comments, introduced by "--", indicate when a value should be yielded, such as to the read-eval-print loop of an interactive session, and where the computation end. The call to a *target* procedure with some argument g consistently and systematically operates on the *trace* of g. Hence, tracing is not explicitly denoted.

the queue is a failure, it is removed from the queue (rule **D**.2). If the expression at the front of the queue is a value, it is removed from the queue as well (rule **D**.3) after being yielded to a consumer, such as the read-eval-print loop of an interpreter. Finally, if no previous case applies, the expression e at the front of the queue is passed to procedure **N**, which executes some steps of e (we will show a finite number) and returns whether the result should be either discarded or put back at the end of the queue (rule **D**.4). A result is discarded when it cannot be derived to a value. If the argument of **D** is the empty queue, the computation halts (rule **D**.5).

Procedure **N** either executes steps (of constructor-rooted expression), or invokes **S**. These steps do not depend on any specific operation of the *source* program. Like the other *target* procedures, the steps executed by **N** update the state of a computation. In addition to the other *target* procedures, **N** also returns a Boolean value. Procedure **N** returns the value "true" if it can determine that its argument cannot be derived to a value, and only in that case. This situation occurs when the argument e of an invocation of **N** is constructor-rooted, and an argument of the root is either a failure or (recursively) it cannot be reduced to a value (rule **N**.1). If an argument of the root of e is a *choice*, then e undergoes a pull-tab step (rule **N**.2). The resulting reduct is a *choice* that procedure **D** will split it into two expressions. If e is constructor-rooted, and neither of the above conditions holds, then **N** is recursively invoked on each argument of the root (rule **N**.3). Finally, if the argument e of an invocation of **N** is operation-rooted, then procedure **S** is invoked on e (rule **N**.4) in hopes that e will be derived to a constructor-rooted expression and eventually one of the previous cases will be executed.

The following example shows why *target* procedure **N** cannot rewrite constructor-rooted expressions to $\bot$. In the following code fragment, e is a constructor-rooted expression that cannot be derived to a value, hence a failing computation, but not a failure in the sense of Def. 3. Let snd be the operation that returns the second component of a pair and consider:

$$t = e \ ? \ \text{snd} \ e \ \text{where} \ e = (\bot, 0) \tag{5}$$

If e is rewritten to $\bot$, for some orders of evaluation t has no values, since snd $\bot$ is a failure. However, 0 is a value of t, since it is also a value of snd $(\bot, 0)$.

Procedure **S** executes a step of an operation-rooted expression. Each operation f of the *source* program contributes a handful of rules defining **S**. We call them **S**$_f$*–rules*. The pattern (in the *target* program) of all these rules is rooted by f. Consequently, the order in which the operations of the *source* program produce **S**-rules is irrelevant. However, the order among the **S**$_f$–rules is relevant. More specific rules are generated first and, as stipulated earlier, prevent the application of less specific rules. Let $\mathcal{T}$ be a definitional tree of f. At least one rule is generated for each node of $\mathcal{T}$. Procedure compile, which generates the **S**$_f$–rules, visits the nodes of $\mathcal{T}$ in post-order. If π is the pattern of a node $\mathcal{N}$ of $\mathcal{T}$, the patterns in the children of $\mathcal{N}$ are instances of π. Hence, rules with more specific patterns textually occur before rules with less specific patterns. In the following account, let e be an f-rooted expression and the argument of an application of an **S**$_f$-rule R and $\mathcal{N}$ the node of the definitional tree of f whose visit by compile produced R. If $\mathcal{N}$ is a *rule* node, then e is a redex and consequently reduced (rule **S**.1). If $\mathcal{N}$ is an *exempt* node, then e is a failure and it is reduced to $\bot$ (rule **S**.2). If $\mathcal{N}$ is a *branch* node, unless p is reduced to a constructor-rooted expression, e cannot be reduced to a constructor-rooted expression. Thus, if p

is a failure, e is a failure as well and consequently is reduced to $\perp$ (rule **S**.3). If p is a *choice*, e undergoes a pull-tab step (rule **S**.4). Finally, if p is operation-rooted, p becomes the argument of a recursive invocation of **S** (rule **S**.5). The last rule, labeled **S**.6, handles situations in which **S** is applied to an expression which is already constructor-rooted. This application occurs only to nodes that are reachable along multiple distinct paths, and originates only from rule **N**.3.

As an example, we show here the result of compiling operation xor, defined above (3), into the $\mathbf{S}_{xor}$ rules. To ease understanding we take some small notational liberties. Let t be an xor-rooted expression argument of **S**. We use pattern matching for the dispatching of cases on t. REWR(u) (resp. PULL(u)) abbreviates the rewrite (resp. pull-tab) step that replaces t by u.

$$
\begin{aligned}
&\mathbf{S}(\texttt{xor True x}) = \text{REWR}(\texttt{not x}); \\
&\mathbf{S}(\texttt{xor False x}) = \text{REWR}(\texttt{x}); \\
&\mathbf{S}(\texttt{xor} \perp \texttt{x}) = \text{REWR}(\perp); \\
&\mathbf{S}(\texttt{xor (x ? y) z}) = \text{PULL}((\texttt{xor x z}) \texttt{ ? } (\texttt{xor y z})); \\
&\mathbf{S}(\texttt{xor x y}) = \mathbf{S}(\texttt{x});
\end{aligned}
\tag{6}
$$

4   Properties

A *call tree* is a possibly infinite, finitely branching tree in which a branch is a call to a *target* procedure whereas a leaf is a step in the *source* program. This concept offers a simple relation between computations in a *source* program and computations in the corresponding *target* program. If e is an expression of the *source* program, a left-to-right traversal of the *call tree* of $\mathbf{D}(e)$ visits the sequence of steps of a computation of e in the *source* program. In this computation, we allow pull-tab steps in addition to rewrite steps, but never apply a rule of *choice*. An example of call tree is presented below.

Definition 7 (Call Tree). *Let S be a* source *program and T the* target *program obtained from S according to the* Fair Scheme. *A call tree* rooted by X, *denoted $\Delta(X)$, is inductively defined as follows: if X is a null action or a rewrite or pull-tab step, we simply let $\Delta(X) = X$. If X is a call to a* target *procedure of T executing a rule with sequence of actions $X_1; \ldots X_n$, then $\Delta(X)$ is the tree rooted by X and whose children are $\Delta(X_1), \ldots \Delta(X_n)$. If e is an expression of S, then a left-to-right traversal of rewrite and pull-tab steps of $\mathbf{D}(e)$ is called the* simulated *computation of e and denoted $\omega(\mathbf{D}(e))$.*

The name "simulated computation" [24, 35] stems from the observation that, under the assumption of Def. 7, $\omega(\mathbf{D}(e))$ is indeed a pull-tabbing computation of e in the *source* program. This will be proved in Cor. 2. We start with some preliminary results. We disregard the fact that pull-tabbing creates inconsistent expressions. Inconsistent expressions should not be passed as arguments to procedures **S** and **N**. We will describe later how to ensure this condition, but for the time being we ignore whether an expression is consistent.

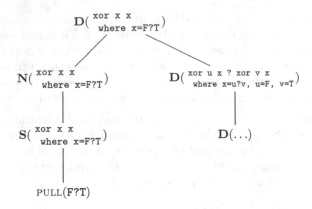

Fig. 3. Topmost portion of the call tree of the expression defined in *(4)*. The syntax of expressions is Curry. The values **False** and **True** are abbreviated by **F** and **T**, respectively.

Theorem 1 (Optimality). *Let S be a* source *program and* **S** *the step procedure of the corresponding* target *program. If e is an operation-rooted expression of S, then:*

1. **S**(e) *executes a replacement at some node n of e,*
2. *node n is needed for e,*
3. *if the step at n is the reduction to* ⊥, *then e is a failure.*

Theorem 1 is significant because it shows that the execution of **S**(e), for any operation-rooted expression e, terminates with a step. If the step is a rewrite to ⊥, then e has no values. This knowledge is important to avoid wasting unproductive computational resources on e. If the step is a rewrite, then that rewrite is unavoidable. More precisely, if e has some value (a fact that generally *cannot* be known *before* obtaining a value), then we have to execute that rewrite to obtain a value of e. In other words, computational resources are used conservatively. If the step is a pull-tab, then reducing the *choice* source of the pull-tab is needed to reduce the redex target of the pull-tab to a constructor-rooted expression. Generally, we cannot know in advance which alternative of the *choice* might produce a value, hence both alternatives must be tried. This is exactly what pull-tabbing provides, without committing to either alternative. In this case, too, no computational resources are wasted.

Below we state some properties of the computation space of the *target* program that culminate in Corollary 2. The correctness of the *Fair Scheme* is a relatively straightforward consequence of this corollary.

Corollary 1 (N Termination). *Let S be a* source *program and* **N** *the normalize procedure of the corresponding* target *program. For any expression e of S, the execution of* **N**(e) *terminates.*

Lemma 1 (Space). *Let S be a* source *program, $\mathbf{D}$ the dispatch procedure of the corresponding* target *program, and e an expression of S. If $\Delta(\mathbf{D}(e))$ is infinite, then:*

1. *$\Delta(\mathbf{D}(e))$ has exactly one infinite path, say B;*
2. *B is rightmost in $\Delta(\mathbf{D}(e))$;*
3. *B contains all and only the applications of $\mathbf{D}$ in $\Delta(\mathbf{D}(e))$;*
4. *rule $\mathbf{D}.4$ is applied an infinite number of times in B.*

Lemma 2 (State Subexpressions). *Let S be a* source *program, $\mathbf{D}$ the dispatch procedure of the corresponding* target *program, and e an expression of S. If $\mathbf{D}(L_0), \mathbf{D}(L_1), \ldots$ is the (finite or infinite) rightmost path of $\Delta(\mathbf{D}(e))$, then for every L_i in the path, the elements of L_i are subexpressions of a state of the computation of e.*

Corollary 2 (Simulation). *Let S be a* source *program, $\mathbf{D}$ the dispatch procedure of the corresponding* target *program, and e an expression of S. $\omega(\mathbf{D}(e))$ is a pull-tabbing derivation of e.*

Corollary 2 shows that a computation in the *target* program can be seen as a pull-tabbing computation in the *source* program. Each element in the queue argument of $\mathbf{D}$ is a subexpression s of a state of a computation t of the top-level expression e. Expression t is not explicitly represented. Every node in the path from the root of t to s, excluding the root of s, is labeled by the *choice* symbol. Hence, any value of s is a value of e. Furthermore, s can be evaluated independently of any other element of the queue argument of $\mathbf{D}$, though it may share subexpressions with them, which both improves efficiency and simplifies computing the values of e.

As presented in Fig. 2, the queue argument of $\mathbf{D}$ may contain unintended expressions originating from pull-tab steps. The following statement characterizes all and only the intended values. A simple modification of $\mathbf{D}$, discussed shortly, avoids creating these unintended expressions in the *target* program. A *consistent computation*, formally defined in [7, Def. 4], avoids combining the left and right alternatives of the clones of a same *choice* produced by pull-tab steps.

Theorem 2 (Correctness). *Let S be a* source *program, $\mathbf{D}$ the dispatch procedure of the corresponding* target *program, e an expression of S, and $\omega(\mathbf{D}(e)) = t_0 \rightarrow t_1 \rightarrow \ldots$ the simulated computation of e. Modulo a renaming of nodes: (1) if $e \xrightarrow{*} v$ in S, for some value v of S, and t_k is an element of $\omega(\mathbf{D}(e))$, for some $k \geqslant 0$, then $t_k \xrightarrow{*} v$, for some consistent computation in S; and (2) if t_k is an element of $\omega(\mathbf{D}(e))$, for some $k \geqslant 0$, and $t_k \xrightarrow{*} v$ is a consistent computation in S, for some value v of S, then $e \xrightarrow{*} v$ in S.*

Given an expression e of the *source* program, we evaluate $\mathbf{D}(e)$ in the *target* program. From any state of the computation of e, through consistent computations, we find all and only the values of e in S. Point (1) ensures a weak form of *completeness*—from any state of the computation of e in *target* program it is possible to produce any value of e. Point (2) ensures the *soundness* of the *fair scheme*—the *target* program does not produce any value of e that would not be produced in the *source* program. We will address the weakness of our completeness statement shortly.

The consistent computations sought for obtaining the values of e come almost for free with the *fair scheme*. A simple modification of $\mathbf{D}$ eliminates inconsistencies so that only intended values are produced. A *fingerprint* [19] is a finite set $\{(c_1, a_1), \ldots, (c_j, a_j)\}$, where c_i is a *choice* identifier [7, Def. 1] and $a_i \in \{1, 2\}$. A fingerprint is associated with a path in an expression. Given an expression e and a path $p = n_0, n_1, \ldots$ in e starting at the root of e, the fingerprint of p in e, denoted $F_e(p)$, is defined by induction on the length of p as follows. Base case: $F_e(n_0) = \varnothing$. Ind. case: Let $f = F_e(n_0, n_1, \ldots n_k)$, for $k \geqslant 0$. If n_k is labeled by the *choice* symbol and has *choice* identifier i, then $F_e(n_0, n_1, \ldots n_{k+1}) = f \cup \{(i, h)\}$, where $h = 1$, resp. $h = 2$, iff n_{k+1} is the first, resp. second, successor of n_k. Otherwise, n_k is not labeled by the *choice* symbol, and $F_e(n_0, n_1, \ldots n_{k+1}) = f$. A fingerprint f is *inconsistent* iff for some *choice* identifier i both $(i, 1)$ and $(i, 2)$ are in f.

An implementation associates a fingerprint to each expression in the queue argument of $\mathbf{D}$. Expressions with consistent fingerprints are evaluated as discussed earlier whereas expressions with inconsistent fingerprints are removed from the queue

```
D(g; Ḡ) =
   if fingerprint(g) is consistent
      then case g of
         ... rules as in Fig. 2 ...
      else D(Ḡ);
```

Fig. 4. Refinement of the dispatch procedure to avoid evaluating inconsistent expressions

5 Strong Completeness

The completeness statement of Th. 2 is weak since, e.g., any hypothetical *target* program that keeps rewriting any expression to itself satisfies the same completeness statement. Of course, rewriting any expression to itself is useless, whereas our *target* program rewrites only needed redexes (Th. 1.2). In orthogonal term rewriting systems, this suffices to compute the normal form of an expression, when it exists [32, Th. 3.26]. Our systems are not orthogonal, hence we cannot apply this result. The reason why the theorem of [32] does not extend to *LOIS* systems is the *choice* operation. Consider the rewrite rule:

$$f(n) \rightarrow f(n+1) ? n \tag{7}$$

The (infinite) derivation $f(0) \xrightarrow{+} f(1) ? 0 \xrightarrow{+} f(2) ? 1 ? 0 \xrightarrow{+} \ldots$ makes only steps without which some value of $f(0)$ could not be reached. Hence, in an intuitive sense these steps are needed for those values. Yet the derivation does not end in a normal form of $f(0)$ (nor does it end at all).

Ideally, we would like to state that if S is a *source* program and T is the program obtained from S according to the *Fair Scheme*, then if a computation of e in S produces a value v, then a computation of e in T produces v as well. Formalizing this statement is complicated by the fact that the computation of e in T may not terminate, yet still produce the value v. These considerations suggest to formulate the (strong) completeness of our scheme as follows.

Statement (Strong Completeness). *Let S be a* source *program, $\mathbf{D}$ the dispatch procedure of the corresponding* target *program, and e an expression of S. If $e \xrightarrow{*} v$ in S, for some value v, then $\Delta(\mathbf{D}(e))$ has a node $\mathbf{D}(v; \bar{G})$ for some, possibly empty, sequence of expressions $\bar{G}$.*

The above statement is exactly what we need in practice. If e has value v, node $\mathbf{D}(v; \bar{G})$ of $\Delta(\mathbf{D}(e))$ is where v becomes available for consumption.

We have a solid argument supporting the validity of the above statement, based on a construction showing that in $\Delta(\mathbf{D}(e))$, the computation space of e in the *target* program, there is a computation of e that makes the "same choices" as a computation of e in *source* program, and consequently these computations produce the same value.

Our argument is rigorous, but it assumes that some facts about term rewriting systems cross over to graph rewriting systems of the same class, e.g., repeatedly reducing needed redexes in *admissible graphs* [22, Def. 18] in inductively sequential systems is a normalizing strategy. While a result of this kind seems relatively easy to prove by reduction to term rewriting, we are not aware of any published proof. For this reason, we do not present the strong completeness statement as a theorem.

6   Related Work

Our work principally relates to the implementation of functional logic languages [18, 20, 21, 23, 29, 40]. This is a long-standing and active area of research whose difficulties originate from the combination of laziness, non-determinism and sharing [37].

The 90's saw various implementations, such as PAKCS [29] and $\mathcal{TOY}$ [21], in which Prolog is the target language. This target environment provides built-in logic variables, hence sharing, and non-determinism through backtracking. The challenge of these approaches is the implementation in Prolog of lazy functional computations [26].

The following decade saw the emergence of virtual machines [14, 31, 40], with a focus on operational completeness and/or multithreading. In some very recent implementations [18, 23] Haskell is the target language. This target environment provides lazy functional computations and to some extent sharing. The challenge of these approaches is the implementation of non-determinism in Haskell.

Our approach follows that of Antoy and Peters [16], which relies less on the peculiarities of the target environment than most previous approaches. The *target* procedures, being abstract, can be mapped to a variety of programming languages and paradigms. For example, one implementation [16] maps to OCaml using its functional, but not its object-oriented, features.

Our work extends the *Basic Scheme* [16]. The *Basic Scheme* is also defined by three procedures, but a direct procedure-wise comparison would not be fruitful. The *Basic Scheme* has neither a queue of subexpressions of the state of the computation nor does it handle explicit failures. It defines a procedure, **H**, obtained by compiling definitional trees, which is similar to **S**, but contrary to **S**, **H** does not return until its argument has been derived to a constructor-rooted expression. The *Fair Scheme* changes this condition to ensure fairness in the sense that any subexpression of a state of a computation which could produce a result is eventually reduced with a needed step. Fairness ensures that, given enough computational resources, all the values of any expression are eventually produced, a very desirable property of computations in any declarative language. We showed that achieving fairness is conceptually simple, the complexities of the definitions of *Fair* and *Basic Scheme* are comparable. One major contribution of the *Fair Scheme* is its provability. No proof of optimality is given in the presentation of the *Basic Scheme*, and it is not strongly complete.

A strategy for the same class of *source* programs accepted by our compiler was presented by one of us long before the *Basic Scheme* [3]. This strategy executes rewrite (and narrowing) steps, but not pull-tabs, and is non-deterministic, i.e., it assumes that a *choice* is *always* reduced to the "appropriate" alternative to produce a result, when there exists such a result. This assumption is obviously unrealistic. In practice, all implementations of this approach resolve the non-determinism in one way or another, but without any guarantees. By contrast, the *Fair Scheme* strategy is deterministic and its essential properties are well-understood and provable.

7   Conclusion

We presented the design of a compiler for functional logic programming languages. Our compiler is abstract and general in the sense that both *source* programs input to the compiler and *target* programs output from the compiler are encoded in intermediate languages. This separation greatly contributes to the flexibility of our compilation scheme. A *source* program is a graph rewriting system obtainable from a program in a concrete syntax such as that of Curry or $\mathcal{TOY}$. A *target* program consists of three procedures that make recursive calls and rewriting and pull-tab steps. From these procedures, it is easy to obtain concrete code in any number of programming languages.

Our compiler is remarkably simple—it is described by the 15 rules presented in Fig. 2. The simplicity of the compiler description enables us to prove properties of the compilation to a degree unprecedented for a work of this kind. We showed both correctness and optimality. Loosely speaking *correctness* means that the *target* code produces all and only the results produced by the *source* code, and *optimality* means that the *target* program makes only steps that the *source* program must make to obtain a result.

The focus of this paper has been formalizing the *Fair Scheme* and discovering and proving some of its fundamental properties. Future work will focus on the implementation. The presentation of the *Fair Scheme* in Fig. 2 is conceptually simple and suitable to prove various properties of the computations of the *target* program. This presentation is not intended as a faithful or complete blueprint of an implementation.

The *Fair Scheme* is the only deterministic strategy for non-deterministic functional logic computations with a proof of optimality and correctness.

Acknowledgments. We thank the anonymous reviewers for constructive suggestions.

References

1. Albert, E., Hanus, M., Huch, F., Oliver, J., Vidal, G.: Operational semantics for declarative multi-paradigm languages. Journal of Symbolic Computation **40**(1), 795–829 (2005)
2. Antoy, S.: Definitional trees. In: Kirchner, H., Levi, G. (eds.) Algebraic and Logic Programming. LNCS, vol. 632, pp. 143–157. Springer, Heidelberg (1992)
3. Antoy, S.: Optimal non-deterministic functional logic computations. In: Hanus, M., Heering, J., Meinke, K. (eds.) ALP 1997 and HOA 1997. LNCS, vol. 1298, pp. 16–30. Springer, Heidelberg (1997). Extended version at http://cs.pdx.edu/ antoy/homepage/publications/alp97/full.pdf
4. Antoy, S.: Constructor-based conditional narrowing. In: Proc. of the 3rd International Conference on Principles and Practice of Declarative Programming (PPDP 2001), pp. 199–206. ACM, Florence, Italy (September 2001)
5. Antoy, S.: Evaluation strategies for functional logic programming. Journal of Symbolic Computation **40**(1), 875–903 (2005)
6. Antoy, S.: Programming with narrowing. Journal of Symbolic Computation **45**(5), 501–522 (2010)
7. Antoy, S.: On the correctness of pull-tabbing. TPLP **11**(4–5), 713–730 (2011)
8. Antoy, S., Brown, D., Chiang, S.: Lazy context cloning for non-deterministic graph rewriting. In: Proc. of the 3rd International Workshop on Term Graph Rewriting, Termgraph 2006, Vienna, Austria, pp. 61–70 (April 2006)
9. Antoy, S., Brown, D.W., Chiang, S.-H.: On the correctness of bubbling. In: Pfenning, F. (ed.) RTA 2006. LNCS, vol. 4098, pp. 35–49. Springer, Heidelberg (2006)
10. Antoy, S., Hanus, M.: Functional logic design patterns. In: Hu, Z., Rodríguez-Artalejo, M. (eds.) FLOPS 2002. LNCS, vol. 2441, pp. 67–87. Springer, Heidelberg (2002)
11. Antoy, S., Hanus, M.: Overlapping rules and logic variables in functional logic programs. In: Etalle, S., Truszczyński, M. (eds.) ICLP 2006. LNCS, vol. 4079, pp. 87–101. Springer, Heidelberg (2006)

12. Antoy, S., Hanus, M.: Set functions for functional logic programming. In: Proceedings of the 11th ACM SIGPLAN International Conference on Principles and Practice of Declarative Programming (PPDP 2009), Lisbon, Portugal, pp. 73–82 (September 2009)
13. Antoy, S., Hanus, M.: Functional logic programming. Comm. of the ACM **53**(4), 74–85 (2010)
14. Antoy, S., Hanus, M., Liu, J., Tolmach, A.: A virtual machine for functional logic computations. In: Grelck, C., Huch, F., Michaelson, G.J., Trinder, P. (eds.) IFL 2004. LNCS, vol. 3474, pp. 108–125. Springer, Heidelberg (2005)
15. Antoy, S., Jost, A.: Compiling a functional logic language: The fair scheme. In: 23rd Int'nl Symp. on Logic-based Program Synthesis and Transformation (LOPSTR 2013), pp. 129–143. Dpto. de Systems Informaticos y Computation, Universidad Complutense de Madrid, TR-11-13, Madrid, Spain (Sept. 2013). Extended version available at http://web.cecs.pdx.edu/~antoy/homepage/publications/lopstr13/long.pdf
16. Antoy, S., Peters, A.: Compiling a functional logic language: *The basic scheme*. In: Schrijvers, T., Thiemann, P. (eds.) FLOPS 2012. LNCS, vol. 7294, pp. 17–31. Springer, Heidelberg (2012)
17. Brassel, B.: Implementing Functional Logic Programs by Translation into Purely Functional Programs. PhD thesis, Christian-Albrechts-Universität zu Kiel (2011)
18. Braßel, B., Hanus, M., Peemöller, B., Reck, F.: KiCS2: A new compiler from Curry to Haskell. In: Kuchen, H. (ed.) WFLP 2011. LNCS, vol. 6816, pp. 1–18. Springer, Heidelberg (2011)
19. Brassel, B., Huch, F.: On a tighter integration of functional and logic programming. In: Shao, Z. (ed.) APLAS 2007. LNCS, vol. 4807, pp. 122–138. Springer, Heidelberg (2007)
20. Brassel, B., Huch, F.: The Kiel Curry System KiCS. In: Seipel, D., Hanus, M. (eds.) Preproceedings of the 21st Workshop on (Constraint) Logic Programming (WLP 2007), Würzburg, Germany, pp. 215–223 (October 2007). Technical report 434
21. Caballero, R., Sánchez, J. (eds.) TOY: A Multiparadigm Declarative Language (version 2.3.1) (2007). http://toy.sourceforge.net
22. Echahed, R., Janodet, J.C.: On constructor-based graph rewriting systems. Technical report 985-I, IMAG (1997). ftp://ftp.imag.fr/pub/labo-LEIBNIZ/OLD-archives/PMP/c-graph-rewriting.ps.gz
23. Fischer, S., Kiselyov, O., Chieh Shan, C.: Purely functional lazy nondeterministic programming. J. Funct. Program. **21**(4–5), 413–465 (2011)
24. Fokkink, W., van de Pol, J.: Simulation as a correct transformation of rewrite systems. In: Privara, I., Ružička, P. (eds.) MFCS 1997. LNCS, vol. 1295, pp. 249–258. Springer, Heidelberg (1997)
25. González Moreno, J.C., López Fraguas, F.J., Hortalá González, M.T., Rodríguez Artalejo, M.: An approach to declarative programming based on a rewriting logic. The Journal of Logic Programming **40**, 47–87 (1999)
26. Hanus, M.: Efficient translation of lazy functional logic programs into Prolog. In: Proietti, M. (ed.) LOPSTR 1995. LNCS, vol. 1048, pp. 252–266. Springer, Heidelberg (1996)
27. Hanus, M. (ed.) Curry: An Integrated Functional Logic Language (Vers. 0.8.2) (2006). http://www-ps.informatik.uni-kiel.de/currywiki/
28. Hanus, M.: Multi-paradigm declarative languages. In: Dahl, V., Niemelä, I. (eds.) ICLP 2007. LNCS, vol. 4670, pp. 45–75. Springer, Heidelberg (2007)

29. Hanus, M. (ed.) PAKCS 1.9.1: The Portland Aachen Kiel Curry System (2008). http://www.informatik.uni-kiel.de/~pakcs

30. Hanus, M., Kuchen, H., Moreno-Navarro, J.J.: Curry: A truly functional logic language. In: Proceedings of the ILPS 1995 Workshop on Visions for the Future of Logic Programming, Portland, Oregon, pp. 95–107 (1995)

31. Hanus, M., Sadre, R.: An abstract machine for Curry and its concurrent implementation in Java. Journal of Functional and Logic Programming **1999**(Special Issue 1), 1–45 (1999)

32. Huet, G., Lévy, J.-J.: Computations in orthogonal term rewriting systems. In: Lassez, J.-L., Plotkin, G. (eds.) Computational Logic: Essays in Honour of Alan Robinson. MIT Press, Cambridge, MA (1991)

33. Hussmann, H.: Nondeterministic algebraic specifications and nonconfluent rewriting. Journal of Logic Programming **12**, 237–255 (1992)

34. Johnsson, T.: Lambda lifting: Transforming programs to recursive equations. In: Jouannaud, J.-P. (ed.) FPCA 1985. LNCS, vol. 201, pp. 190–203. Springer, Heidelberg (1985)

35. Kamperman, J.F.T., Walters, H.R.: Simulating TRSs by minimal TRSs a simple, efficient, and correct compilation technique. Technical report CS-R9605, CWI (1996)

36. López-Fraguas, F.J., de Dios-Castro, J.: Extra variables can be eliminated from functional logic programs. Electron. Notes Theor. Comput. Sci. **188**, 3–19 (2007)

37. López-Fraguas, F.J., Martin-Martin, E., Rodríguez-Hortalá, J., Sánchez-Hernández, J.: Rewriting and narrowing for constructor systems with call-time choice semantics. In: Theory and Practice of Logic Programming, pp. 1–49 (2012)

38. López-Fraguas, F.J., Rodríguez-Hortalá, J., Sánchez-Hernández, J.: A simple rewrite notion for call-time choice semantics. In: PPDP 2007: Proceedings of the 9th ACM SIGPLAN international conference on Principles and practice of declarative programming, pp. 197–208. ACM, New York, NY, USA (2007)

39. Fraguas, F.J.L., Hernández, J.S.: TOY: A multiparadigm declarative system. In: Narendran, P., Rusinowitch, M. (eds.) RTA 1999. LNCS, vol. 1631, pp. 244–247. Springer, Heidelberg (1999)

40. Lux, W. (ed.) The Münster Curry Compiler (2012). http://danae.uni-muenster.de/~lux/curry/

41. Warren, D.H.D.: Higher-order extensions to PROLOG: are they needed? Machine Intelligence **10**, 441–454 (1982)

Generating Specialized Interpreters for Modular Structural Operational Semantics

Casper Bach Poulsen[✉] and Peter D. Mosses

Department of Computer Science, Swansea University, Swansea SA2 8PP, UK
{cscbp,p.d.mosses}@swansea.ac.uk

Abstract. Modular Structural Operational Semantics (MSOS) is a variant of Structural Operational Semantics (SOS). It allows language constructs to be specified independently, such that no reformulation of existing rules in an MSOS specification is required when a language is extended with new constructs and features.

Introducing the Prolog MSOS Tool, we recall how to synthesize executable interpreters from small-step MSOS specifications by compiling MSOS rules into Prolog clauses. Implementing the transitive closure of compiled small-step rules gives an executable interpreter in Prolog. In the worst case, such interpreters traverse each intermediate program term in its full depth, resulting in a significant overhead in each step.

We show how to transform small-step MSOS specifications into corresponding big-step specifications via a two-step specialization by internalizing the rules implementing the transitive closure in MSOS and 'refocusing' the small-step rules. Specialized specifications result in generated interpreters with significantly reduced interpretive overhead.

Keywords: Interpreter generation · Structural operational semantics · Modular SOS · Specialization · Partial evaluation · Program derivation · Refocusing

1 Introduction

Background. Structural operational semantics (SOS) [21] provides a simple and direct method for specifying the semantics of programming language constructs and process algebras. The behaviour of constructs defined by an SOS is modelled by a labelled transition system whose transition relation is defined by a set of inference rules and axioms. For programming language semantics, the configurations of the transition system are typically given by terms and auxiliary entities, such as stores (recording the values of imperative variables before and after each transition step) and environments (determining the bindings of identifiers). In conventional SOS, auxiliary entities are explicit in all rules. This gives rise to the modularity problem with SOS: language extensions involving new auxiliary entities require reformulation of existing rules. Modular SOS (MSOS) [16] solves the modularity problem in SOS by implicitly propagating all unmentioned auxiliary entities.

© Springer International Publishing Switzerland 2014
G. Gupta and R. Peña (Eds.): LOPSTR 2013, LNCS 8901, pp. 220–236, 2014.
DOI: 10.1007/978-3-319-14125-1_13

Besides propagating auxiliary entities, small-step (M)SOS rules relate terms to partly evaluated terms. Evaluation in small-step (M)SOS specifications is given by a sequence of transition steps that eventually reaches a final state. In contrast, big-step (M)SOS rules relate terms directly to final states. As illustrated elsewhere [2,16], small-step rules are typically more concise than big-step rules for programming languages with abrupt termination and/or divergence.

The *PLanCompS*[1] project is developing an open-ended set of reusable *fundamental constructs* (or *funcons*), whose dynamic semantics is given by small-step MSOS rules[2]. Translating concrete constructs of a programming language into fundamental constructs gives a *component-based semantics*. MSOS rules provide a basis for verification, using, e.g., bisimulation [5,18] or structural induction on the underlying MSOS rules [16], and prototype interpreter generation. In this paper we focus on generating prototype interpreters in Prolog.

Contribution. It is well-known that big-step SOS rules can be compiled into Prolog clauses [7], and compilation of small-step MSOS rules into Prolog clauses has been utilized and hinted at in earlier publications [5,15–17]. The present paper presents the first systematic account of how to synthesize executable interpreters in Prolog from small-step MSOS specifications. We also assess and show how to reduce interpretive overhead in these interpreters.

The efficiency of generated interpreters is significantly improved by adapting *refocusing* [9] to MSOS. This is achieved by specializing a *refocusing rule* wrt an MSOS specification. The specialization forces evaluation of sub-terms, effectively transforming small-step rules into big-step rules. Compiling these big-step rules gives interpreters that avoid the computational overhead of decomposing the program term in each intermediate step, which previous interpreters generated from small-step MSOS specifications [3,5,16,17] have suffered from.

Through a subsequent specialization step, called *striding*, a small-step specification is transformed into its corresponding big-step counterpart by compressing corridor transitions, in a similar style to [8]. By *left-factoring* [1,20] the resulting big-step specification, back-tracking in generated interpreters is minimized.

We demonstrate and illustrate our techniques on MSOS specifications due to the pragmatic advantages of MSOS over SOS, but expect that the techniques are straightforward to extend to SOS.

Related work. The Maude MSOS Tool [3] executes MSOS specifications encoded as rewriting logic rules in Maude [6]. It allows for elegant representation of MSOS rules utilizing Maude features such as sorts and records. The approach to interpreting MSOS specifications is essentially similar to that of the Prolog MSOS Tool, where evaluation is implemented by sequences of small-step transitions, resulting in a significant overhead in each step.

The refocusing rule that we introduce is inspired by the work on refocusing by Danvy et al. [8,9]. That work is based on program transformations applied to

---

[1] Programming Language Components and Specifications: www.plancomps.org.
[2] In fact, funcons are specified using Implicitly Modular SOS [19], a variant of MSOS with syntax closer to SOS.

functional programs implementing reduction semantics. In contrast, the specialization we present here applies directly to MSOS rules, and is based on simple rule unfolding.

Partial evaluation in logic programming [10,13] has been extensively studied as a means of compiling programs and speeding up interpreters based on binding time analyses. The specializations that we consider here correspond to partial evaluations of the refocusing rule wrt to small-step inference rules.

Horn logical semantics [11] uses Horn clauses to relate terms to values or denotations. The big-step style inherent to Horn logical semantics makes specification of control instructions challenging, as witnessed by Wang et al.'s suggestion of using Horn logical continuation-based semantics [22] to handle abrupt termination: in the continuation-based approach each predicate is parameterized over terms, semantic domains, control stacks, and continuations. In contrast, small-step MSOS can deal with abrupt termination without parameterizing and modifying existing rules. This paper uses small-step MSOS for specification, and describes how to systematically derive a corresponding big-step specification by specialization.

Refocused rules bear a striking resemblance to Charguéraud's pretty-big-step rules [4]. As demonstrated in [2], pretty-big-step rules can be derived from small-step rules by unfolding refocused rules.

Outline. Section 2 reviews MSOS. Section 3 recalls how the Prolog MSOS Tool compiles MSOS rules into Prolog clauses. Section 4 shows how to improve the efficiency of generated interpreters by refocusing. Section 5 introduces the striding transformation, which unfolds refocused rules into classic big-step rules. The efficiency of generated naïve, refocused, and striding interpreters is assessed in Sect. 6. Section 7 concludes and suggests further lines of research.

2   Modular Structural Operational Semantics

This section outlines the main features of MSOS by comparing it with SOS.

2.1   An Example SOS

SOS rules define possible transitions between configurations in an underlying labelled transition system. In SOS, a configuration γ can make a transition to γ' if: (1) γ matches the *conclusion source* of an SOS rule

$$\frac{C_1 \quad \cdots \quad C_n}{\gamma \xrightarrow{\alpha} \gamma'}$$

where $\gamma \xrightarrow{\alpha} \gamma'$ is the rule conclusion, α is a (possibly empty) *transition label*, and C_i are the *premises* (e.g., transition steps or side-conditions) of the rule; and (2) using only SOS rules, for each premise C_i we can construct an upwardly branching derivation tree whose leaves are axioms, i.e., rules with empty premises and satisfied side-conditions[3]. For a more detailed introduction to (M)SOS, the reader is referred to [16, 21].

The following SOS rules define the applicative constructs $\mathsf{let}(id, e_1, e_2)$ and $\mathsf{bound}(id)$. We let ρ range over environments, id over identifiers, e over expressions, and v over values. The formula $\rho \vdash \gamma \to \gamma'$ asserts that γ makes a transition to γ' under environment ρ. $\rho[id \mapsto v]$ returns an environment ρ' where $\rho'(id) = v$ and $\rho'(id') = \rho(id')$ for $id' \neq id$.

$$\frac{\rho \vdash e_1 \to e_1'}{\rho \vdash \mathsf{let}(id, e_1, e_2) \to \mathsf{let}(id, e_1', e_2)} \; [\text{LET1-SOS}] \qquad \frac{\rho[id \mapsto v] \vdash e_2 \to e_2'}{\rho \vdash \mathsf{let}(id, v, e_2) \to \mathsf{let}(id, v, e_2')} \; [\text{LET2-SOS}]$$

$$\frac{}{\rho \vdash \mathsf{let}(id, v_1, v_2) \to v_2} \; [\text{LET3-SOS}] \qquad \frac{\rho(id) = v}{\rho \vdash \mathsf{bound}(id) \to v} \; [\text{BOUND-SOS}]$$

We now turn our attention to a semantics for sequential composition, $\mathsf{seq}(e_1, e_2)$, variable assignment, $\mathsf{assign}(ref, e)$, and variable dereferencing, $\mathsf{deref}(ref)$. We let σ range over stores, ref over references, and skip is a value. The formula $\langle e, \sigma \rangle \to \langle e', \sigma' \rangle$ asserts that the configuration given by term e and store σ can make a transition to the configuration given by term e' and store σ'.

$$\frac{\langle e_1, \sigma \rangle \to \langle e_1', \sigma' \rangle}{\langle \mathsf{seq}(e_1, e_2), \sigma \rangle \to \langle \mathsf{seq}(e_1', e_2), \sigma' \rangle} \; [\text{SEQ1-SOS}] \qquad \frac{}{\langle \mathsf{seq}(\mathsf{skip}, e_2), \sigma \rangle \to \langle e_2, \sigma \rangle} \; [\text{SEQ2-SOS}]$$

$$\frac{\langle e_1, \sigma \rangle \to \langle e_1', \sigma' \rangle}{\langle \mathsf{assign}(ref, e_1), \sigma \rangle \to \langle \mathsf{assign}(ref, e_1'), \sigma' \rangle} \; [\text{ASN1-SOS}]$$

$$\frac{\sigma' = \sigma[ref \mapsto v]}{\langle \mathsf{assign}(ref, v), \sigma \rangle \to \langle \mathsf{skip}, \sigma' \rangle} \; [\text{ASN2-SOS}] \qquad \frac{\sigma(ref) = v}{\langle \mathsf{deref}(ref), \sigma \rangle \to \langle v, \sigma \rangle} \; [\text{DEREF-SOS}]$$

Combining the constructs let, bound, seq, assign, and deref in SOS requires that we reformulate all rules: the rules for let and bound must propagate a store σ; similarly, seq, assign, and deref must propagate an environment ρ. We refrain from this tedious reformulation, and use MSOS instead.

2.2   Modular SOS

Like in SOS, MSOS rules define transition steps; i.e., a configuration makes a transition if we can construct a derivation tree using the rules defining the transition relation. Auxiliary entities in MSOS are encoded in the label of the transition relation, and are only explicitly mentioned when required. For example, in Fig. 1 the [LET1] rule makes no explicit mention of auxiliary entities, since they are not explicitly used by that rule. Crucially, computations in MSOS require labels on consecutive transitions to be *composable*. The remainder of this section defines MSOS labels and label composition.

---

[3] This notion of transition is based on positive SOS specifications. Rules with negative premises are not considered here.

$$\frac{e_1 \xrightarrow{\{\ldots\}} e_1'}{\mathsf{let}(id, e_1, e_2) \xrightarrow{\{\ldots\}} \mathsf{let}(id, e_1', e_2)} \text{ [LET1]} \qquad \frac{e_2 \xrightarrow{\{\mathbf{env}=\rho[id \mapsto v_1],\ldots\}} e_2'}{\mathsf{let}(id, v_1, e_2) \xrightarrow{\{\mathbf{env}=\rho,\ldots\}} \mathsf{let}(id, v_1, e_2')} \text{ [LET2]}$$

$$\frac{}{\mathsf{let}(id, v_1, v_2) \xrightarrow{\{-\}} v_2} \text{ [LET3]} \qquad \frac{\rho(id) = v}{\mathsf{bound}(id) \xrightarrow{\{\mathbf{env}=\rho,-\}} v} \text{ [BOUND]}$$

$$\frac{e_1 \xrightarrow{\{\ldots\}} e_1'}{\mathsf{seq}(e_1, e_2) \xrightarrow{\{\ldots\}} \mathsf{seq}(e_1', e_2)} \text{ [SEQ1]} \qquad \frac{}{\mathsf{seq}(\mathsf{skip}, e_2) \xrightarrow{\{-\}} e_2} \text{ [SEQ2]}$$

$$\frac{e_1 \xrightarrow{\{\ldots\}} e_1'}{\mathsf{assign}(ref, e_1) \xrightarrow{\{\ldots\}} \mathsf{assign}(ref, e_1')} \text{ [ASN1]}$$

$$\frac{\sigma' = \sigma[ref \mapsto v]}{\mathsf{assign}(ref, v) \xrightarrow{\{\mathbf{sto}=\sigma,\mathbf{sto}'=\sigma',-\}} \mathsf{skip}} \text{ [ASN2]} \qquad \frac{\sigma(ref) = v}{\mathsf{deref}(ref) \xrightarrow{\{\mathbf{sto}=\sigma,-\}} v} \text{ [DEREF]}$$

Fig. 1. MSOS rules for example constructs

Definition 1 (MSOS Label). *An* MSOS label L *is an unordered set of* label components, *where each* label component $\mathbf{ix} = E$ *consists of a distinct label index* $\mathbf{ix}$ *and an auxiliary entity* E *such that each index is either unprimed (e.g.,* $\mathbf{env}$*) meaning the label is* readable, *or primed (e.g.,* $\mathbf{sto}'$*) meaning the label is* writable.

Label variables *refer to sets of label components. The label variable '—' ranges over sets of* unobservable *label components, while label variables '...', X, Y, etc. refer to sets of arbitrary label components.*

Informally, a label component is *observable* if it exhibits side effects. An example of an observable label component is the component pair $\mathbf{sto} = \sigma$, $\mathbf{sto}' = \sigma'$ such that $\sigma \neq \sigma'$. The change from $\mathbf{sto}$ to $\mathbf{sto}'$ is an observable side effect. Another example of an observable component is illustrated by the print construct:

$$\frac{}{\mathsf{print}(v) \xrightarrow{\{\mathbf{out}'=[v],-\}} \mathsf{skip}} \text{ [PRINT]}$$

The $\mathbf{out}'$ component represents an output channel. An output channel may emit observable output several times during program execution. The observable output of evaluating the print construct above is the single element list $[v]$. The label component is unobservable when it contains the empty list, i.e., $\mathbf{out}' = [\,]$.

Environments, stores, and output channels each exemplify a distinct category of label components. These categories define how information is propagated between consecutive transition labels (i.e., how labels *compose*). Labels are defined by arrows in a category. The category gives the semantics of label composition [14, 16]. For the purpose of this paper, the following definition of a label composition operator '∘' suffices:

– Read-only label components (e.g., environments) remain unchanged between consecutive transition steps; e.g., $\{\mathbf{env} = \rho\} \circ \{\mathbf{env} = \rho\} = \{\mathbf{env} = \rho\}$.

- Read-write label components (e.g., stores) compose like binary relations; e.g., $\{\mathbf{sto}=\sigma', \mathbf{sto'}=\sigma''\} \circ \{\mathbf{sto}=\sigma, \mathbf{sto'}=\sigma'\} = \{\mathbf{sto}=\sigma, \mathbf{sto'}=\sigma''\}$.
- Write-only label components (e.g., output channels) are monoidal, generating lists of observable outputs; e.g., $\{\mathbf{out'}=l_2\} \circ \{\mathbf{out'}=l_1\} = \{\mathbf{out'}=l_1 \cdot l_2\}$, where '$\cdot$' is list concatenation, and l_1, l_2 are lists.

The formula $\mathsf{assign}(\mathit{ref}, v) \xrightarrow{\{\mathbf{sto}=\sigma, \mathbf{sto'}=\sigma', -\}} \mathsf{skip}$ says that $\mathsf{assign}(\mathit{ref}, v)$ makes a transition to skip under the label where the readable label component $\mathbf{sto}$ is σ and the writable label component $\mathbf{sto'}$ is σ'. It also says that no observable side effects occur in the remaining label components. Label composition in MSOS propagates the written σ' entity to the $\mathbf{sto}$ label component in the next transition. The following consecutive steps illustrate this propagation:

$$\mathsf{seq}(\mathsf{assign}(\mathit{ref}, v), \mathsf{skip}) \xrightarrow{\{\mathbf{sto}=\sigma, \mathbf{sto'}=\sigma', -\}} \mathsf{seq}(\mathsf{skip}, \mathsf{skip}) \xrightarrow{\{\mathbf{sto}=\sigma', \mathbf{sto'}=\sigma', -\}} \mathsf{skip}$$

The second transition has σ' in both $\mathbf{sto}$ and $\mathbf{sto'}$; i.e., no unobservable side effects occur on the $\mathbf{sto}, \mathbf{sto'}$ label components. Since no observable side effects occur in the second label, it could alternatively be written as $\{-\}$.

3 Generating MSOS Interpreters

This section describes how the Prolog MSOS Tool synthesizes interpreters in Prolog from MSOS specifications.

3.1 From MSOS Rule to Prolog Clause

MSOS terms are compiled as summarized in Table 1. Table 2 shows the compiled Prolog clauses for the seq construct. Solving a goal step(_,_,_) in Prolog using the compiled clauses corresponds to checking that the step is valid relative to the MSOS rules. Using the clauses in Table 2, we can check that the term seq(seq(skip, skip), skip) can make a transition step:

```
?- init_label(L), step(seq(seq(v(skip),v(skip)),v(skip)), L, X).
L = [env=map_empty, sto=map_empty, sto+=map_empty, out+=[]],
X = seq(v(skip), v(skip))
```

Here, init_label initializes MSOS labels with initial label components; in this case, env=map_empty, sto=map_empty, sto+=Sigma_, and out+=Out. Solving this goal executes the second Prolog clause in Table 2 which by label_instance(L,unobs) unifies sto=map_empty with sto+=Sigma_, and out+=Out with the unobservable output out+=[].

3.2 Implementing the Transitive Closure in Prolog

The steps predicate[4] generates the transitive closure of the transition relation:

---

[4] This predicate is not tail-recursive. It is, however, possible to construct a tail-recursive version: post_comp accumulates sequences of emitted write-only data. If this data were to be emitted as it is generated, the call to post_comp could be removed.

Table 1. Compilation of MSOS terms into Prolog predicates

	MSOS term	Prolog predicate

Rule	$\left[\begin{array}{c} C_1 \quad \cdots \quad C_n \\ \hline \gamma \xrightarrow{L} \gamma' \end{array}\right]$	step($[\![\gamma]\!]$,L,$[\![\gamma']\!]$) :-
		    label_instance(L,$[\![L]\!]$),
		    $[\![C_1]\!]$,$\cdots$,$[\![C_n]\!]$.

Transition step	$[\![\gamma \xrightarrow{L} \gamma']\!]$	step($[\![\gamma]\!]$,$[\![L]\!]$,$[\![\gamma']\!]$)
Readable label	$[\![\{\mathbf{ix}=E,X\}]\!]$	[$[\![\mathbf{ix}]\!]$=$[\![E]\!]$	$[\![X]\!]$]
Writable label	$[\![\{\mathbf{ix}'=E,X\}]\!]$	[$[\![\mathbf{ix}]\!]$+=$[\![E]\!]$	$[\![X]\!]$]
Unobservable label	$[\![-]\!]$	unobs
Map (e.g., $\rho,\sigma,\ldots$)	$[\![\,[x_1 \mapsto v_1,\ldots,x_n \mapsto v_n]\,]\!]$	[$[\![x_1]\!]$+>$[\![v_1]\!]$,$\ldots$,$[\![x_n]\!]$+>$[\![v_n]\!]$]
Terms, values, and label indices	$[\![t]\!]$	Prolog atoms, annotated with v(_) for values.
Variables	$[\![x]\!]$; $[\![x_1]\!]$; $[\![x']\!]$	X; X1; X_

Table 2. Compiled Prolog clauses for the seq construct

MSOS rule	Prolog clause

$\dfrac{e_1 \xrightarrow{\{\ldots\}} e_1'}{\mathsf{seq}(e_1,e_2) \xrightarrow{\{\ldots\}} \mathsf{seq}(e_1',e_2)}$	step(seq(E1,E2),L,seq(E1_,E2)) :-
	    label_instance(L,Dots),
	    step(E1,Dots,E1_).

$\mathsf{seq}(\mathsf{skip},e_2) \xrightarrow{\{-\}} e_2$	step(seq(v(skip),E2),L,E2) :- label_instance(L,unobs).

```
steps(T1,L,T3) :-
    pre_comp(L,L1), step(T1,L1,T2), mid_comp(L1,L2),
    steps(T2,L2,T3), post_comp(L1,L2,L).
steps(v(V),L,v(V)) :-
    label_instance(L,unobs).
```

These clauses are mutually exclusive; i.e., values are final terms for which no further transition is possible. pre_comp, mid_comp, and post_comp propagate readable and writable label components as described in Sect. 2.2.

```
?- init_label(L), steps(seq(seq(v(skip),v(skip)),v(skip)), L, X).
L = [env=map_empty, sto=map_empty, sto+=map_empty, out+=[]],
X = v(skip)
```

Prolog fails if no sequence of steps exists that yields a value:

```
?- init_label(L), steps(seq(seq(v(0),v(skip)),v(skip)), L, X).
false.
```

Figure 2 summarizes the number of inferences required for interpreters generated by the Prolog MSOS Tool to reduce terms of the structure[5]:

$$\underbrace{seq(seq(\cdots seq(skip, skip) \cdots, skip), skip)}_{n}$$

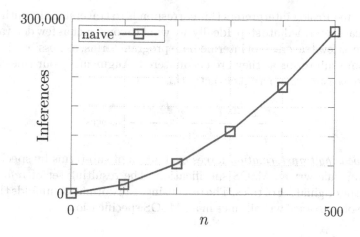

Fig. 2. Naïve evaluation of deeply nested seq terms

Since each step occurs on the outermost program term, the Prolog interpreter traverses the term in its full depth in each step, i.e., each step uses $O(n)$ inferences. It takes n steps to evaluate a seq term of depth n, hence evaluation of deeply nested seq terms uses $O(n^2)$ inferences. We next demonstrate how *refocusing* reduces the number of required inferences to $O(n)$.

4 Refocused MSOS Interpreters

The transitive closure implemented by the **steps** predicate in Prolog is straightforwardly internalized in MSOS by the $\rightarrow^*$ relation defined by the rules:

$$\frac{x \xrightarrow{L_1} y \qquad y \xrightarrow{L_2}{}^* z}{x \xrightarrow{L_2 \circ L_1}{}^* z} \text{ [TRANS]} \qquad \frac{}{v \xrightarrow{\{-\}}{}^* v} \text{ [REFL-V]}$$

Evaluating a term s using these rules proceeds by constructing an upwardly branching derivation tree, if one exists, from the root formula $s \xrightarrow{L}{}^* t$. Using $\Gamma, \Delta, \ldots$ to

---

[5] Right-nested seq terms do not suffer from runtime overhead. This is not the case, however, for deeply right-nested arithmetic expressions or λ-applications. We use left-nested seq terms here for simplicity of exposition.

refer to instances of $\rightarrow^*$ and $A, B, \ldots$ to refer to instances of $\rightarrow$, derivation trees have the structure:

$$
\cfrac{\cfrac{\vdots}{A} \quad \cfrac{\cfrac{\vdots}{B} \quad \cfrac{\cfrac{\vdots}{C} \quad \cfrac{\cdot\cdot}{}}{\Psi}}{\Delta}}{\Gamma}
$$

In generated Prolog interpreters this corresponds to traversing the entire program term in each intermediate step. Ideally, we want to construct as few derivation trees, and have as few traversals of intermediate program terms, as possible; i.e., we want to evaluate sub-terms as they are encountered. Augmenting our rules by the following *refocusing rule* permits exactly this:

$$
\frac{x \xrightarrow{L_1} y \qquad y \xrightarrow{L_2}{}^* z}{x \xrightarrow{L_2 \circ L_1} z} \; [\textsc{refocus}]
$$

The *refocusing transformation* forces evaluation of sub-terms by specializing the refocusing rule wrt an MSOS specification. The resulting set of *refocused rules* replace the original set of rules. The refocusing transformation unfolds the leftmost premise of [REFOCUS] wrt all rules in an MSOS specification:

$$
\frac{\cfrac{C \quad D}{B}\,[d_1] \quad \Gamma}{A}\;[\textsc{refocus}] \qquad \Longrightarrow \qquad \frac{C \quad D \quad \Gamma}{A}\;[d_1\text{-}\textsc{refocus}]
$$

Using refocused rules changes the structure of derivation trees:

$$
\cfrac{\cfrac{\cfrac{\vdots}{B}}{\cfrac{\vdots \quad}{A}} \quad \cfrac{\cfrac{C \quad \vdots}{\Gamma}}{\Psi}}{\Delta}
$$

For example, refocusing [SEQ1] (from Fig. 1, page 224) gives:

$$
\frac{\cfrac{e_1 \xrightarrow{L_1} e_1'}{\mathsf{seq}(e_1, e_2) \xrightarrow{L_1} \mathsf{seq}(e_1', e_2)}\;[\textsc{seq}1] \quad \mathsf{seq}(e_1', e_2) \xrightarrow{L_2}{}^* z}{\mathsf{seq}(e_1, e_2) \xrightarrow{L_2 \circ L_1} z}\;[\textsc{refocus}]
$$

$$
\Longrightarrow \qquad \frac{e_1 \xrightarrow{L_1} e_1' \quad \mathsf{seq}(e_1', e_2) \xrightarrow{L_2}{}^* z}{\mathsf{seq}(e_1, e_2) \xrightarrow{L_2 \circ L_1} z}\;[\textsc{seq}1\text{-}\textsc{refocus}]
$$

Unfolding [SEQ2] and applying the [REFL-V] rule trivially gives an identical rule. Thus the refocused rules for seq are:

$$
\frac{e_1 \xrightarrow{L_1} e_1' \quad \mathsf{seq}(e_1', e_2) \xrightarrow{L_2}{}^* z}{\mathsf{seq}(e_1, e_2) \xrightarrow{L_2 \circ L_1} z}\;[\textsc{seq}1\text{-}\textsc{refocus}] \qquad \frac{}{\mathsf{seq}(\mathsf{skip}, e_2) \xrightarrow{\{-\}} e_2}\;[\textsc{seq}2\text{-}\textsc{refocus}]
$$

Figure 3 summarizes the number of inferences the interpreter generated from the refocused MSOS specification uses to evaluate deeply nested seq terms. In contrast to naïve evaluation, the number of inferences increases linearly, since each sub-term is reduced when it is first encountered: evaluation uses $O(n)$ inferences.

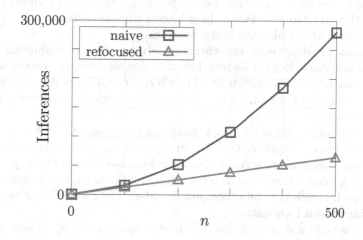

Fig. 3. Refocused and naïve evaluation of deeply nested seq terms

Introducing the refocusing rule permits sub-terms to be evaluated locally in derivations. Specializing the refocusing rule wrt an MSOS specification produces a specialized interpreter which forces evaluation of all sub-terms. However, forcing evaluation of sub-terms is not semantically sound in the presence of abrupt termination.

4.1 Refocusing and Abrupt Termination

Consider the language given by the following add, blocking, block, and loop constructs, where $+_i$ is integer addition, and **block**$'$ is a write-only label component:

$$\frac{}{\text{block} \xrightarrow{\{\text{block}'=1,-\}} \text{stuck}} \text{[BLOCK]} \qquad \frac{e \xrightarrow{\{\text{block}'=1,...\}} e'}{\text{blocking}(e) \xrightarrow{\{\text{block}'=0,...\}} \text{skip}} \text{[BLOCKING1]}$$

$$\frac{e \xrightarrow{\{\text{block}'=0,...\}} e'}{\text{blocking}(e) \xrightarrow{\{\text{block}'=0,...\}} \text{blocking}(e')} \text{[BLOCKING2]} \qquad \frac{}{\text{loop} \xrightarrow{\{-\}} \text{loop}} \text{[LOOP]}$$

$$\frac{}{\text{blocking}(v) \xrightarrow{\{-\}} v} \text{[BLOCKING3]} \qquad \frac{v = v_1 +_i v_2}{\text{add}(v_1, v_2) \xrightarrow{\{-\}} v} \text{[ADD1]}$$

$$\frac{e_1 \xrightarrow{\{...\}} e_1'}{\text{add}(e_1, e_2) \xrightarrow{\{...\}} \text{add}(e_1', e_2)} \text{[ADD2]} \qquad \frac{e_2 \xrightarrow{\{...\}} e_2'}{\text{add}(e_1, e_2) \xrightarrow{\{...\}} \text{add}(e_1, e_2')} \text{[ADD3]}$$

If a block term is evaluated inside a blocking term, evaluation terminates and produces the value skip. Evaluating the loop-construct results in divergence.

Under ordinary small-step evaluation of the term blocking(add(block, loop)) we have the two possible outcomes of evaluation: either evaluation terminates with the value skip, or it diverges. If block is evaluated, the **block'** = 1 label component is matched by the [BLOCKING1] rule for the blocking term, which terminates the program with value skip. Otherwise, the sub-term loop is evaluated, which results in a program term identical to the initial program.

Refocused evaluation, on the other hand, always diverges: evaluating block gives the term add(stuck, loop). This term has an evaluable sub-term, namely loop. Refocused evaluation forces evaluation of this term, resulting in divergence. In other words, adding the refocusing rule to a semantics with abrupt termination is not correct by default.

The issue of dealing with abrupt termination is symptomatic for big-step rules. In the presence of abrupt termination, one typically needs extra rules propagating the abruptly terminated term [4]. We show how to circumvent the problem with abrupt termination in refocused and big-step MSOS rules in a generic way: we introduce a special read-write label component, labeled by ε and ε', representing a flag indicating abrupt termination.

First, we add a single reflexive rule that propagates abruptly terminated configurations[6] ($\varepsilon = 1$), and update our existing evaluation rules to indicate that they apply only to configurations that are not abruptly terminated ($\varepsilon = 0$):

$$\frac{x \xrightarrow{\{\varepsilon=0, X_1\}} y \qquad y \xrightarrow{L_2}{}^{*} z}{x \xrightarrow{L_2 \circ \{\varepsilon=0, X_1\}}{}^{*} z} \text{[TRANS-}\varepsilon] \qquad \frac{x \xrightarrow{\{\varepsilon=0, X_1\}} y \qquad y \xrightarrow{L_2}{}^{*} z}{x \xrightarrow{L_2 \circ \{\varepsilon=0, X_1\}} z} \text{[REFOCUS-}\varepsilon]$$

$$\frac{}{v \xrightarrow{\{\varepsilon=0, —\}}{}^{*} v} \text{[REFL-V-}\varepsilon] \qquad \frac{}{x \xrightarrow{\{\varepsilon=1, —\}}{}^{*} x} \text{[REFL-}\varepsilon]$$

Second, MSOS specifications must explicitly indicate abrupt termination in rules. For example, rules that are sensitive to the behaviour of their sub-terms, such as [BLOCKING1] which inspects the writable **block'** component during evaluation of its sub-term, must explicitly indicate abruptly terminating steps via $\varepsilon, \varepsilon'$:

$$\frac{}{\text{block} \xrightarrow{\{\mathbf{block'}=1, \varepsilon'=1, —\}} \text{stuck}} \text{[BLOCK-}\varepsilon] \qquad \frac{e \xrightarrow{\{\mathbf{block'}=1, \varepsilon=0, \varepsilon'=1, ...\}} e'}{\text{blocking}(e) \xrightarrow{\{\mathbf{block'}=0, \varepsilon=0, \varepsilon'=0, ...\}} \text{skip}} \text{[BLOCKING1-}\varepsilon]$$

Using this alternative set of rules, refocused evaluation has the same possible outcomes for the example term blocking(add(block, loop)) as small-step evaluation.

Refocusing is a simple specialization which significantly reduces overhead compared to traditional small-step MSOS rules. However, it requires explicit specification of abrupt termination and of rules for constructs whose behaviour is sensitive to the behaviour of their sub-terms. It is ongoing work to identify syntactic

---

[6] We refer to configurations as being *abruptly terminated* rather than *stuck*, since the terms in the configuration may have computational behaviour. E.g., the add(stuck, loop) term is not stuck in a strict sense, since it has evaluable sub-terms.

constraints which uniquely distinguish abruptly terminating constructs and constructs that are sensitive to the number of steps their sub-terms can make. Such constraints would enable automatic insertion of $\varepsilon, \varepsilon'$ label components.

5   Big-Step MSOS Interpreters

A small-step transition relation relates terms to other partly evaluated terms. Under refocused evaluation, the transition relation relates terms directly to values or abruptly terminated terms. Refocused rules are therefore in big-step style. However, refocused rules may use several intermediate inferences to map a term to a value. The *striding transformation* specializes refocused rules to remove the extra overhead. The resulting rules are similar to classic big-step rules.

5.1   The Striding Transformation

The striding transformation has the effect of compressing 'corridor' transitions [8], i.e., transitions for which a unique further transition exists. The striding transformation specializes a refocused rule, $[d_1\text{-REFOCUS}]$, wrt another refocused rule, $[d_2\text{-REFOCUS}]$. The result is a big-step style rule, $[d_1\text{-}d_2\text{-STRIDING}]$:

$$\cfrac{C \quad D \quad \cfrac{E \qquad \Delta}{\Gamma}\ [d_2\text{-REFOCUS}]}{A}\ [d_1\text{-REFOCUS}] \quad\Longrightarrow\quad \cfrac{C \quad D \quad E \quad \Delta}{A}\ [d_1\text{-}d_2\text{-STRIDING}]$$

The striding transformation generates the set of all possible combinations of rule unfoldings. To filter semantically equivalent rules resulting from the transformation, we use *formal hypothesis simulation* (*fh-simulation*) [18]. For example, specializing the [SEQ1-REFOCUS] rule wrt itself gives:

$$\cfrac{e_1 \xrightarrow{L_1} e_1' \qquad e_1' \xrightarrow{L_2} e_1'' \qquad \mathsf{seq}(e_1'', e_2) \xrightarrow{L_3}{}^* z}{\mathsf{seq}(e_1, e_2) \xrightarrow{L_3 \circ L_2 \circ L_1} z}\ [\text{SEQ1-SEQ1-STRIDING}]$$

However, every possible step this rule can make can be matched by [SEQ1-REFOCUS]. Hence, we omit this rule from the set of striding rules. Specializing [SEQ1-REFOCUS] wrt [SEQ2-REFOCUS] gives the rule:

$$\cfrac{e_1 \xrightarrow{\{\ldots\}} \mathsf{skip}}{\mathsf{seq}(e_1, e_2) \xrightarrow{\{\ldots\}} e_2}\ [\text{SEQ1-SEQ2-STRIDING}]$$

By the MSOS rules for the seq construct, substituting the $\rightarrow$ with $\rightarrow^*$ is equivalent:

$$\cfrac{e_1 \xrightarrow{\{\ldots\}}{}^* \mathsf{skip}}{\mathsf{seq}(e_1, e_2) \xrightarrow{\{\ldots\}} e_2}\ [\text{SEQ1-SEQ2-STRIDING}^*]$$

This rule matches all steps that can be made using the [SEQ2] rule. There are no rules which can match all possible steps that the [SEQ1-SEQ2-STRIDING*] rule can make. Any subsequent rule unfoldings can be shown to be equivalent to the current set of rules by fh-similarity. Thus the set of rules resulting from applying the striding transformation to the seq construct are:

$$\frac{e_1 \xrightarrow{\{\dots\}}{}^* \text{skip}}{\text{seq}(e_1, e_2) \xrightarrow{\{\dots\}} e_2} \text{[SEQ1-SEQ2-STRIDING*]} \qquad \frac{e_1 \xrightarrow{L_1} e_1' \quad \text{seq}(e_1', e_2) \xrightarrow{L_2}{}^* z}{\text{seq}(e_1, e_2) \xrightarrow{L_2 \circ L_1} z} \text{[SEQ1-REFOCUS]}$$

5.2 Left-Factoring

The [SEQ1-REFOCUS] rule relates a seq term to a result, which is characteristic of big-step rules. While big-step derivation trees contain fewer inferences, compiled big-step Prolog clauses potentially give rise to non-determinism and back-tracking during proof search. For example, the conclusions of both [SEQ1-SEQ2-STRIDING*] and [SEQ1-REFOCUS] match arbitrary seq terms. In the worst case, this non-determinism leads to back-tracking, which would increase the number of inferences required to evaluate terms that do not yield values.

Left-factoring [1, 20] is a simple clause transformation which improves the determinacy of Prolog clauses generated from big-step style rules:

$$\begin{array}{l} H \leftarrow A \wedge B \\ H \leftarrow A \wedge C \end{array} \implies H \leftarrow A \wedge (B \vee C)$$

Using this simple idea, Prolog clauses are transformed to obtain specialized interpreters without the back-tracking penalty incurred by compiling big-step style rules into Prolog clauses. Figure 4 summarizes the reduction in the number of inferences resulting from striding and left-factoring when evaluating deeply nested seq terms.

6 Benchmark Experiments

We assess the viability of the specializations proposed in previous sections by considering a variant of a larger MSOS example semantics [5] with function closures and imperative state.

Figure 5 summarizes the number of Prolog inferences used to calculate the factorial of n, the nth Fibonacci number, and the greatest common divisor of the nth and $n + 1$st Fibonacci numbers using Euclid's algorithm. Each program is implemented[7] in two ways: applicatively, based on recursive unfolding; and imperatively, based on assignment and a while loop construct.

The refocusing rule introduces extra label composition operations in generated Prolog clauses for refocused rules. For deeply nested program terms this saves having to re-traverse the term in the next step. However, it entails redundant computations for values. This explains both the encouraging speed-ups in the applicative

---

[7] Benchmark code, generated interpreters, and details about the Prolog system used are available online: http://cs.swansea.ac.uk/~cscbp/lopstr13.

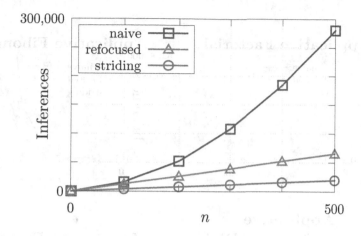

Fig. 4. Striding, refocused, and naïve evaluation of deeply nested seq terms

benchmarks (which unfold function closures to form deeply nested terms), and the slight overhead that refocusing and striding introduces for shallowly nested terms, such as the imperative factorial and Fibonacci benchmarks.

We emphasize that our specialization significantly reduces overhead in 4 out of 6 benchmarks, where the number of inferences is reduced by 4 times or more. Evaluating shallowly nested terms using big-step rules compared to small-step entails a relatively modest overhead of around 1.3 times more inferences.

7   Conclusion and Further Work

We have described how to generate interpreters from MSOS specifications and how such interpreters can be encoded in Prolog. After assessing the overhead of interpreters generated from small-step rules, we applied refocusing and striding to derive their big-step counterparts. The resulting generated interpreters significantly reduced the number of inferences used to evaluate deeply nested program terms.

Label composition is computationally expensive in generated interpreters as Fig. 5 illustrates. Our label composition strategy alleviates the need to re-compile rules as new constructs are added to languages, but requires us to traverse the Prolog list representation of label components multiple times (in the worst case) in each step. One could use a partial evaluator, such as LOGEN [12], to unfold label composition predicates. This would correspond to compiling an MSOS specification into an SOS specification, similar to compiling generalized transition systems (underlying MSOS) to labelled transition systems (underlying SOS), as described in [16]. Unfolding label composition predicates in generated Prolog interpreters should decrease the number of inferences required to evaluate terms.

Applicative Factorial

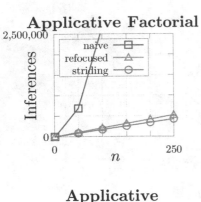

Applicative Fibonacci

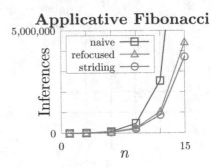

Applicative Greatest Common Divisor

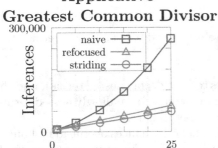

Imperative Factorial

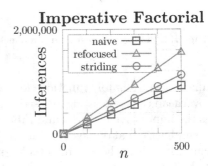

Imperative Fibonacci

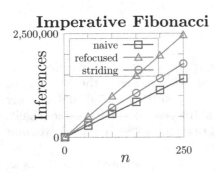

Imperative Greatest Common Divisor

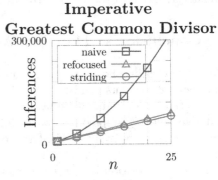

Fig. 5. Benchmark inference graphs

The refocusing rule requires MSOS rules to be explicit about abruptly terminating constructs and constructs that are sensitive to the number of steps their sub-terms make. It should be possible to specify a rule format for conservatively identifying abruptly terminating constructs. This would enable automatic annotation of MSOS rules with $\varepsilon, \varepsilon'$ label components.

Striding requires filtering specialized rules that are equivalent to existing ones. We suggested using fh-simulation [18] for this. For the purposes of this paper, these proofs were constructed manually. While bisimulation is undecidable in general, it should be possible to automate proofs for at least some constructs.

Acknowledgments. Thanks to Paolo Torrini, Martin Churchill, Ferdinand Vesely, and the anonymous referees for their useful comments. This work was supported by an EPSRC grant (EP/I032495/1) to Swansea University in connection with the *PLanCompS* project (www.plancomps.org).

References

1. Aho, A.V., Ullman, J.D.: The theory of parsing, translation, and compiling. Prentice-Hall, Inc. (1972)
2. Bach Poulsen, C., Mosses, P.D.: Deriving pretty-big-step semantics from small-step semantics. In: Shao, Z. (ed.) ESOP 2014 (ETAPS). LNCS, vol. 8410, pp. 270–289. Springer, Heidelberg (2014)
3. Chalub, F., Braga, C.: Maude MSOS tool. ENTCS **176**(4), 133–146 (2007)
4. Charguéraud, A.: Pretty-big-step semantics. In: Felleisen, M., Gardner, P. (eds.) Programming Languages and Systems. LNCS, vol. 7792, pp. 41–60. Springer, Heidelberg (2013)
5. Churchill, M., Mosses, P.D.: Modular bisimulation theory for computations and values. In: Pfenning, F. (ed.) FOSSACS 2013 (ETAPS 2013). LNCS, vol. 7794, pp. 97–112. Springer, Heidelberg (2013)
6. Clavel, M., Durán, F., Eker, S., Lincoln, P., Martí-Oliet, N., Meseguer, J., Talcott, C.: Maude manual (version 2.6) (2008). http://maude.cs.uiuc.edu/maude2-manual/
7. Clement, D., Despeyroux, J., Despeyroux, T., Hascoet, L., Kahn, G.: Natural semantics on the computer. Research Report RR-0416, INRIA (1985)
8. Danvy, O.: From reduction-based to reduction-free normalization. In: Koopman, P., Plasmeijer, R., Swierstra, D. (eds.) AFP 2008. LNCS, vol. 5832, pp. 66–164. Springer, Heidelberg (2009)
9. Danvy, O., Nielsen, L.R.: Refocusing in reduction semantics. BRICS Research Series RS-04-26, Dept. of Computer Science, Aarhus University (2004)
10. Gallagher, J.P.: Tutorial on specialisation of logic programs. In: PEPM 1993, pp. 88–98. ACM (1993)
11. Gupta, G.: Horn logic denotations and their applications. In: Apt, K.R., Marek, V.W., Truszczynski, M., Warren, D.S. (eds.) The Logic Programming Paradigm. Artificial Intelligence, pp. 127–159. Springer, Heidelberg (1999)
12. Leuschel, M., Jorgensen, J., Vanhoof, W., Bruynooghe, M.: Offline specialisation in Prolog using a hand-written compiler generator. TPLP **4**(1), 139–191 (2004)
13. Lloyd, J.W., Shepherdson, J.C.: Partial evaluation in logic programming. J. Log. Program. **11**(3–4), 217–242 (1991)
14. Mosses, P.D.: Foundations of Modular SOS. BRICS Research Series RS-99-54, Dept. of Computer Science, Aarhus University (1999)
15. Mosses, P.D.: Pragmatics of Modular SOS. In: Kirchner, H., Ringeissen, C. (eds.) AMAST 2002. LNCS, vol. 2422, pp. 21–40. Springer, Heidelberg (2002)
16. Mosses, P.D.: Modular structural operational semantics. J. Log. Algebr. Program. **60–61**, 195–228 (2004)

17. Mosses, P.D.: Teaching semantics of programming languages with Modular SOS. In: Boca, P., Bowen, J.P., Duce, D.A. (eds.) TFM 2006. Electr. Workshops in Comput, BCS (2006)
18. Mosses, P.D., Mousavi, M.R., Reniers, M.A.: Robustness of equations under operational extensions. In: Fröschle, S.B., Valencia, F.D. (eds.) EXPRESS 2010. EPTCS, vol. 41, pp. 106–120 (2010)
19. Mosses, P.D., New, M.J.: Implicit propagation in structural operational semantics. ENTCS **229**(4), 49–66 (2009)
20. Pettersson, M. (ed.): Compiling Natural Semantics. LNCS, vol. 1549. Springer, Heidelberg (1999)
21. Plotkin, G.D.: A structural approach to operational semantics. J. Log. Algebr. Program. **60–61**, 17–139 (2004)
22. Wang, Q., Gupta, G., Leuschel, M.: Towards provably correct code generation via horn logical continuation semantics. In: Hermenegildo, M.V., Cabeza, D. (eds.) PADL 2004. LNCS, vol. 3350, pp. 98–112. Springer, Heidelberg (2005)

Author Index

Printed in the United States
By Bookmasters